INTERNATIONAL SERIES OF MONOGRAPHS IN

NATURAL PHILOSOPHY

GENERAL EDITOR: D. TER HAAR

VOLUME 23

QUANTUM FIELD THEORY

OTHER TITLES IN THE SERIES IN NATURAL PHILOSOPHY

Vol. 1. DAVYDOV—Quantum Mechanics

Vol. 2. FOKKER—Time and Space. Weight and Inertia

Vol. 3. KAPLAN—Interstellar Gas Dynamics

Vol. 4. ABRIKOSOV, GOR'KOV and DZYALOSHINSKII—Quantum Field Theoretical Methods in Statistical Physics

Vol. 5. OKUN'—Weak Interaction of Elementary Particles

Vol. 6. SHKLOVSKII—Physics of the Solar Corona

Vol. 7. AKHIEZER *et al.*—Collective Oscillations in a Plasma

Vol. 8. KIRZHNITS—Field Theoretical Methods in Many-body Systems

Vol. 9. KLIMONTOVICH—The Statistical Theory of Non-equilibrium Processes in a Plasma

Vol. 10. KURTH—Introduction to Stellar Statistics

Vol. 11. CHALMERS—Atmospheric Electricity (2nd edition)

Vol. 12. RENNER—Current Algebras and their Applications

Vol. 13. FAIN and KHANIN—Quantum Electronics, Vol. 1. Basic Theory

Vol. 14. FAIN and KHANIN—Quantum Electronics, Vol. 2. Maser Amplifiers and Oscillators

Vol. 15. MARCH—Liquid Metals

Vol. 16. HORI—Spectral Properties of Disordered Chains and Lattices

Vol. 17. SAINT JAMES, THOMAS and SARMA—Type II Superconductivity

Vol. 18. MARGENAU and KESTNER—Theory of Intermolecular Forces

Vol. 19. JANCEL—Foundations of Classical and Quantum Statistical Mechanics

Vol. 20. TAKAHASHI—An Introduction to Field Quantization

Vol. 21. YVON—Correlations and Entropy in Classical Statistical Mechanics

Vol. 22. PENROSE—Foundations of Statistical Mechanics

QUANTUM FIELD THEORY

IN TWO VOLUMES

VOLUME 1

by

A. VISCONTI

Faculté des Sciences d'Aix-Marseille

TRANSLATED BY

H. BACRY AND C. M. HANCOCK

THE QUEEN'S AWARD TO INDUSTRY 1966

PERGAMON PRESS

OXFORD · LONDON · EDINBURGH · NEW YORK
TORONTO · SYDNEY · PARIS · BRAUNSCHWEIG

Pergamon Press Ltd., Headington Hill Hall, Oxford
4 & 5 Fitzroy Square, London W.1

Pergamon Press (Scotland) Ltd., 2 & 3 Teviot Place, Edinburgh 1

Pergamon Press Inc., Maxwell House, Fairview Park, Elmsford, New York 10523

Pergamon of Canada Ltd., 207 Queen's Quay West, Toronto 1

Pergamon Press (Aust.) Pty. Ltd., 19a Boundary Street, Rushcutters Bay, N.S.W. 2011, Australia

Pergamon Press S.A.R.L., 24 rue des Écoles, Paris 5e

Vieweg & Sohn GmbH, Burgplatz 1, Braunschweig

First English edition 1969

Library of Congress Catalog Card No. 66-18237

PRINTED IN HUNGARY

08 011821 6

Contents of Volume 1

Contents of Volume 2

Preface to the English Edition

I HAVE taken advantage of the English translation of my book to correct misprints and errors of the French edition, and to rewrite completely some of its paragraphs.

For practical reasons the translation will appear in two volumes, but the division between the first and second volumes does not correspond to the distinction between free and interacting fields; rather, the first volume is devoted to the study of the lagrangian and hamiltonian formulations of Quantum Field Theory and of the free fields.

I wish to thank my translators Professor H. Bacry and Dr. C. M. Hancock for the fine job they have done, and the staff of Pergamon Press for their care in preparing this book.

Introduction

THIS book is a revised and elaborated version of a number of lectures delivered at various universities, including the Science Faculty of Aix-Marseilles. Its level corresponds to that of a post-graduate student.

To give a clear picture of its scope, I shall reproduce the programme of a year's course in quantum field theory at the University of Aix-Marseilles: a course which is divided into six parts, the approximate duration of which is reckoned in weeks, each week comprising three hours of lecturing.

I. QUANTIZATION OF FIELDS

(a) The state vector and field variables.
(b) The Heisenberg–Pauli quantization method.
(c) The three fundamental pictures.
(d) Brief investigation of the Schwinger–Feynman action principle.

The basis for these lectures is essentially Chap. II: such questions as action formulation and the structure of the state vector equation or the field equations can be more or less emphasized according to the students' level; but the Heisenberg–Pauli quantization method (§§ 6, 7, 8 and 10) needs to be examined thoroughly; § 9 may be left until later. It is also advisable to introduce and examine, if only briefly, the notion of quantum pictures (§§ 11 and 12).

Since it is generally found necessary to refresh students' memories by recapitulating notions assumed to be known (Chap. I is designed to meet this need), we advise lecturers to devote three weeks to this first group of lessons.

II. FIELD OBSERVABLES

The contents of Chap. III form the subject matter of this second group of lessons. Sections 1, 2 and 3 should be studied *in toto*, except perhaps for the somewhat abstract considerations concerning the Lie relations at the end of § 2, which may be replaced by a study of the note at the end of that section. The following sections (as far as § 9) are concerned with building, with the help of the Schwinger–Feynman action principle, the observables inferred from the Lorentz group and gauge group. One can, if absolutely necessary, confine oneself at a first reading to applications of Noether's theorem (without using the operator character of field variables), but in that case a number of properties are omitted which should really be covered. Sections 9 and 11 may be studied if need be after group III (free fields).

The applications of the notions presented in this chapter are, it will be noted, extremely important: our entire knowledge of elementary particles is conditioned by them. We have been obliged, however, to relegate them to Chap. VII, §§ 8 and 9; they call for a certain degree of physical knowledge concerning the particles, and this knowledge, although elementary in nature, cannot *a priori* be expected of the reader at this stage. The above sections should, in any case, be read concurrently with Chap. III: they represent its most natural illustration.

III. Free fields

(a) Classification and properties of elementary particles.

(b) Scalar and pseudo-scalar fields.

(c) Dirac field.

(d) Maxwell field.

The classification of elementary particles is studied in Chap. IV, Part 1; lecturers are advised to develop the basic notions found in this part with the help of the works quoted in the notes: a minimum of two weeks seems necessary for this study (except in the frequent cases where the study of particles is the subject of a separate course of lectures). The rest of this chapter can be omitted at a first reading, except for §§ 5 and 7, which apply directly to the Dirac equation.

The Hermitian and non-Hermitian scalar fields must be studied in full detail (Chap. V, Part 1), since they are of basic importance for the quantization of all other fields; from one to two weeks should be devoted to them, depending on the students' previous knowledge. The rest of Chap. V (Parts 2 and 3) dealing with the quantization of the general equation

$$\{\beta_\mu \partial_\mu + m\} Q(x) = 0$$

may also be omitted at a first reading (§ 16 devoted to the explicit representation of the creation and annihilation operators may be studied rapidly). Indeed, although the results of this chapter find their application in the study of the Dirac equation, we have been careful, in Chap. VI, § 1, devoted to that equation, to prove afresh and quite independently all the results obtained in §§ 2 and 3 of Chap. V. We investigate, for instance, independently of the general theories elaborated in the latter chapter, the commutation relations between field variables of spin $\frac{1}{2}$, the splitting of the field variables of spin $\frac{1}{2}$ into creation and annihilation operators, etc.

The study of the Dirac field (Chap. VI, Part 1) can easily take four weeks. As for the Maxwell field (Chap. VI, Part 2), it requires from one to two weeks, depending on the time given to classical electrodynamics (§§ 16 and 17 of this chapter) and the consequences of the Fermi condition (§§ 20 and 21). Part 3 of Chap. VI can be passed over rapidly at a first reading.†

IV. Interacting fields (Heisenberg picture)

The contents of this group of lessons are to be found in Chap. VII, Part 1: the aspects dealt with in this book may be completed if need be, certain interactions being stressed more particularly than others according to the direction the lecturer wishes to give to his course. Two weeks seem necessary for this study.

Part 2 of this chapter can be reserved until later; § 13 (polarization of the vacuum...) may be dealt with in the last lessons on quantum field theory—in other words, be included in group VI.

V. Schrödinger picture

This formulation is of much interest, at least pedagogically, since it permits one to exhibit the matrix elements of the field operators. Part 1 of this chapter may be passed over rapidly, but the utmost attention should be given to §§ 4, 7 and 9, dealing respectively with the $\mathcal{S}$ operator, transition probabilities, cross-sections and a simplified version of quantum electrodynamics. The non-covariant calculation of the Lamb–Retherford effect can be dealt with in the last lessons of the course (group VI). From one to two weeks seem necessary for this study.

† These six chapters make up the first volume; the second volume comprises the five other chapters (VII to XI) and two appendixes.

VI. Interaction picture. Diagrams. Applications

The contents of Chap. IX are central to an introductory course on quantum field theory; they involve a study of the expressions of the elements of the $\mathcal{S}$ operator which enter into the expression of the cross-sections of the scattering phenomena, whose importance cannot, of course, be overstressed. I have adopted for this study the language of Feynman's diagrams because of the intuitive and physical character of that method of exposition. It should be noted, however, that there are two possible approaches to the study of these diagrams. The first approach, an eminently logical one, consists in starting out from first principles, showing that any matrix element of the $\mathcal{S}$ operator may be represented graphically, and finally giving the rules of correspondence between the diagrams and their analytical expressions. This is obviously the appropriate method for a rigorous presentation, and we have discussed it in Chap. IX, Parts 1 and 2.

But another and far more intuitive method exists: the one originally presented by Feynman himself in his well-known paper. It consists of starting out from the concepts of wave mechanics, extending them appropriately, and "inferring" the language of the diagrams from a few examples. We discuss this method in Chap. IX, Part 3.† It can, in my opinion, prove fruitful, for it can serve as a conclusion to the course on quantum mechanics which precedes the one on quantum field theory. An experiment carried out at the University of Aix-Marseilles (following indeed the example of certain American universities) has shown that students at the end of their first year of quantum mechanics are perfectly ready to learn and use the diagram method, even if it is presented just as a "recipe". It has the added advantage of keeping up their interest during the abstract and formal part of quantum field theory, which will lead them to a more rigorous formulation of rules permitting the expression of the matrix elements of the $\mathcal{S}$ operator.

Depending as always on students' receptiveness, at least eight weeks seem necessary for this study.

† We have used in that part the formalism of the evolution operator in the Schrödinger picture of quantum mechanics, at the same time introducing integral densities. In a first reading it seems simpler to verify directly that the function $K(\boldsymbol{x}, t; \boldsymbol{x}', t')$ is the solution of the Schrödinger equation which reduces to $\delta(\boldsymbol{x}-\boldsymbol{x}')$ when $t=t'$. The rest of the exposition then follows that of Part 3 of Chap. IX.

VII. THE BASIC IDEAS UNDERLYING THE RENORMALIZATION OF QUANTUM ELECTRODYNAMICS

(a) Non-relativistic calculation of the Lamb–Retherford shift and indications concerning relativistic calculation.
(b) Anomalous magnetic momentum of the electron.
(c) Fluctuations of the vacuum; renormalization of the charge.
(d) Infinite diagrams; renormalization techniques.

The subjects of this last group of lessons are dealt with respectively in Chap. VIII, § 10; Chap. X, §§ 12d and 12b; Chap. VII, § 13; and finally Chap. X, §§ 2, 6, 7, 8 and 9. This study may take about four weeks.

Lastly, if time permits and students prove sufficiently receptive, some of the questions dealt with in Chap. XI (e.g. §§ 10, 13 and 15) can be discussed. At the Science Faculty of Aix-Marseilles, it has in fact been found profitable to devote a course of further studies to the more intensive examination of the present forms of quantum field theory (Chap. XI and Appendix II) and the theories and techniques of renormalization.

It will be apparent from the above analysis that this book is a "classical" study of quantum field theory in that, broadly speaking, it presents the foundations of this theory within the framework laid down by Heisenberg and Pauli thirty years ago.

I begin with a description of the Hamiltonian formalism, rather than with a statement of the Schwinger–Feynman action principle which could have replaced it, but whose undoubted elegance is counterbalanced by the difficulties involved in its formulation and application. I next investigate free fields, then interacting fields and the $\mathcal{S}$ operator, and only in the last chapter (Chap. XI) do I bring in considerations independent of the existence of an action and which represent the essential feature of present-day attempts at axiomatization.

Another important question should perhaps be mentioned in this connection: what place should mathematical rigour occupy in the arguments of theoretical physics? It is certainly tempting to present quantum field theory according to a rigorously axiomatic method, with concepts and the rules governing their combinations stated and specified once for all: such an approach has in fact been tried out by a number of physicists. But does it really pay? Is it the method that leads most efficiently to results which can be tested by experiment?

Personally, I do not think so. Firstly, because the physicist—even when he is baptized a theorist—works in a different manner from the mathematician, and pursues a different goal. The aim of all physicists is after all the statement of rules whereby the greatest number of experimental facts can be explained and classified, and the writing of formulae, numerical tables and graphs. The method which leads most quickly and easily to this objective (even if it requires some unproved assumptions) is in my view the method to be adopted. The mathematician's attitude is different: the manner of obtaining results is at least as important in his eyes as the results themselves; it is only right that he should devote a large part of his time to classifying, generalizing and reducing to its essentials what the physicist has often perceived only vaguely and applied with little concern either for the conditions of application or the logical coherence of the method adopted.† There has been no lack of examples in recent years of this division of labour between mathematicians and physicists.

In actual fact, of course, between the pure experimental physicist and the pure mathematician there exists a whole range of different types of attitude (the theoretical physicist, for instance, being less of a mathematician than the mathematical physicist). It would be presumptuous in the extreme and contrary to the very principle of freedom of research to wish to pronounce value-judgements establishing a hierarchy among them. It is healthy and natural that such minds should coexist and endeavour to understand each other; all one can wish for is that in any group in which they collaborate a certain balance should prevail.‡

For my own part, I admit that I am far closer to the theoretical physicist who does not shrink from using arguments which are logically insecure but rich in physical consequences; who might even be inclined to distrust the most impeccable physical theories because of the very rigidity of their axiomatization—symptomatic in his eyes almost of sterility. Such an attitude, which I would not for a moment wish to impose as the "ideal" one, may help to explain the choice of certain methods in this book.

In the first place a general work on quantum field theory must necessarily deal with a number of questions whose present form of exposition is far from satisfactory (e.g. the interaction picture, whose existence raises

† The above remarks apply to the *ideal* method of exposition of a physical theory (ideal, that is, from the point of view of logical coherence!): it is the method which, while preserving its intuitive and physical character, can be rendered sufficiently rigorous by a mathematician who confines his role to clarifying a number of subsidiary points, suggesting, for example, slight alterations in the order of arguments or the vocabulary.

Unfortunately, I have learnt from personal experience that to achieve the degree of rigour demanded by the mathematician, such substantial alterations have to be introduced that the intuitive and physical character of the exposition is profoundly modified.

‡ As for the extreme case of those who use "mathematical jargon" with the aim of stating in a new form (without attempting to examine them more deeply) theories that are already perfectly well known, such attempts seem doomed to a certain sterility.

undoubted difficulties but which is nevertheless extremely helpful in dealing with scattering problems) or which may seem untopical (for a time, the study of particles with arbitrary spin seemed purely academic), if only to save newcomers the trouble of gradually finding them out as they go along.

A further question is whether (independently of one's personal viewpoint) one should adopt an axiomatic presentation in line with one of the present tendencies, or whether alternatively one should confine oneself to an exposition in which the assumptions are not always analysed in an exhaustive manner and are only introduced when their need is actually felt. The axiomatic method would certainly have greater weight if examples could be given of fields interacting with an $\mathcal{S}$ operator different from identity. For after all, even if one is sure of having kept back all the necessary properties and only such properties, there always remains the possibility—remote, it is true, but impossible to exclude in strict logic—that the abstraction has been carried too far, or as Wightman wittily puts it: "it is just barely conceivable that the baby has been let out with the bath water."

Furthermore, a course of instruction such as Quantum Field Theory must be above all *open* in character, that is to say it must indicate, so far as possible, numerous lines of approach. It would be harmful indeed if the form of exposition adopted were to dismiss as definitively explored a particular approach which later research revealed to be a fertile line of investigation.

To return to more general questions: in dealing with the Dirac equation, should one describe the spinor theory for its own sake and preferably in an n-dimensional space, or should one study that equation directly? In a work intended for mathematicians the former method would obviously be the appropriate one; for my part, I have confined myself to studying the Dirac equation as a specific case of a particular class of differential matrix equations with constant coefficients. Although the considerations one is obliged to introduce for this purpose lack the rigour of those of the first method, I consider this approach a valid one, since it clearly exhibits a number of physical consequences. It is immediately apparent, for instance, that the charge conjugation matrix C obeys $C^T = -C$ in the case of fermions and $C^T = C$ in the case of bosons.

Finally, as a last example, there was the problem of choosing among the various renormalization methods. Which should one adopt, Källen's or Bogoliubov's, both of these described in excellent books, or Dyson's, which is based on Feynman's diagrams and their analysis? The first two obviously lend themselves to a precise and "sufficiently" rigorous presentation; the third one has the merit of directly interpreting the difficulties which have held up quantum field theory over a period of about twenty years and which inevitably arise when it comes to calculating the radiative corrections of the Coulomb scattering, in other words (in the last analysis) Rutherford's well-known formula.

Concerning the physical applications which readers may legitimately expect to find in such a work, I want to point out that I originally planned this book to form a single volume, but the exigencies of publication compelled me to divide it into two. As a result, many applications found their place in Vol. 2, a disadvantage which had not been apparent in the original plan of the book. They will be found, in all, to be sufficiently numerous: from the restrictions to the structure of the Hamiltonian as a result of the local character of the field equations (Chap. II) to the physical consequences of renormalization; they are particularly numerous in Chaps. VII, VIII, IX and X.

A final word on the English edition of my book: improvements have been made in several places and mistakes and misprints corrected. I want also to thank my translators for their kind collaboration.

Formulae are designated by two groups of figures, the first of which represents the chapter and the second the serial number: e.g. formula (10.15) is formula no. 15 of Chap. X. A peculiarity should be noted in the case of Chap. VI: formula (6.2.15) is formula no. 15 of the second part of Chap. VI.

Notation

WE HAVE made a point of using (with the exception of the Hermitian conjugation which we denote by $\sim$ instead of *, reserving the latter symbol for the complex conjugation) the notation most commonly used by physicists, even in some cases where their choice is not a particularly fortunate one! One of our purposes in writing this book was indeed to place readers rapidly in a position to be able to read original papers for themselves.

The unit system adopted is the natural unit one: $\hbar = 1$, $c = 1$.

X, a vector with components x_j, $j = 1, 2, 3$ of the three-space;

x, a vector with components x_μ, $\mu = 1, 2, 3, 4$ of the four-space; its first three components x_j are real, and its fourth one $x_4 = ix_0 = ict$ is imaginary.

The Latin indices range from 1 to 3; the Greek indices from 1 to 4. Summation occurs over all repeated indices.

$\boldsymbol{ab} = a_j b_j$;

$ab = a_\mu b_\mu = \boldsymbol{ab} - a_0 b_0$;

$d^3\boldsymbol{x} = dx_1\, dx_2\, dx_3$, a volume element of the three–space;

$d^4x = d^3\boldsymbol{x}\, dx_0$, a volume element of space–time;

$Q_A(x)$, a *field variable*, a linear operator operating in a space $\mathcal{R}$ and depending on the parameters x defining a space–time point and A symbolizing a set of discrete parameters (spin, isobaric spin (isospin), etc.); $Q_A(x)$ is called a *wave function* when it is considered as a differentiable function belonging to some normed space;

$[A, B]_{\text{cl}}$, Poisson bracket;

$[A, B] \equiv AB - BA$, a commutator of the operators A and B;

$[A, B]_+ \equiv AB + BA$, an anticommutator of the operators A and B;

$\dfrac{\partial f}{\partial x_\mu} \equiv \partial_\mu f \equiv f_{,\mu}$, the operator ∂_μ has the following components: ∂_j, and $\partial_4 = -\dfrac{i\partial}{\partial x_0}$; $f\overleftarrow{\partial}_\mu = \partial_\mu f$.

$f(x) = (2\pi)^{-4}\int e^{ipx}f(p)d^4p$, the Fourier transform of a function $f(x)$, is generally denoted, following the normal (and unsatisfactory!) usage of physicists, by the same letter f as the function. Only its argument p, instead of x, indicates that $f(p)$ is not a function of p identical to the function $f(x)$ of x.

* (asterisk) indicates the conjugate complex of a c-number;

~ (tilde) indicates the Hermitian conjugate of an operator or a matrix;

T indicates the transpose of a matrix;

I, the identity operator;

$\hat{A} = \gamma_\mu A_\mu$; γ_μ: Dirac matrices; A_μ: a four-vector.

CHAPTER I

General principles

WE SHALL devote this chapter to a number of basic notions on the Lorentz transformations, linear spaces, quantum mechanics, the Fourier transformation, the Dirac function, and, lastly, functional analysis.

It is assumed that the reader is familiar with the techniques here referred to. The following exposition is therefore simply an outline of ideas and formulae we shall be making use of hereafter. It hardly seems necessary to add that the few deductions it contains are by no means claimed to be rigorously exact.

1. Lorentz group

Let us consider a metric four-space endowed with the following metric tensor:†

$$g_{ij} = g_{ji} = \delta_{ij}, \quad g_{\mu_0} = g_{0\mu} = -\delta_{\mu 0} \tag{1.1}$$

where Greek indices run from 0 to 3 and Latin indices from 1 to 3. $\delta_{\alpha\beta}$ is the Kronecker symbol equal to 0, except when $\alpha = \beta$ and in that case equal to 1.

The transformation

$$A = (A_{\mu\nu}), \tag{1.2}$$

a 4×4 matrix with real elements, transforms the vector X represented by a matrix column with four elements into a vector X':

$$X' = AX \tag{1.3a}$$

or

$$X'_\mu = A_{\mu\nu} X_\nu \tag{1.3b}$$

† Among all the books about the Lorentz group, I want to quote the ones I have been using: H. Bacry, *Leçons sur le Théorie des Groupes et les symétries des particules élémentaires*; I. M. Gel'fand, R. A. Minlos and Z. Ya. Shapiro, *Representations of the Rotation and Lorentz Groups and their Applications*; H. Lipkin, *Lie Groups for Pedestrians.*

with summation over the dummy index ν. The components X_μ being chosen to be real, it follows that the components X'_μ are also real.

We shall call the transformation A a homogeneous Lorentz transformation if the scalar product of two vectors X and Y of this space is an invariant, in other words if

$$\tilde{X}gY = \tilde{X}'gY' \tag{1.4a}$$

$\tilde{X}$, $\tilde{X}'$ are the adjoint (single-row) matrices (simply transposed since the components X_μ and X'_μ are real) of the (single-column) matrices representing X and X'. The above formula may be written in the more explicit form:

$$X_\mu g_{\mu\nu} Y_\nu = X'_\mu g_{\mu\nu} Y'_\nu. \tag{1.4b}$$

Expressing in formula (1.4) the transformed vectors in terms of the vectors X and Y, we obtain the necessary and sufficient condition for A to be a Lorentz transformation:

$$\tilde{A}gA = g \tag{1.5}$$

or

$$A_{\varrho\mu} g_{\varrho\sigma} A_{\sigma\nu} = g_{\mu\nu} \tag{1.6a}$$

where we sum over the repeated indices ϱ and σ. We also obtain, since $g_{\mu\nu} = g_{\nu\mu}$:

$$Ag\tilde{A} = g. \tag{1.6b}$$

It then becomes easy to prove that any Lorentz transformation is unimodular; we obtain, in fact, in accordance with (1.5)

$$\det \tilde{A} \cdot \det g \cdot \det A = \det g$$

or

$$|\det A|^2 = 1. \tag{1.7}$$

Let us now take $\mu = \nu = 0$; formula (1.6a) is written

$$A_{\varrho 0} g_{\varrho\sigma} A_{\sigma 0} \equiv \sum_j |A_{j0}|^2 - (A_{00})^2 = g_{00} = -1$$

from which we obtain the condition

$$(A_{00})^2 \geqslant 1. \tag{1.8}$$

As may be easily verified, the Lorentz transformations form a group; formulae (1.7) and (1.8) will enable us to divide this so-called "full group L" into four sheets:

$A_{00} \geq 1$. Orthochronous subgroup $L^{\uparrow}$ of the Lorentz transformations conserving the direction of time.

- det $A = +1 : L^{\uparrow}_{+}$, the special Lorentz group, the elements of which are the "proper Lorentz transformations". This group contains in particular the rotations of the three-dimensional Euclidean space.
- det $A = -1 : L^{\uparrow}_{-}$, contains in particular the reflections of the Euclidean space:

$$x'_j = -x_j, \qquad x'_0 = x_0$$

with $x' = \sigma x$, where

$$\sigma = \begin{pmatrix} -1 & 0 & 0 & 0 \\ 0 & -1 & 0 & 0 \\ 0 & 0 & -1 & 0 \\ 0 & 0 & 0 & 1 \end{pmatrix} = -g.$$

$A_{00} \leq -1$. Antichronous transformations $L^{\downarrow}$ reversing the direction of time.

- det $A = +1 : L^{\downarrow}_{+}$, the total reflection $x'_j = -x_j$, $x'_0 = -x_0$ belongs to this sheet.
- det $A = -1 : L^{\downarrow}_{-}$, contains in particular the time reflections τ

$$x'_j = x_j, \qquad x'_0 = -x_0$$

with $x' = \tau x$, where

$$\tau = \begin{pmatrix} 1 & 0 & 0 & 0 \\ 0 & 1 & 0 & 0 \\ 0 & 0 & 1 & 0 \\ 0 & 0 & 0 & -1 \end{pmatrix} = g.$$

These four sheets are disconnected and the sum of $L^{\uparrow}_{+}$ with one of the other three sheets forms a group.

The transformation

$$\sigma\tau = \tau\sigma$$

is a total reflection in the four-space: it changes, in fact, x into $-x$ and belongs to $L^{\downarrow}_{+}$.

On the other hand, from the relations

$$L = L^{\uparrow} + L^{\downarrow} = L^{\uparrow} + \tau L^{\uparrow} = (1+\tau)L^{\uparrow}, \tag{1.9}$$

$$L^{\uparrow} = L^{\uparrow}_{+} + L^{\uparrow}_{-} = L^{\uparrow}_{+} + \sigma L^{\uparrow}_{+} = (1+\sigma)L^{\uparrow}_{+} \tag{1.10}$$

which may be easily verified, we infer that the full Lorentz group L may be generated solely with the aid of the special group $L^{\uparrow}_{+}$ and the two transformations σ and τ as follows:

$$L = (1+\tau)(1+\sigma)L^{\uparrow}_{+}. \tag{1.11}$$

Let us finally consider a point of coordinates $(\mathbf{x}, x_0 = ct)$; the fundamental quadratic form

$$g_{\mu\nu}x_\mu x_\nu \equiv \boldsymbol{x}^2 - x_0^2 \tag{1.12}$$

divides the space into two domains which are invariant under the full group L: the inside and outside of the "light cone":

$$g_{\mu\nu}x_\mu x_\nu = 0. \tag{1.13}$$

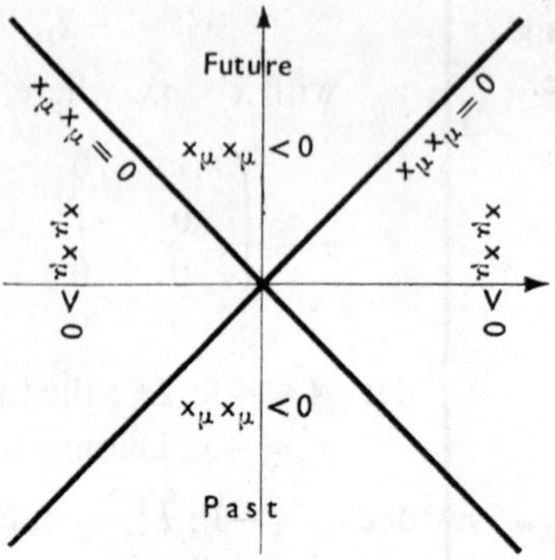

FIG. 1.1.

The inside of this cone is in turn divided into two domains invariant under the subgroup $L^\dagger$: the past which corresponds to $x_0 < 0$ and the future which corresponds to $x_0 > 0$.

Instead of the metric tensor $g_{\mu\nu}$, the tensor $\delta_{\mu\nu}$ will be used hereafter. To this end, the fourth component of a vector X which we shall denote by X_4, instead of X_0, will be purely imaginary:

$$X_4 = iX_0 \tag{1.14}$$

and the index μ takes, therefore, the values of 1 to 4.

The vector space thus defined is pseudo-Euclidean and the scalar product of two vectors X and Y is written

$$XY = X_\mu Y_\mu = \boldsymbol{XY} - X_0 Y_0. \tag{1.15}$$

It is easy to verify that with this metric, the matrix L of a Lorentz transformation is written in terms of A as follows:

$$L = \begin{pmatrix} A_{11} & A_{12} & A_{13} & -iA_{10} \\ A_{21} & A_{22} & A_{23} & -iA_{20} \\ A_{31} & A_{32} & A_{33} & -iA_{30} \\ iA_{01} & iA_{02} & iA_{03} & A_{00} \end{pmatrix} \tag{1.16a}$$

and that in accordance with (1.6) we have

$$L^{-1} = L^T, \tag{1.16b}$$

the superscript T denoting the transposed matrix.

Relative to their norms, we can distinguish three kinds of four-vectors: space-like vectors such as

$$X_\mu X_\mu > 0;$$

time-like vectors such as

$$X_\mu X_\mu < 0;$$

and isotropic vectors with zero norms†:

$$X_\mu X_\mu = 0.$$

It is clear that the isotropic vectors belong to the light cone. The distinction between space-like vectors and time-like vectors is particularly important: let there be two points A and B of the four-space and let $AB = x$ be the four-vector which joins them. Let us assume x to be space-like:

$$x_\mu x_\mu = \boldsymbol{x}^2 - c^2t^2 > 0,$$

in this case the velocity of propagation

$$v = \frac{|x|}{t}$$

of a signal starting from A at the time 0 and arriving at B at the time t will be greater, in virtue of the preceding inequality, than the velocity of light c:

$$v > c.$$

It may therefore be stated in accordance with the general principle of special relativity, that no signal coming from A will be able to arrive at B or that the events which occur at A and B are completely independent of each other. It is this basic property which distinguishes space-like vectors from time-like ones and explains the role of the space-like surfaces as will be made clear in the following section.

We shall finally be led to consider mathematical entities (covariants) with well-defined variances under Lorentz transformations. The simplest of these are scalars: under a Lorentz transformation, a scalar φ is turned into φ' ($\varphi \to \varphi'$) such that

$$\varphi' = \varphi.$$

† Note also that each kind of four-vector is invariant for any L transformation and that an $L^{\dagger}$ transformation does not change the sign of the time component of a time-like vector, but that this is no longer the case for a space-like vector.

The vector v_μ is such that a Lorentz transformation transforms $\boldsymbol{v}$ into v' with

$$v'_\alpha = L_{\alpha\nu} v_\nu$$

a tensor t of nth rank is such that

$$t'_{\alpha_1 \ldots \alpha_n} = L_{\alpha_1 \nu_1} \ldots L_{\alpha_n \nu_n} t_{\nu_1 \ldots \nu_n}.$$

From Chap. III onwards we shall introduce entities with more complicated variances such as spinors; at the beginning of the same chapter we shall also examine infinitesimal Lorentz transformations and their representations.†

2. Space-like surfaces

It is particularly important for some problems in quantum field theory to bring out clearly the relativistic variance of certain formulae and equations. For instance, such virtually meaningless results as the "infinite" proper energy of an electron at rest can be interpreted to some degree provided one can determine at each stage of the calculation, the relativistic variance of the expression obtained. The introduction of space-like surfaces and the "covariant formalism" satisfy this requirement.‡

We specify as in the preceding section a point of space-time by its coordinates:

$$\boldsymbol{x},\ x_4 = ix_0 = ict.$$

The expression "measurement at a given time" only has a precise meaning if one specifies the frame of reference in which this measurement is carried out, and this remark, in a slightly different form, amounts to recognizing that a family of equitemporal planes, $t =$ constant, is not covariant under a Lorentz transformation: this transformation does not transform it into a new family of equitemporal planes.

Let us therefore consider a family of space-like surfaces σ, i.e. surfaces, with a positive linear element:

$$ds^2 = dx_\mu dx_\mu = d\boldsymbol{x}^2 - c^2\, dt^2 > 0. \tag{1.17}$$

We saw at the end of the preceding section that these surfaces are such that two events which occur on one of them are necessarily independent of each other. We may therefore fix on such a surface σ, exactly as we would do on an equitemporal plane, a number of arbitrary initial data.

† To the scalar, vector, tensor correspond the pseudo-scalar, the pseudo-vector, the pseudo-tensor whose definitions imply an extra factor (det L). For instance, the pseudo-vector w_α is transformed as follows:

$$w'_\alpha = (\det L) L_{\alpha\nu} w_\nu.$$

‡ S. Tomonaga, *Progr. Theor. Phys.*, **1**, 27 (1946).

These surfaces form, moreover—as we can easily verify—a covariant family: a Lorentz transformation turns a given space-like surface into another of the same kind.

We are therefore going to attribute to such surfaces the part played by the time variable, and a number of notions must accordingly be made clear.

Let us note first that any surface with a time-like normal is a space-like surface. Let n_μ be this normal vector; according to its definition, it obeys the relations

$$\left.\begin{aligned} n_\mu n_\mu &\equiv \boldsymbol{n}^2 - n_0^2 < 0, \\ n_\mu dx_\mu &= \boldsymbol{n}\, d\boldsymbol{x} - n_0\, dx_0 = 0. \end{aligned}\right\} \tag{1.18}$$

By introducing into the expression of the ds^2 the value of dx_0 taken from (1.18) we have

$$ds^2 = dx_\mu dx_\mu = d\boldsymbol{x}^2 - \frac{\boldsymbol{n}^2\, d\boldsymbol{x}^2 \cos^2\theta}{n_0^2} = \frac{d\boldsymbol{x}^2}{n_0^2}(n_0^2 - \boldsymbol{n}^2 \cos^2\theta) > 0;$$

θ is the angle between the vectors $\boldsymbol{n}$ and $d\boldsymbol{x}$.

The surfaces σ will hereafter by definition be those whose normal satisfies

$$n_\mu n_\mu = -1, \tag{1.19}$$

and an example of such a family of surfaces is the family of hyperplanes depending on the parameter τ:

$$n_\mu x_\mu - \tau = 0. \tag{1.20}$$

In particular, for an equitemporal plane the vector $\boldsymbol{n}$ has the following components:

$$\boldsymbol{n} = (0, 0, 0, -i). \tag{1.21}$$

We shall define a surface element $d\sigma_\mu$ directed according to the normal

$$d\sigma_\mu = n_\mu d\sigma$$

by its components:

$$(dx_2 dx_3 dx_0,\ dx_1 dx_3 dx_0,\ dx_1 dx_2 dx_0,\ -i dx_1\, dx_2\, dx_3). \tag{1.22}$$

The surface element of an equitemporal plane has therefore the following components:

$$(0, 0, 0, -i d^3\boldsymbol{x} \equiv -i dx_1\, dx_2\, dx_3). \tag{1.23}$$

When a surface element $d\sigma_\mu$ is subject to a displacement δx_μ it generates the volume element

$$\delta\omega = d\sigma_\mu \delta x_\mu \tag{1.24}$$

in such a way that if $d\sigma_\mu$ is a plane surface element

$$\delta\omega = \delta x_0 d^3\boldsymbol{x} = c d^3\boldsymbol{x}\, dt \equiv d^4 x. \tag{1.25}$$

From the relation $d\sigma = -n_\mu d\sigma_\mu$, we infer that for a plane

$$d\sigma = d^3\boldsymbol{x}. \tag{1.26}$$

Note, finally, two other definitions: the "normal derivation" and the "tangential derivation", which are defined as follows:

$$D_{(n)} \equiv -n_\mu \frac{\partial}{\partial x_\mu}, \tag{1.27}$$

$$D_\mu \equiv \frac{\partial}{\partial x_\mu} - n_\mu D_{(n)} = (\delta_{\mu\lambda} + n_\mu n_\lambda)\frac{\partial}{\partial x_\lambda}. \tag{1.28}$$

When the surface on which one differentiates is a plane, these definitions are reduced to the following ones:

$$D_{(n)} = \frac{\partial}{\partial x_0}, \tag{1.29}$$

$$D_j = \frac{\partial}{\partial x_j}, \qquad D_4 = 0. \tag{1.30}$$

Note that the vectors n_μ and D_μ are orthogonal:

$$n_\mu D_\mu = n_\mu \frac{\partial}{\partial x_\mu} + D_{(n)} = 0,$$

so that the vector $\frac{\partial}{\partial x_\mu}$ is decomposed into its tangential and normal components

$$\frac{\partial}{\partial x_\mu} = D_\mu + n_\mu D_{(n)}. \tag{1.31}$$

We may also verify

$$n_\nu D_\mu - n_\mu D_\nu = n_\nu \partial_\mu - n_\mu \partial_\nu. \tag{1.32}$$

3. The natural unit system

Let us begin by recalling a few elementary principles of relativistic and wave mechanics.

Considering a particle of mass m and speed $\boldsymbol{v}$, we have†

$$\left.\begin{aligned} \boldsymbol{p} = m\boldsymbol{v}\gamma = \frac{E}{c^2}\boldsymbol{v} \quad E = mc^2\gamma, \quad \beta = \frac{|\boldsymbol{v}|}{c}, \quad \gamma = \frac{1}{\sqrt{1-\beta^2}}, \\ |\boldsymbol{p}| = \beta\frac{E}{c} = \beta\gamma mc \end{aligned}\right\} \tag{1.33}$$

† It will be noted that the momentum $\boldsymbol{k}$ of a particle of mass 0, and speed c (photon), is to be defined from the formula $\boldsymbol{k} = \frac{E}{c^2}\boldsymbol{v}$ which gives $|\boldsymbol{k}| = \frac{E}{c}$.

and the law of force in special relativity is written

$$\boldsymbol{F} = \frac{d\boldsymbol{p}}{dt}. \tag{1.34}$$

We now obtain the theorem of kinetic energy as a first approximation of relativistic dynamics:

$$E - E_0 = E - mc^2 = \frac{1}{2}\,mv^2\left(1 + \frac{3}{4}\,\beta^2 + \ldots\right) \tag{1.35}$$

and $E_0 = mc^2$ (Einstein formula) is the energy of the particle at rest.

Note also that $c\boldsymbol{p}$, E define a four-vector, since

$$c^2\boldsymbol{p}^2 - E^2 = -m^2c^4. \tag{1.36}$$

Consider on the other hand a plane wave of frequency ν; its angular frequency is $\omega = 2\pi\nu$ and the Planck and de Broglie relations can be written respectively

$$E = h\nu = \hbar\omega, \quad \hbar = \frac{h}{2\pi}; \tag{1.37}$$

$$\lambda = 2\pi\,\frac{\hbar}{|\boldsymbol{p}|} = \frac{2\pi}{|\boldsymbol{k}|} \tag{1.38}$$

where $\boldsymbol{p} = \hbar\boldsymbol{k}$ is the momentum of the particle accompanying the wave. Let V be the propagation speed of the wave; from the relation

$$\lambda = \frac{2\pi V}{\omega}$$

we get

$$|\boldsymbol{p}| = \frac{\hbar\omega}{V} = \frac{E}{V}, \quad |\boldsymbol{k}| = \frac{\omega}{V}. \tag{1.39}$$

The vector $\boldsymbol{k}$ is called the wave propagation vector; since, if $\boldsymbol{n}$ is a unit vector normal to the wave surface (direction of propagation), the wave accompanying the particle of mass m, momentum $\boldsymbol{p}$, and energy E is described by the expression

$$u(\boldsymbol{x}, t) = Ae^{i\omega\left(t - \frac{\boldsymbol{n}.\boldsymbol{x}}{v}\right)} = Ae^{-\frac{i}{\hbar}(\boldsymbol{p}.\boldsymbol{x} - Et)} = Ae^{-i(\boldsymbol{k}.\boldsymbol{x} - \omega t)}. \tag{1.40}$$

Finally, the Compton wavelength of a particle of mass m is

$$\lambda\!\!\!^{-}_{\mathrm{C}} = \frac{\hbar}{mc}. \tag{1.41}$$

This wavelength gives an idea of the extent of the particle in accordance with a well-known train of reasoning based on the uncertainty relation $\Delta x \Delta p \approx \hbar$.

In the C.G.S. system the dimensions of $\hbar$ may be computed from (1.37): ω having the dimensions of the inverse of a time (T^{-1}), the dimensions of $\hbar$ are those of an energy multiplied by a time, namely

$$ML^2T^{-1},$$

and c obviously has LT^{-1} as dimensions.

The quantum field theory formulae are simplified to a marked degree when we adopt the "natural unit system" according to which

$$\hbar = 1, \qquad c = 1.$$

The unit of length remains to be fixed. We should, in fact, have a natural unit of length if we were able to construct a theory based on a certain quantum of length. Attempts to work out such a theory having produced no precise results, we shall retain the centimetre as unit of length.

The Compton formula (1.41) enables us to express the mass in terms of $\hbar$, c and of $\lambda\!\!\!^{-}_c$ which in both systems is measured in centimetres. Writing this formula successively in the C.G.S. and "natural" (n.u.) systems, we find

$$\frac{m_g}{m_{\text{n.u.}}} = \frac{\hbar}{c}, \tag{1.41a}$$

where $\hbar$ and c are approximately equal to 10^{-27} and 3×10^{10}.

Finally, by expressing $\lambda\!\!\!^{-}_c$ in terms of the period $T[\lambda\!\!\!^{-}_c = (2\pi)^{-1}cT]$, we can find the ratio of a time measured in seconds to a time measured in n.u.:

$$T_{\text{n.u.}} = cT_{\text{sec}}. \tag{1.41b}$$

With the help of the Planck and Einstein formulae, the reader will easily prove that the dimensions of an energy, of an angular frequency ω and of a mass are all equal to L^{-1}.

Similarly, we can verify that the natural unit of charge equals $(\hbar c)^{\frac{1}{2}}$ electrostatic C.G.S. units. In order to avoid the factors π in the Maxwell equations, we shall take as the charge of the electron $\sqrt{4\pi}$ times its charge in the non-rationalized n.u. system.

With the above values of $\hbar$ and c, the charge of the electron has the magnitude†

$$(e)_{\text{n.u.}} = \left(\sqrt{\frac{4\pi e^2}{\hbar c}}\right)_{\text{C.G.S.}} = \sqrt{\frac{4\pi}{137}}. \tag{1.42}$$

Let us recall that in the natural unit system, the "classical radius" of

† The system used here is the C.G.S. e.s.u. system and formula (1.42) is derived by calculating the electrostatic energy of the electron in both systems.

the electron ($r_0 = e^2/mc$) and the Compton wave length which is associated with it are respectively

$$r_0 = 2{\cdot}8\times 10^{-13}\ \text{cm}, \quad \lambda\!\!\!^{-}_{C} = \left(\frac{1}{m_e}\right)_{\text{n.u.}} = 3{\cdot}85\times 10^{-11}\ \text{cm}.$$

The mass of the electron is therefore $\left(\frac{1}{\lambda\!\!\!^{-}_{c}}\right)_{\text{n.u.}}$. We shall finally summarize the above indications in the following table:

CONVERSION TABLE

Physical quantities in the n.u. system	Physical quantities in the C.G.S. system
$m_{\text{n.u.}}$	$\frac{c}{\hbar} m$
$t_{\text{n.u.}}$	ct_{sec}
$E_{\text{n.u.}}$	$\frac{1}{\hbar c} E_{\text{ergs}}$
$v_{\text{n.u.}}$	$\frac{1}{c} v_{\text{C.G.S.}}$
$\omega_{\text{n.u.}}$	$\frac{1}{c} \omega_{\text{C.G.S.}}$
$k_{\text{n.u.}}$	$k_{\text{C.G.S.}}$
$p_{\text{n.u.}}$	$\frac{1}{\hbar} p_{\text{C.G.S.}}$
$e_{\text{n.u.}}$	$\sqrt{\frac{4\pi}{\hbar c}}\, e_{\text{C.G.S.}}$
$(\sqrt{k^2+m^2})$ n.u.	$\left(\sqrt{k^2+\frac{c^2}{\hbar^2} m^2}\right)_{\text{C.G.S.}} = \frac{1}{c\hbar} E_{\text{C.G.S.}}$
	$c \approx 3\times 10^{10}$ cm/sec, $\hbar \approx 10^{-27}$ g. cm²/sec, $\alpha = \left(\frac{e^2}{\hbar c}\right)_{\text{C.G.S.}} = \left(\frac{e^2}{4\pi}\right)_{\text{n.u.}} \approx \frac{1}{137}$, $\lambda\!\!\!^{-}_{c} = 3{\cdot}85\times 10^{-11}$ cm.

Note. We sometimes wish to express in the C.G.S. system a transition probability per unit time computed in the n.u. system: according to (8·97*b*) (noting that $W_{\text{II, I}}$ is a number without dimensions), we have

$$(w_{\text{II, I}})_{\text{C.G.S.}} = c(w_{\text{II, I}})_{\text{n.u.}}.$$

On the other hand, in (8·98), we note that H' has the dimensions of an energy and we express the argument $E_{\text{II}}-E_{\text{I}}$ also in C.G.S. units. We then have, in a first approximation, for example:

$$(w_{\text{II, I}})_{\text{C.G.S.}} = \frac{2\pi}{\hbar}(|\langle II\,|\,H'\,|\,I\rangle|^2\,\delta(E_{\text{II}}-E_{\text{I}}))_{\text{C.G.S.}}.$$

The cross-sections having L^2 as dimensions are the same in both systems (1 barn$=10^{-24}$ cm^2).

4. Notation concerning linear spaces†

We shall consider linear spaces (with a finite or infinite number of dimensions), in other words such that if $|u\rangle$ and $|v\rangle$ are two ket vectors (or vectors) of such a space, the vector

$$|w\rangle = \lambda|u\rangle+\mu|v\rangle \qquad (\lambda,\ \mu,\ \text{complex numbers}).$$

is also a vector of this same space.

Note that we shall sometimes write u instead of $|u\rangle$.

We shall consider in a general manner *Hilbert* vector spaces, in other words such that to the space of the ket vectors there corresponds a dual space, that of the bra vectors (or covectors) $\langle u|$, such that the scalar product $\langle u|v\rangle$ of two ket vectors is a number with the following properties.‡

$$\langle u|(|v\rangle+|w\rangle) = \langle u|v\rangle+\langle u|w\rangle, \tag{1.43a}$$

$$\langle u|(\lambda|v\rangle) = \lambda\langle u|v\rangle \qquad (\lambda,\ \text{complex number}), \tag{1.43b}$$

$$\langle u|v\rangle = \langle v|u\rangle^*, \tag{1.43c}$$

$$\langle u|u\rangle \geqslant 0. \tag{1.43d}$$

In particular, the vectors $|u\rangle$ and $|v\rangle$ will be called orthogonal if $\langle u|v\rangle = 0$; moreover, when equality occurs in (1.43d) the vector $|u\rangle$ is said to be null.

† For a detailed study, cf. P. A. M. Dirac, *The Principles of Quantum Mechanics*, Oxford, 1958, p. 14 *et seq.*; D. Kastler, *Introduction à l'Electrodynamique quantique*, chap. I, Dunod, 1960; A. Lichnerowicz, *Algèbre et Analyse linéaires*, Masson, 1947 or *Eléments de Calcul Tensoriel*, Armand Colin, 1950; J. M. Souriau, *Calcul linéaire*, Presses Universitaires de France, 1960.

‡ A bra is therefore a linear form on the space of the kets; the set of the bras is the dual space. The scalar product is thus the product of two kets and can be obtained by making a bra act on a ket.

The sign * denotes the complex conjugate of a complex number (c-number).

Any linear mapping of this space is described by a linear operator Ω which by definition satisfies the following equalities:

$$\Omega\{|u\rangle+|v\rangle\} = \Omega|u\rangle+\Omega|v\rangle, \qquad (1.44a)$$

$$\Omega\{\lambda|u\rangle\} = \lambda\Omega|u\rangle; \qquad (1.44b)$$

since $\Omega|u\rangle$ is a ket vector, we may define the number

$$\langle v|(\Omega|u\rangle) \equiv \langle v|\Omega|u\rangle. \qquad (1.45)$$

The Hermitian conjugate operator $\tilde{\Omega}$ of Ω is defined by the relation

$$\langle u|\Omega|v\rangle = \langle v|\tilde{\Omega}|u\rangle^* \qquad (1.46a)$$

and the bra vector corresponding to the ket vector $\Omega|u\rangle$ is

$$\langle u|\tilde{\Omega}. \qquad (1.46b)$$

The inverse Ω^{-1} of an operator Ω is defined, when it exists, by the relation

$$\Omega\Omega^{-1} = \Omega^{-1}\Omega = I, \qquad (1.47)$$

where I is the identity operator.

An operator Ω is called Hermitian if

$$\Omega = \tilde{\Omega}, \qquad (1.48)$$

and unitary if†

$$\Omega^{-1} = \tilde{\Omega}. \qquad (1.49)$$

When we can find a number ω_n and a vector $|\omega_n\rangle$ such that

$$\Omega|\omega_n\rangle = \omega_n|\omega_n\rangle, \qquad (1.50)$$

we say that ω_n and $|\omega_n\rangle$ are respectively eigenvalue and eigenvector of Ω. It is convenient to denote an eigenvector corresponding to the eigenvalue ω_n by $|\omega_n\rangle$ (supposing that the eigenvalue under consideration is not degenerate).

Concerning Hermitian and unitary operators, the following two theorems may be stated:

THEOREM 1. *Eigenvalues of a Hermitian operator are real; two eigenvectors corresponding to different eigenvalues are orthogonal.*

To prove this, let us first form

$$\langle\omega_n|\Omega|\omega_m\rangle = \omega_m\langle\omega_n|\omega_m\rangle,$$

then using (1.46*a*) and (1.48), let us write the left side of the above equation

$$\langle\omega_n|\Omega|\omega_m\rangle = \langle\omega_m|\Omega|\omega_n\rangle^* = \omega_n^*\langle\omega_n|\omega_m\rangle.$$

† Strictly speaking, one should have $\Omega\tilde{\Omega} = I$ and $\tilde{\Omega}\Omega = I$, see (2.21) and (2.22) of p. 49.

We have therefore, finally,

$$(\omega_m - \omega_n^*)\langle \omega_n | \omega_m \rangle = 0$$

if $n \neq m$: $\langle \omega_n | \omega_m \rangle = 0$, if $n = m$, then $\omega_n = \omega_n^*$ provided we take (1.43*d*) into consideration, hence ω_n is a real number.†

The following theorem may be proved with the same facility:

THEOREM 2. *The eigenvectors of a unitary operator (corresponding to non-degenerate eigenvalues) are orthogonal; its eigenvalues have modulus one.*

Theorem 1 enables us therefore to consider sets of orthogonal vectors $|\omega_m\rangle$:

$$\langle \omega_n | \omega_m \rangle = \delta_{nm} \tag{1.51a}$$

and the matrix corresponding to the Hermitian operator Ω in such a basis is then diagonal:

$$\langle \omega_n | \Omega | \omega_m \rangle = \omega_n \delta_{nm}. \tag{1.51b}$$

We know that when we have a set of commuting (Hermitian) operators‡ $\Omega, \Omega', \Omega'', \ldots$; these operators may be diagonalized simultaneously. It is convenient to write their eigenvectors in the form

$$|\omega, \omega', \omega'', \ldots\rangle,$$

with

$$\left.\begin{aligned} \Omega\,|\omega, \omega', \omega'', \ldots\rangle &= \omega\;\,|\omega, \omega', \omega'', \ldots\rangle, \\ \Omega'|\omega, \omega', \omega'', \ldots\rangle &= \omega'|\omega, \omega', \omega'', \ldots\rangle. \end{aligned}\right\} \tag{1.52}$$

Such a set of operators is called complete, if none of the vectors $|\omega, \omega', \omega'', \ldots\rangle$ is degenerate, that is to say if to each set of eigenvalues there corresponds a uniquely determined vector. Such sets enable us to define the state of a quantum system.

When $\omega_1, \omega_2, \ldots$, are the elements of a discrete set, we say that the spectrum of Ω is a discrete spectrum, otherwise it is continuous. We normalize the eigenvectors of a continuous spectrum by writing

$$\langle \omega | \omega' \rangle = \delta(\omega - \omega'); \tag{1.53}$$

† In the case of degenerate eigenvalues, the reader is referred to the books quoted at the beginning of this section.

‡ More explicitly, let these operators be Ω_N; they will be called commuting if $[\Omega_N, \Omega_{N'}] = 0$, whatever N and N' are. For instance, the components L_1, L_2, L_3 of an angular momentum $\boldsymbol{L}$ each commute with $\boldsymbol{L}^2$; these components taken separately may each be diagonalized with $\boldsymbol{L}^2$. But the operators $\boldsymbol{L}^2, L_1, L_2, L_3$ cannot be diagonalized simultaneously.

$\delta(\omega)$ is the Dirac function (cf. § 6 of this chapter). It should be noted that in the case of a continuous spectrum, an operator is diagonal when its matrix elements are of the form

$$\omega\delta(\omega-\omega'), \tag{1.54}$$

for in order to conserve the commuting property of the diagonal matrices, it is essential to restrict the definition of the diagonal matrices to those matrices whose elements are of this form. For instance, the matrix whose elements are

$$\omega\delta'(\omega-\omega'),$$

where $\delta'(\omega)$ is the derivative of the Dirac function $\delta(\omega)$, is not a diagonal matrix.†

Finally, we shall assume that the eigenvectors of the Hermitian operators we shall be considering, form not only a set of orthonormal vectors but also a complete set, that is to say that any vector of this space is a linear combination of vectors of this set.

If therefore $|v\rangle$ is any vector, and $|\omega_n\rangle$, $|\omega\rangle$ a complete set of orthonormal vectors:

$$|v\rangle = \sum_n |\omega_n\rangle\, c_n + \int |\omega\rangle\, d\omega c(\omega),$$

the discrete summation concerns the discontinuous spectrum, while the integral concerns the continuous spectrum of Ω. Since the basis is orthonormal, the coefficients c_n and $c(\omega)$ take the values:

$$c_n = \langle\omega_n|v\rangle, \qquad c(\omega) = \langle\omega|v\rangle;$$

so that

$$|v\rangle = \sum_n |\omega_n\rangle\langle\omega_n|v\rangle + \int |\omega\rangle\, d\omega\,\langle\omega|v\rangle. \tag{1.55}$$

Any other operator O of this space may be written

$$\begin{aligned} O = &\sum_n\sum_m |\omega_n\rangle\langle\omega_n|O|\omega_m\rangle\langle\omega_m| \\ &+\iint |\omega\rangle\, d\omega\,\langle\omega|O|\omega'\rangle\, d\omega'\,\langle\omega'|, \end{aligned} \tag{1.56a}$$

in particular the identity operator takes the form

$$I = \sum_n |\omega_n\rangle\langle\omega_n| + \int |\omega\rangle\, d\omega\,\langle\omega|. \tag{1.56b}$$

† Cf. P. A. M. Dirac, *op. cit.*, p. 70.

By using the preceding formula, we may form projection operators $\mathcal{P}_N$ (idempotent Hermitian operators) on the subspace spanned by the vectors $|\omega_1\rangle, \ldots, |\omega_N\rangle$ as follows:

$$\mathcal{P}_N = \sum_{n=1}^{N} |\omega_n\rangle\langle\omega_n|. \tag{1.57}$$

Note also that all the summations used may be replaced by Stieltjes integrals:

$$|v\rangle = \int |\omega\rangle \, d\varphi(\omega) \, \langle\omega|v\rangle, \tag{1.58a}$$

where $\varphi(\omega)$ is equal to ω for all the vectors of the continuous spectrum and

$$\varphi(\omega) = \sum_{n=1} \int_{-\infty}^{\omega} \delta(\omega_n - \omega') \, d\omega' \tag{1.58b}$$

for the discrete spectrum of the eigenvalues ω_n.

We may readily generalize these results to the vectors identified by several parameters $|\omega, \omega', \ldots\rangle$ by replacing simple integrations and summations by multiple integrations and summations. We leave it to the reader to write these formulae explicitly.

To obtain an example of a linear space consider the square-integrable functions $u(x)$, where x is a point of space R_N:$x = (x_1, \ldots, x_N)$: these functions form the space L^2 which is a Hilbert space.

Let us consider the complete set of orthonormal vectors $|x\rangle$ such that

$$\langle x'|x\rangle = \delta(x'-x) \equiv \prod_{j=1}^{N} \delta(x'_j - x_j) \tag{1.59}$$

and let us denote the components of a vector u:

$$\langle x|u\rangle = u(x), \tag{1.60}$$

since $u(x)$ is a function of x. The scalar product of two vectors $|u\rangle$ and $|v\rangle$ is then

$$\langle u|v\rangle = \int \langle u|x\rangle\langle x|v\rangle \, dx = \int u^*(x)v(x) \, dx. \tag{1.61}$$

Denoting by x the operator whose matrix element is by definition

$$\langle x''|x|x'\rangle = x'' \, \delta(x''-x'), \tag{1.62a}$$

we see that an eigenvector of this operator is precisely the vector $|x'\rangle$ corresponding to the value x':

$$x|x'\rangle = x'|x'\rangle. \tag{1.62b}$$

Similarly, by writing

$$\left\langle x'' \left| \frac{\partial}{\partial x_n} \right| x' \right\rangle = \frac{\partial \delta(x''-x')}{\partial x''_n}, \tag{1.63}$$

we see that the components of the vector $\frac{\partial}{\partial x_n} |u\rangle$ are in fact $\frac{\partial u(x)}{\partial x_n}$.

The definitions of the Hermitian and unitary operators are well known. Let us prove, for example, that in Euclidean space, the operator $i\nabla$ with components $i\frac{\partial}{\partial x_j}$ is Hermitian:

$$\left\langle u \left| i\frac{\partial}{\partial x_j} \right| v \right\rangle = \int u^*(x) i \frac{\partial v(x)}{\partial x_j} \, d^3x = \int \left(i \frac{\partial u(x)}{\partial x_j} \right)^* v(x) \, d^3x$$
$$= \left\{ \int v^*(x) i \frac{\partial u(x)}{\partial x_j} \, d^3x \right\}^* = \left\langle v \left| i \frac{\partial}{\partial x_j} \right| u \right\rangle^* = \left\langle u \left| \left(i \frac{\partial}{\partial x_j} \right)^\sim \right| v \right\rangle. \tag{1.64}$$

The definition of the elementary solution or Green function of a linear operator $\mathscr{L}$ is an interesting application of the above considerations. We propose to calculate the components $u(x)$ of the vector $|u\rangle$, verifying

$$\mathscr{L}|u\rangle = |b\rangle, \tag{1.65a}$$

where $\mathscr{L}$ is a linear operator. For this purpose, we shall see that it is sufficient to know the elementary solution $|G(\xi)\rangle$ of

$$\tilde{\mathscr{L}}|G(\xi)\rangle = |\xi\rangle. \tag{1.65b}$$

We have, indeed, on the one hand

$$\langle u|\tilde{\mathscr{L}}|G(\xi)\rangle = \langle u|\xi\rangle = u^*(\xi),$$

and on the other

$$\langle u|\tilde{\mathscr{L}}|G(\xi)\rangle = \langle G(\xi)|\mathscr{L}|u\rangle^* = \langle G(\xi)|b\rangle^*,$$

from which it follows that

$$u(\xi) = \langle G(\xi)|b\rangle = \int G(x, \xi)^* b(x) \, dx. \tag{1.66a}$$

Generally $G(x, \xi)^* = G(\xi, x)$, so that we find ourselves with the usual formula

$$u(\xi) = \int G(\xi, x) b(x) \, dx. \tag{1.66b}$$

We shall now consider canonical transformations: by definition such a transformation is generated by a unitary operator U and any vector

$|v\rangle$ is transformed into a vector $|v'\rangle$ according to

$$|v'\rangle = U|v\rangle. \tag{1.67}$$

In order to conserve the matrix elements of the operators in such a transformation, i.e.

$$\langle u|A|v\rangle = \langle u'|UAU^{-1}|v'\rangle = \langle u'|A'|v'\rangle,$$

we must have

$$A' = UAU^{-1} \tag{1.68}$$

and it is easy to prove that the eigenvalues of A' are those of A.†

We also know that the canonical transformed operator of an analytical function of an operator A is obtained by taking the same function, but changing its argument A into A'; if indeed

$$F(A) = \sum a_n A^n \qquad (a_n,\ c\text{-number}),$$

we have

$$UF(A)U^{-1} = \sum a_n(UAU^{-1})^n = F(A'). \tag{1.69}$$

Finally, if U is a unitary operator infinitely close to identity (we shall call it infinitesimal), depending on a parameter ε, U may be written down in the form

$$U = I + i\varepsilon T, \tag{1.70a}$$

where T is Hermitian as may be easily verified. We then have

$$A' = UAU^{-1} = (I+i\varepsilon T)A(I-i\varepsilon T) = A + i\varepsilon[T, A]. \tag{1.70b}$$

When U is not infinitesimal, the Lie formula enables us to express the expansion of A'. Since the operator U is unitary, it may be expressed in the form e^{iT}, where T is a Hermitian operator; consequently,

$$U = \sum_0 \frac{(i)^n}{n!} T^n = \lim_{n=\infty} \left(1 + \frac{i}{n} T\right)^n,$$

from which it follows that

$$A' = UAU^{-1} = A + i[T, A] + \frac{(i)^2}{2!}[T, [T, A]] + \ldots = \sum_{n=0} \frac{(i)^n}{n!}[T, A]^{(n)}, \tag{1.71}$$

† In a more general way, let U be any regular operator. Consider as in (1.67) the transformation $|v'\rangle = U|v\rangle$: it turns the vector $|w\rangle = A|v\rangle$ into

$$|w'\rangle = UA|v\rangle = UAU^{-1}|v'\rangle = A'|v'\rangle.$$

The matrix element $\langle u'|A'|v'\rangle = \langle u|(U^{-1})^{\sim}U^{-1}A|v\rangle$ is not generally equal to $\langle u|A|v\rangle$.

with

$$[T, A]^{(n)} = [T \ldots [T, A] \ldots] \qquad \text{(where } n \text{ is equal to the number of commutators),}$$

$$[T, A]^{(0)} = A.$$

The commutators have well-known properties:

$$[A_1, A_2] = -[A_2, A_1], \tag{1.72a}$$

$$[\alpha_1 A_1 + \alpha_2 A_2, A_3] = \alpha_1 [A_1, A_3] + \alpha_2 [A_2, A_3] \qquad (\alpha_1, \alpha_2, c\text{-numbers}), \tag{1.72b}$$

$$[A_1, [A_2, A_3]] + [A_2, [A_3, A_1]] + [A_3, [A_1, A_2]] = 0; \tag{1.72c}$$

the last relation is known as the Jacobi identity.

It is interesting to define implicitly by formulae (1.72) a particular type of product as follows: let $A_1, A_2, A_3, \ldots,$ be "vectors" of a given vector space; we shall define the product $B = [A_1, A_2]$ as another "vector" of this same space obeying the relations (1.72): these vectors form therefore a kind of algebra (N.B. the product $[\ldots, \ldots]$, is not associative), the so-called Lie algebra.

Such a product may be realized in various ways: A_1, A_2 may be operators and this product is then the commutator of A_1 and A_2: $[A_1, A_2]_Q$.

A_1 and A_2 may be functions $f(p, q)$ and $g(p, q)$ of a set of $2N$ coordinates p_k and q_k and the product may be defined as follows:

$$[f, g]_{\text{cl}} = \sum_k \frac{\partial f}{\partial q_k}\frac{\partial g}{\partial p_k} - \frac{\partial f}{\partial p_k}\frac{\partial g}{\partial q_k},$$

This product is the Poisson bracket which plays a fundamental part in Hamiltonian formalism.

Finally, A_1 and A_2 may be vectors of Euclidean space and it will then be verified that $[A_1, A_2]$ may be defined as the vector product $\boldsymbol{A}_1 \wedge \boldsymbol{A}_2$.

On the other hand, both Poisson brackets and commutators satisfy the following relation:

$$[A_1, A_2 A_3] = [A_1, A_2]A_3 + A_2[A_1, A_3]. \tag{1.72d}$$

Let us add this rule to rules (1.72), which is equivalent, in the abstract language we have adopted, to defining together with the product $[A_1, A_2]$ a new associative product $A_1 \cdot A_2$ (which we shall also write in the form $A_1 A_2$), distributive with respect to the addition and which moreover satisfies (1.72*d*).†

† From (1.72*d*) we infer

$$[A_1 A_2, A_3] = -[A_3, A_1 A_2] = A_1[A_2, A_3] + [A_1, A_3]A_2.$$

The following theorem is basic to quantum mechanics:
If two "vectors" u_k and v_k are such that

$$[u_k, u_{k'}] = [v_k, v_{k'}] = 0, \quad [u_k, v_{k'}] = \alpha\delta_{kk'} \quad (\alpha, \text{c-number}), \tag{1.73}$$

then for any polynomial or analytical function $f(u, v)$ defined by means of the signs + and ., we have

$$[u_k, f(u, v)] = \alpha \frac{\partial f}{\partial v_k}, \tag{1.74a}$$

$$[v_k, f(u, v)] = -\alpha \frac{\partial f}{\partial u_k}. \tag{1.74b}$$

The demonstration is carried out by induction. We verify the formulae for $f = u_j$ and $f = v_j$, then we prove that if they are valid for two functions f_1 and f_2, they remain so for $f = f_1 f_2$ and $f = f_1 + f_2$. These formulae are therefore valid for any polynomial in p and q and also for any analytical function.†

In particular, if we choose for $f(u, v)$:

$$\mathcal{N} = \sum v_k u_k,$$

the preceding formulae give

$$[u_p, \mathcal{N}] = \alpha u_p, \tag{1.75a}$$
$$[v_p, \mathcal{N}] = -\alpha v_p, \tag{1.75b}$$

relations which in quantum mechanics define the number operator together with the creation and annihilation operators.

We shall now consider the following case: let there be a system with N degrees of freedom whose coordinates and their conjugate momenta are: $q_1, q_2, \ldots, q_N; p_1, p_2, \ldots, p_N$ and which is described by the Hamiltonian $H(p, q)$.

(a) By setting $u_k = q_k$, $v_k = p_k$, we have

$$[q_k, q_{k'}]_{\text{cl}} = [p_k, p_{k'}]_{\text{cl}} = 0, \quad [q_k, p_{k'}]_{\text{cl}} = \delta_{kk'} \tag{1.76}$$

and the Hamilton equations are

$$[q_k, H(p, q)]_{\text{cl}} = \frac{\partial H}{\partial p_k} = \frac{dq_k}{dt}, \qquad [p_k, H(p, q)]_{\text{cl}} = -\frac{\partial H}{\partial q_k} = \frac{dp_k}{dt}. \tag{1.77}$$

(b) Let us suppose that the variables q_k and p_k may be formally considered as operators on a given Hilbert space. We may then define beside

† The derivatives $\frac{\partial f}{\partial u_k}$ and $\frac{\partial f}{\partial v_k}$ are well defined once one restricts oneself to forming increments δu_k and δv_k which commute (with respect to the operation denoted by the sign .) with all the u_j and v_j. N.B. Do not forget that the order of the variables u and v in $f(u, v)$ is essential.

the classical Poisson brackets, a commutator denoted by $[\ldots]_Q$. We suppose furthermore that q_k and p_k may be chosen such that this commutator obeys the relations

$$[q_k, q_{k'}]_Q = [p_k, p_{k'}]_Q = 0; \qquad [q_k, p_{k'}]_Q = \alpha\, \delta_{kk'}; \tag{1.78}$$

the above theorem enables us to write†

$$\alpha^{-1}[q_k, H(p, q)]_Q = \frac{\partial H}{\partial p_k}, \qquad \alpha^{-1}[p_k, H(p, q)]_Q = -\frac{\partial H}{\partial q_k}.$$

We thus obtain a second way of writing down the Hamilton equations:

$$\frac{dq_k}{dt} = \alpha^{-1}[q_k, H(p, q)]_Q, \tag{1.79a}$$

$$\frac{dp_k}{dt} = \alpha^{-1}[p_k, H(p, q)]_Q. \tag{1.79b}$$

This transition from one formulation of the Hamilton equations to another constitutes the quantization of the mechanical system under consideration. It is the fundamental operation of quantum mechanics, the principles of which will be recalled in the next section, and its extension to systems with an infinite number of degrees of freedom is the basis of quantum field theory (cf. in particular, Chap. II, § 7).

Note also that the choice of the constant α in (1.78) is not completely arbitrary: if, indeed, one requires this relation to be invariant by Hermitian conjugation $\sim$, it is necessary (since this operation, because of the Hermitian character of p_k and q_k, merely changes the sign of the commutator) that $\alpha^* = -\alpha$, in other words that α should be purely imaginary.‡ It is in fact taken to be equal to $i\hbar$ so that when $\hbar \to 0$, we find ourselves back with a c-number theory (as required by the correspondence principle).

The reader will also have noted the importance of the relations (1.75): as we have said, they enable us to define the number operator in quantum mechanics; indeed, introducing the operators

$$a_k = \frac{1}{\sqrt{-2\alpha^2}}(q_k + \alpha p_k), \tag{1.80a}$$

$$a_k^* = \frac{1}{\sqrt{-2\alpha^2}}(q_k - \alpha p_k) \tag{1.80b}$$

which satisfy

$$[a_k, a_{k'}^*] = \delta_{kk'} \tag{1.80c}$$

† Poisson brackets will hereafter be denoted by $[\ldots]_{\text{cl}}$ and commutators simply by $[\ldots]$.

‡ Note that, in more general cases, when p and q are not Hermitian, but are mutually Hermitian conjugate, α is a real number.

we can define $\mathcal{N}$ as follows

$$\mathcal{N} = \sum a_k^* a_k \tag{1.81}$$

and we note that if $|n\rangle$ is an eigenvector of $\mathcal{N}$ corresponding to the eigenvalue n, formulae (1.75) applied to $|n\rangle$ enable us to write

$$\mathcal{N} a_p |n\rangle = (n-1) a_p |n\rangle, \qquad \mathcal{N} a_p^* |n\rangle = (n+1) a_p^* |n\rangle. \tag{1.82}$$

The Hermitian character of $\mathcal{N}$ implies that it is positive definite; it enables us to state that there exists an eigenvector $|0\rangle$ corresponding to the eigenvalue 0, that its eigenvalues are all the positive integers and that its eigenvectors are obtained by applying a certain number of operators a_p^* to the vector $|0\rangle$.†

To conclude this section, let us recall a few formulae concerning tensor products.

To the vectors $u^{(1)}$, $u^{(2)}$, ..., corresponds the vector $u^{(1)} \otimes u^{(2)} \otimes, \ldots$ of the space which is the direct product of the spaces to which $u^{(1)}, \ldots,$ belong, and whose components are

$$\{u^{(1)} \otimes u^{(2)} \otimes \ldots \otimes u^{(N)}\}_{i_1 \ldots i_N} = u^{(1)}_{i_1} u^{(2)}_{i_2} \ldots u^{(N)}_{i_N}. \tag{1.83}$$

The outer product has by definition the following components:

$$\{u^{(1)} \wedge u^{(2)} \wedge \ldots \wedge u^{(N)}\}_{i_1 \ldots i_N} = \sum_{\text{perm.}} \chi u^{(1)}_{i_1} \ldots u^{(N)}_{i_N}, \tag{1.84}$$

summation being made over all permutations $i_1, \ldots i_N$ and χ denoting the signature of the permutation expressed. It will be verified that we obtain in this way the components of the vector product $\boldsymbol{a} \wedge \boldsymbol{b}$.

We know how the tensor product of two operators is defined; in particular, the tensor product of two matrices A and B is the matrix (whose elements are themselves matrices)‡

$$A \otimes B = \begin{pmatrix} Ab_{11} & Ab_{12} & \ldots \\ Ab_{21} & Ab_{22} & \ldots \\ \ldots & \ldots & \ldots \end{pmatrix}. \tag{1.85}$$

We verify that we also obtain for the operators $\Omega_1, \Omega_2, \ldots$

$$(\Omega_1 \otimes \Omega_2)(\Omega_3 \otimes \Omega_4) = (\Omega_1 \Omega_3) \otimes (\Omega_2 \Omega_4), \tag{1.86a}$$

$$(\Omega_1 \otimes \Omega_2)^{\sim} = \tilde{\Omega}_1 \otimes \tilde{\Omega}_2. \tag{1.86b}$$

† Any operator of the form $\tilde{A}A$ is positive definite (therefore with positive eigenvalues). Indeed, for any vector $|u\rangle$:

$$\langle u | \tilde{A}A | u \rangle = \| A | u \rangle \|^2 > 0.$$

‡ The matrix elements of $A \otimes B$ are therefore:

$$(A \otimes B)_{i_1 i_2, j_1 j_2} = (A b_{i_2 j_2})_{i_1 j_1} = a_{i_1 j_1} b_{i_2 j_2}$$

so that to the vector $u \otimes v$ corresponds the vector $u' \otimes v'$ such that:

$$u' \otimes v' = (A \otimes B)(u \otimes v) = Au \otimes Bv.$$

5. Quantum mechanics

It is a well-known fact that quantum mechanics originally presented itself in two forms: wave mechanics and matrix mechanics. It was the latter form which gave rise to quantum field theory.

We are therefore going to devote this section to a brief survey of matrix mechanics. Reference must be made to the particularly fruitful attempts in recent years to interpret scattering processes in terms of wave functions of wave mechanics: we shall devote the last section of Chap. IX to this matter.†

Let us consider a conservative mechanical system with N degrees of freedom; it is described by a set of N functions of t:

$$q_1(t),\ q_2(t),\ \ldots,\ q_N(t)$$

and it is defined by a Lagrangian

$$L = L(q_1(t), \ldots, q_N(t);\ \dot{q}(t), \ldots, \dot{q}_N(t)).$$

By requiring that the action $\mathcal{A} = \int L\,dt$ be stationary, we obtain the equations of the motion of the system in Lagrangian form:

$$\frac{d}{dt}\frac{\partial L}{\partial \dot{q}_k} - \frac{\partial L}{\partial q_k} = 0 \qquad (k = 1, \ldots, N). \tag{1.87}$$

Let us assume furthermore that by means of the equations

$$p_k = \frac{\partial L}{\partial \dot{q}_k} \qquad (p_k = \text{momentum canonically conjugate to } q_k) \tag{1.88}$$

we are able to express the $\dot{q}_k$'s in terms of the p_j's; we may then define a Hamiltonian

$$H = \sum_1^N p_k \dot{q}_k - L, \tag{1.89}$$

H is expressed solely in terms of the q_k's and p_k's:

$$H \equiv H(p, q).$$

Finally, the Lagrange equations are replaced by Hamilton equations:

$$\dot{q}_k = \frac{\partial H}{\partial p_k}, \qquad \dot{p}_k = -\frac{\partial H}{\partial q_k}. \tag{1.90}$$

† The reader is assumed to be familiar with the concepts and methods of wave mechanics and matrix mechanics. In view of the large number of works on this subject, we shall quote only the classical: P. A. M. Dirac, *The Principles of Quantum Mechanics*, Oxford, Clarendon Press; and A. Messiah, *Mécanique quantique*, Vols. I and II, Dunod, 1960.

In the course of this work we shall not touch on the question of the interpretation of quantum mechanics; we shall remain within the framework of the statistical interpretation.

Since any mechanical quantity G of this system may be expressed in terms of the conjugate variables p, q, we may consider the problem as entirely solved, once the Hamilton equations have been integrated. Taking these equations into account, we also obtain

$$\frac{dG(p, q)}{dt} = \sum_k \left(\frac{\partial G}{\partial q_k} \frac{\partial H}{\partial p_k} - \frac{\partial G}{\partial p_k} \frac{\partial H}{\partial q_k} \right) \equiv [G, H]_{cl}, \tag{1.91}$$

where the derivative of the left-hand side is a total derivative. This equation defines the classical Poisson brackets introduced in the preceding section.

Let us now suppose that the system defined by the Hamiltonian (1.89) has atomic or subatomic dimensions. We know that the above mechanical interpretation is no longer valid: classical mechanics must be replaced by quantum mechanics which postulates the existence of classes of incompatible measurements. The link between these two mechanics, or in other words the link between the macroscopic and microscopic universes, is assured by Bohr's correspondence principle which states that the results of quantum mechanics coincide with those of classical mechanics when we take into consideration the expectation values of the observables.

The remarkable property of the quantum systems is that a physical quantity such as $G(p, q)$ can no longer take one single and well-defined value but can take a set of possible values, each having a well-defined probability. Let us try to state this property more precisely: in quantum mechanics, each physical quantity is replaced by a Hermitian operator or *observable* by means of p and q which cease to be commuting numbers (c-numbers) and are defined by the commutation relations

$$[q_k(t), p_l(t)] \equiv q_k(t)p_l(t) - p_l(t)q_k(t) = i\hbar\, \delta_{kl}, \tag{1.92}$$

relations which, as we saw at the end of the preceding section, enable us to express the Hamilton equations in two ways (cf. (1.77) and (1.79)).†

$G(p, q)$ having been replaced by a Hermitian operator, possesses eigenvalues and eigenvectors (vectors of Hilbert space):

$$G(p, q)\,|\,G'_n\rangle = G'_n\,|\,G'_n\rangle, \tag{1.93}$$

a first fundamental postulate of physical interpretation is to assume that the result of a precise measurement concerning the observable G can only

† The determination of the algebra of the p_k, q_k's is based on the definition of the quantum Poisson brackets: cf. Chap. II, § 6, or P. A. M. Dirac, *op. cit.*, p. 86. It is shown moreover that the algebra generated by such commutation relations is uniquely determined, that is to say such that if p'_l and q'_k satisfy the same commutation relations as p_l, q_k we have $p'_l = Up_lU^{-1}$, $q'_k = Uq_kU^{-1}$, where U is a unitary operator. For a proof, cf. P. A. M. Dirac, *op. cit.*, p. 106.

be one of the eigenvalues of that operator. The state of the system is then by definition described by the corresponding eigenvector.

In a general manner, in order to define a state of the system we are interested in, we need merely consider "a complete set of commuting observables", such that there is one and only one eigenvector corresponding to each set of eigenvalues; a simultaneous precise measurement of the observables of such a system is called a "maximal measurement".† Consequently, let such an eigenvector Ψ be chosen as the state vector of the system; the second postulate of physical interpretation will be as follows: the probability of a precise measurement of G being G'_n is

$$P_n = |\langle \Psi | G'_n \rangle|^2. \tag{1.94}$$

The expectation value of G is then

$$\begin{aligned}\langle G \rangle &= \sum_n |\langle \Psi | G'_n \rangle|^2 G'_n = \sum_n \langle \Psi | G'_n \rangle G'_n \langle G'_n | \Psi \rangle \\ &= \sum_{m,n} \langle \Psi | G'_m \rangle \langle G'_m | G | G'_n \rangle \langle G'_n | \Psi \rangle = \langle \Psi | G | \Psi \rangle.\end{aligned} \tag{1.95}$$

At the beginning of Chap. II we shall see how this framework has to be extended in order to take into account the new circumstances met with in quantum field theory.

In this rapid survey we have adopted the Heisenberg picture which consists in giving the $q_k(t), p_k(t)$ and the state vector Ψ independent of t. Now, it may be noted that the Hamilton equations are written from formulae (1.79):

$$\frac{dq_k}{dt} = \frac{1}{i\hbar}[q_k, H], \quad \frac{dp_k}{dt} = \frac{1}{i\hbar}[p_k, H] \tag{1.96}$$

and may be very readily solved: their solutions, which reduce respectively to q_k and p_k for $t = t_0$ are, as we may easily verify:

$$q_k(t) = e^{\frac{i}{\hbar}(t-t_0)H} q_k e^{-\frac{i}{\hbar}(t-t_0)H}, \tag{1.97a}$$

$$p_k(t) = e^{\frac{i}{\hbar}(t-t_0)H} p_k e^{-\frac{i}{\hbar}(t-t_0)H}. \tag{1.97b}$$

Let us now suppose that we carry out a transformation of the axes of the Hilbert space of the system, a transformation which we define by the unitary operator

$$\mathcal{U}(t-t_0) = e^{-\frac{i}{\hbar}(t-t_0)H}, \tag{1.98}$$

† Cf. P. A. M. Dirac, *op. cit.*, p. 57.

the q_k's and the p_k's then become independent of t, but the state vector Ψ is replaced by the following vector depending on t:

$$\Psi^{(S)}(t) = \mathcal{U}(t-t_0)\Psi, \tag{1.99a}$$

and satisfying the equation

$$i\hbar \frac{\partial \Psi^{(S)}(t)}{\partial t} = H\Psi^{(S)}(t). \tag{1.99b}$$

The above equation is the Schrödinger equation and the new picture is the Schrödinger picture. The fundamental part played by this equation in the formulation of wave mechanics is well known. We shall assume that the reader is familiar with these questions, and confine ourselves to comparing the Schrödinger and Heisenberg pictures in the table below.

Comparative Table of Quantum Pictures

Schrödinger Picture	Heisenberg Picture
System with N degrees of freedom	
$\Psi^{(S)}(t) = \mathcal{U}(t-t_0)\Psi(t_0)$, $\quad i\hbar\dfrac{\partial \mathcal{U}}{\partial t} = H^{(S)}\mathcal{U}$.	To each physical quantity G of the Schrödinger picture corresponds the following transform: $G^{(H)} = \mathcal{U}^{-1}(t-t_0)G\,\mathcal{U}(t-t_0)$.
$H^{(S)}$ Hermitian operator function of the q's and p's. $\mathcal{U}(0) = 1$. $\Psi(t_0)$ eigenvector common to a complete set of observables. Expectation value of the observable $G^{(S)}$:	Expectation value of $G^{(H)}$:
$\langle G\rangle = \langle \Psi^{(S)}(t) \| G^{(S)} \| \Psi^{(S)}(t)\rangle = \langle \Psi(t_0) \| \mathcal{U}^{-1}(t-t_0)G^{(S)} \times \mathcal{U}(t-t_0) \| \Psi(t_0)\rangle$.	$\langle G\rangle = \langle \Psi(t_0) \| G^{(H)} \| \Psi(t_0)\rangle$.
Derivative of $\langle G\rangle$:	Derivative of $\langle G\rangle$:
$\dfrac{d\langle G\rangle}{dt} = \langle \Psi^{(S)}(t) \| \dot{G}^{(S)} \| \Psi^{(S)}(t)\rangle$, with $\dot{G}^{(S)} = -\dfrac{i}{\hbar}[G^{(S)}, H^{(S)}]$.	$\dfrac{d\langle G\rangle}{dt} = \left\langle \Psi(t_0) \middle\| \dfrac{dG^{(H)}}{dt} \middle\| \Psi(t_0)\right\rangle$, $\quad \dfrac{dG^{(H)}}{dt} = -\dfrac{i}{\hbar}[G^{(H)}, H^{(H)}]$.

Commutation relations:

$$p_k^{(S)} = -i\hbar \frac{\partial}{\partial q_k^{(S)}},$$

$$[q_k^{(S)}, p_{k'}^{(S)}] = i\hbar\, \delta_{kk'}.$$

The Hamilton equations are:

$$\dot{q}_k^{(S)} = -\frac{i}{\hbar}[q_k^{(S)}, H^{(S)}] = \frac{\partial H^{(S)}}{\partial p_k^{(S)}},$$

$$\dot{p}_k^{(S)} = -\frac{i}{\hbar}[p_k^{(S)}, H^{(S)}] = -\frac{\partial H^{(S)}}{\partial q_k^{(S)}},$$

with

$$\frac{d\langle q_k\rangle}{dt} = \langle \Psi^{(S)}(t)|\dot{q}_k^{(S)}|\Psi^{(S)}(t)\rangle,$$

$$\frac{d\langle p_k\rangle}{dt} = \langle \Psi^{(S)}(t)|\dot{p}_k^{(S)}|\Psi^{(S)}(t)\rangle.$$

Commutation relations:

$$[q_k^{(H)}(t), p_{k'}^{(H)}(t)] = i\hbar\, \delta_{kk'}.$$

The Hamilton equations are:

$$\frac{dq_k^{(H)}}{dt} = -\frac{i}{\hbar}[q_k^{(H)}, H^{(H)}] = \frac{\partial H^{(H)}}{\partial p_k},$$

$$\frac{dp_k^{(H)}}{dt} = -\frac{i}{\hbar}[p_k^{(H)}, H^{(H)}] = -\frac{\partial H^{(H)}}{\partial q_k}.$$

6. Fourier transformation and Dirac function

The few proofs which we will come across in this section are presented, needless to say, without any claim to mathematical rigour.†

Let $f(x)$ be a square-integrable function of the four-dimensional variable x; we shall denote its Fourier transform by the same letter (wherever there is no risk of confusion):

$$\left.\begin{aligned} f(x) &= (2\pi)^{-4}\int e^{ikx}f(k)\, d^4k, \qquad d^4k = d^3\mathbf{k}\, dk_0, \\ f(k) &= \int e^{-ikx}f(x)\, d^4x. \end{aligned}\right\} \qquad (1.100)$$

It must be noted that $f(x)$ and $f(k)$ are not the same function of x or k.

The Dirac function $\delta(x)$, which is indeed a measure, is naively defined as follows:

$$\delta(x) = (2\pi)^{-4}\int e^{ikx}\, d^4k \qquad (1.101)$$

† Introduced by Dirac (cf. P. A. M. Dirac, *op. cit.*, p. 58), these "functions" did not obtain proper recognition from mathematicians until the studies of L. Schwartz, who defined them with greater precision and extended their application in an appropriate manner. See Appendix I.

and we have the formula

$$\int \delta(x-\xi)f(\xi)d^4\xi = (2\pi)^{-4}\int e^{ik(x-\xi)}f(\xi)\,d^4\xi\,d^4k$$
$$= (2\pi)^{-4}\int d^4k e^{ikx}\int d^4\xi e^{-ik\xi}f(\xi) \quad (1.102)$$
$$= (2\pi)^{-4}\int e^{ikx}f(k)\,d^4k = f(x).$$

In the case of a one-dimensional variable (which is generally the component x_0 of x), we shall define the Fourier transform by

$$\left.\begin{aligned} f(x_0) &= (2\pi)^{-1}\int_{-\infty}^{+\infty} e^{-ik_0x_0}f(k_0)\,dk_0, \\ f(k_0) &= \int_{-\infty}^{+\infty} e^{ik_0x_0}f(x_0)\,dx_0, \end{aligned}\right\} \quad (1.103)$$

in such a way as to agree with (1.100).

The Dirac function with one-dimensional argument has a number of properties, the most important of which are the following:

$$\delta(x) = \delta(-x), \quad (1.104a)$$
$$\delta'(x) = \delta'(-x), \quad (1.104b)$$
$$f(x)\,\delta(x) = 0 \quad \text{if} \quad f(0) = 0, \quad (1.104c)$$
$$x\,\delta'(x) = -\delta(x), \quad (1.104d)$$
$$\delta(ax) = \frac{1}{|a|}\,\delta(x), \quad (1.104e)$$
$$\delta(x^2-a^2) = \frac{1}{2|a|}\,[\delta(x-a)+\delta(x+a)], \quad (1.104f)$$
$$\int \delta(a-x)\,\delta(x-b)\,dx = \delta(a-b), \quad (1.104g)$$
$$f(x)\,\delta(x-a) = f(a)\,\delta(x-a), \quad (1.104h)$$
$$\delta(f(x)) = \frac{\delta(x-a)}{|f'(a)|} = \frac{\delta(x-a)}{|f'(x)|}\left\{\begin{array}{l}\text{if } a \text{ is a simple and real} \\ \text{root of } f(x).\dagger\ddagger\|\end{array}\right. \quad (1.104i)$$

† The derivative $\delta'(x)$ of $\delta(x)$ is defined by an integration by parts:

$$\int f(x)\,\delta'(x)\,dx = -\int \frac{df}{dx}\,\delta(x)\,dx = -\left.\frac{df}{dx}\right]_{x=0}.$$

Formula (1.104*d*) is proved by differentiating $x\delta(x)=0$.

‡ The reader may prove, taking formulae (1.104) into account, that

$$\delta(\boldsymbol{x}) \equiv \delta(x_1)\,\delta(x_2)\,\delta(x_3) = -\frac{4\pi^2}{r}\frac{\partial}{\partial r}\,\delta(r). \quad \text{where} \quad r=|\boldsymbol{x}|.$$

(For footnote ‖ see page 290.)

As an important application of the above formula, let us compute the Fourier transform of the convolution product of two functions $f(x)$ and $g(x)$;

$$\begin{aligned}\int f(x-\xi)g(\xi)\,d^4\xi &= (2\pi)^{-8}\int e^{ik(x-\xi)}f(k)e^{ik'\xi}g(k')\,d^4k\,d^4k'\,d^4\xi\\ &= (2\pi)^{-4}\int e^{ikx}\,\delta(k-k')f(k)g(k')\,d^4k\,d^4k' \qquad (1.105)\\ &= (2\pi)^{-4}\int e^{ikx}f(k)g(k)\,d^4k,\end{aligned}$$

hence the Fourier transform of $\int f(x-\xi)g(\xi)\,d^4\xi$ is $f(k)g(k)$.

We verify, similarly,

$$\begin{aligned}\int f(x)g(x)\,d^4x &= (2\pi)^{-8}\int e^{i(p+q)x}f(p)g(q)\,d^4p\,d^4q\,d^4x\\ &= (2\pi)^{-4}\int f(p)g(-p)\,d^4p. \qquad (1.106)\end{aligned}$$

It is helpful to introduce the following three step functions (x, a one-dimensional variable):

$$\theta_+(x) = \begin{cases}1 & \text{if } x>0,\\ 0 & \text{if } x<0;\end{cases} \qquad (1.107a)$$

$$\theta_-(x) = \begin{cases}0 & \text{if } x>0,\\ 1 & \text{if } x<0;\end{cases} \qquad (1.107b)$$

$$\varepsilon(x) = \left\{\begin{matrix}1 & \text{if } x>0\\ -1 & \text{if } x<0\end{matrix}\right\} = \frac{x}{|x|} = \frac{1}{\pi}\int_{-\infty}^{+\infty}\frac{\sin x\lambda}{\lambda}\,d\lambda; \qquad (1.107c)$$

$$\theta_+(x)+\theta_-(x) = 1; \qquad (1.107d)$$

$$\theta_+(x)-\theta_-(x) = \varepsilon(x). \qquad (1.107e)$$

‖ Let us consider the integral:

$$I = \int \frac{d^3\boldsymbol{k}}{2\omega(\boldsymbol{k})}f(\boldsymbol{k}),$$

where $f(\boldsymbol{k})$ is a given function and $\omega(\boldsymbol{k}) = +\sqrt{(\boldsymbol{k}^2+m^2)}$, m being a "mass". By a straightforward application of (1.104f), I can be transformed into an integral over the whole space–time:

$$I = \int d^4k\,\theta_+(k_0)\,\delta(k^2+m^2)f(\boldsymbol{k}),$$

where $k^2 = \boldsymbol{k}^2 - k_0^2$. The former formula will be used extensively in Chap. XI because of its Lorentz invariance properties: when k is space-like (and then $\theta_+(k_0)$ is not an invariant), $\delta(k^2+m^2)$ is zero. One may also note that $\theta_+(k_0)\,\delta(k^2+m^2)$ is, up to a factor, the Fourier transform of $\Delta^+(x)$ (cf. p. 163).

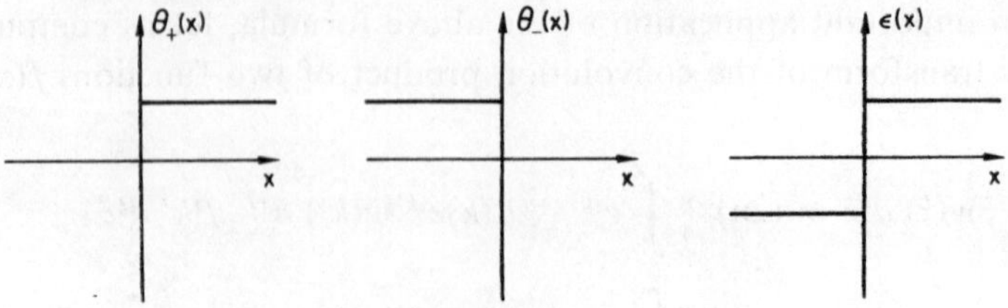

FIG. 1.2.

The step of these functions at the origin enables us to write

$$\frac{d}{dx}\theta_{\pm}(x) = \pm\,\delta(x); \qquad \frac{d}{dx}\varepsilon(x) = 2\,\delta(x). \tag{1.108}$$

With x still denoting a one-dimensional variable, we may obtain a further expression of $\delta(x)$ as follows:

$$\delta(x) = \frac{1}{2\pi}\int_{-\infty}^{+\infty} e^{ikx}\,dk = \frac{1}{\pi}\int_{0}^{\infty}\cos kx\,dk = \frac{1}{\pi}\lim_{K=\infty}\frac{\sin Kx}{x}. \tag{1.109}$$

The Fourier transforms of the step functions $\theta_+(x)$ and $\theta_-(x)$ will enable us to define two new distributions, namely

$$\delta_+(x) \equiv \frac{1}{2\pi}\int_{-\infty}^{+\infty} e^{ikx}\theta_+(k)\,dk = \frac{1}{2\pi}\lim_{\varepsilon=0}\int_{0}^{\infty} e^{ik(x+i\varepsilon)}\,dk \tag{1.110}$$

$$= \frac{1}{2\pi}\lim_{\varepsilon=0}\left.\frac{e^{ik(x+i\varepsilon)}}{i(x+i\varepsilon)}\right]_{k=0}^{+\infty} = -\frac{1}{2i\pi}\lim_{\varepsilon=0}\frac{1}{x+i\varepsilon},$$

$$\delta_-(x) \equiv \frac{1}{2\pi}\int_{-\infty}^{+\infty} e^{ikx}\theta_-(k)\,dk = \frac{1}{2i\pi}\lim_{\varepsilon=0}\frac{1}{x-i\varepsilon}. \tag{1.111}$$

The relation

$$\delta_+(x)+\delta_-(x) = \delta(x) \tag{1.112}$$

which follows readily from the above definitions enables us to obtain a further expression of $\delta(x)$:

$$\delta(x) = \frac{1}{2i\pi}\lim_{\varepsilon=0}\left\{-\frac{1}{x+i\varepsilon}+\frac{1}{x-i\varepsilon}\right\} = \frac{1}{\pi}\lim_{\varepsilon=0}\frac{\varepsilon}{x^2+\varepsilon^2}. \tag{1.113}$$

Let us now evaluate

$$\delta_+(x)-\delta_-(x) = -\frac{1}{i\pi}\lim_{\varepsilon=0}\frac{x}{x^2+\varepsilon^2} = -\frac{1}{i\pi}\frac{P}{x}, \tag{1.114}$$

where P denotes a principal value,

$$P\int_{-\infty}^{+\infty} f(x)\,dx = \lim_{\varepsilon=0}\left\{\int_{-\infty}^{X-\varepsilon}+\int_{X+\varepsilon}^{+\infty} f(x)\,dx\right\} \quad \text{with } X \text{ a singular point of } f(x),$$

and in virtue of the definitions (1.110) and (1.111),

$$\frac{1}{-i\pi}\frac{P}{x} = \frac{1}{2\pi}\int_{-\infty}^{+\infty} e^{ikx}(\theta_+(k)-\theta_-(k))dk = \frac{1}{2\pi}\int_{-\infty}^{+\infty} e^{ikx}\varepsilon(k)\,dk,$$

from which it follows that

$$\frac{P}{x} = -\frac{i}{2}\int_{-\infty}^{+\infty} e^{ikx}\varepsilon(k)\,dk. \tag{1.115}$$

If we add and subtract (1.112) and (1.114) we obtain two new expressions of the distributions $\delta_+(x)$ and $\delta_-(x)$:†

$$\delta_+(x) = -\frac{1}{2i\pi}\frac{P}{x}+\frac{1}{2}\,\delta(x), \tag{1.116a}$$

$$\delta_-(x) = \frac{1}{2i\pi}\frac{P}{x}+\frac{1}{2}\,\delta(x). \tag{1.116b}$$

Formulae (1.110), (1.111) and (1.115) may be reversed:

$$\theta_\pm(k) = \int_{-\infty}^{+\infty} e^{-ikx}\delta_\pm(x)\,dx, \qquad \varepsilon(k) = -\frac{1}{i\pi}\int_{-\infty}^{+\infty} e^{-ikx}\frac{P}{x}\,dx. \tag{1.117}$$

Consider finally

$$e^{-i\alpha x}\,\delta_+(x) = \frac{1}{2\pi}\lim_{\varepsilon=0}\int_0^{\infty} e^{i(k-\alpha)x}e^{-\varepsilon k}\,dk$$

and change the variable:

$$e^{-i\alpha x}\,\delta_+(x) = \frac{1}{2\pi}\lim_{\varepsilon=0} e^{-\varepsilon\alpha}\int_{-\alpha}^{+\infty} e^{iKx}e^{-\varepsilon K}\,dK.$$

Let α now go to $-\infty$; we obtain

$$\lim_{\alpha=-\infty} e^{-i\alpha x}\,\delta_+(x) = 0. \tag{1.118a}$$

If, furthermore, we suppose that for K going to $-\infty$, εK and $\varepsilon\alpha$ both go to zero, we also obtain

$$\lim_{\alpha=+\infty} e^{-i\alpha x}\,\delta_+(x) = \frac{1}{2\pi}\int_{-\infty}^{+\infty} e^{iKx}\,dK = \delta(x). \tag{1.118b}$$

Similarly, we can establish that

$$\lim_{\alpha=\pm\infty} e^{-i\alpha x}\,\delta_-(x) = \begin{cases} 0, \\ \delta(x). \end{cases} \tag{1.119}$$

† Note that we have, for instance,

$$\delta_-(x^2-a^2) = \frac{1}{2a}\,[\delta_-(x-a)-\delta_-(-x-a)].$$

In each particular case it will be necessary to study carefully the uniform convergence of formulae (1.118) and (1.119) and the validity both of the changes of variables and of the interchanges of limits.

We note in concluding that we shall later be led to consider step functions of the four-dimensional variable x. These functions will still be defined by formulae (1.107) whose argument will no longer be x, but x_0. Thus, for instance,

$$\varepsilon(x) = \begin{cases} 1 & \text{if } x_0 > 0, \\ -1 & \text{if } x_0 < 0. \end{cases} \tag{1.120}$$

Formulae (1.108) can be easily extended. We shall obtain, for instance,

$$\frac{\partial}{\partial x_\mu} \varepsilon(x) = -2i\delta(x_0)\,\delta_{\mu 4}. \tag{1.121}$$

7. Basic notions of functional analysis

A functional

$$F = F[y]$$

is a correspondence between a set of numbers and a set of functions; F belongs to the set of numbers and $y(x)$ to the set of functions. In other words, with each curve $y(x)$ (supposing x to be a one-dimensional variable), defined in an interval $a \leqslant x \leqslant b$, the functional $F[y]$ associates a uniquely determined number.

Let us assume that the functions $y(x)$ under consideration belong to a Hilbert space and let $\omega_n(x)$ be a complete orthonormal basis:

$$y(x) = \lim_{N=\infty} \sum_1^N y_n \omega_n(x)$$

and one may write

$$F = \lim_{N=\infty} F\left[\sum_1^N y_n \omega_n(x)\right] \equiv \lim_{N=\infty} f(y_1, \ldots, y_N); \tag{1.122}$$

the right side of (1.122) is a functional, and therefore independent of the argument x, but a function of the N parameters $y_1, y_2, \ldots, y_N$. We may then approach the functional $F[y]$ through a set of functions of several variables $f(y_1, y_2, \ldots, y_N)$, the number of these variables increasing indefinitely.†

† It goes without saying that the arguments set forth make no claim to mathematical rigour; for a more detailed study, the reader is referred to the following works: V. Volterra and J. Pérès, *Théorie générale des fonctionnelles*, Gauthier-Villars, Paris; P. Lévy, *Problèmes concrets d'Analyse fonctionnelle*, Gauthier-Villars, Paris.

The functional derivative of $F[y]$ at the point x which we shall denote as

$$\frac{\delta F[y]}{\delta y(x)}$$

may be defined as follows: consider two infinitely close curves:

$$y(x) \quad \text{and} \quad y(x)+\delta y(x),$$

where $\delta y(x)$ is an arbitrary infinitesimal function of x. Let there be in the neighbourhood of a point ξ the element of area $\delta\omega(\xi)$ (the contour of which is shown in the figure) enclosed between the curves

$$y(x) \quad \text{and} \quad y(x)+\delta\eta(x).$$

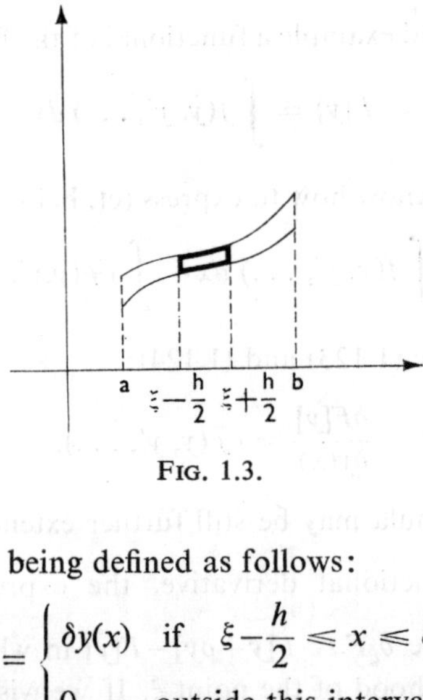

FIG. 1.3.

The function $\delta\eta(x)$ being defined as follows:

$$\delta\eta(x) = \begin{cases} \delta y(x) & \text{if} \quad \xi-\dfrac{h}{2} \leqslant x \leqslant \xi+\dfrac{h}{2}, \\ 0 & \text{outside this interval;} \end{cases}$$

he defini tion of the functional derivative at the point ξ is then

$$\frac{\delta F[y]}{\delta y(\xi)} = \lim_{h,\,|\delta\eta|=0} \frac{F[y+\delta\eta]-F[y]}{\delta\omega(\xi)}. \tag{1.123}$$

The functional derivative at the point ξ is then a functional which is also a function of the variable ξ. It will be noted moreover that if we make appropriate assumptions of continuity

$$\delta\omega(\xi) = \int_{\xi-\frac{h}{2}}^{\xi+\frac{h}{2}} \delta y(\xi')\, d\xi' = h\,\delta\eta(\xi). \tag{1.124}$$

Consider as a first example the following linear functional:

$$K[y|x] = \int K(x, x')y(x')\,dx', \tag{1.125a}$$

its derivative is, as may be easily verified using (1.123):

$$\frac{\delta K[y|x]}{\delta\, y(\xi)} = K(x, \xi). \tag{1.125b}$$

The preceding formula enables us to define

$$\frac{\delta y(x)}{\delta y(\xi)} = \frac{\delta}{\delta y(\xi)} \int \delta(x-x')y(x')\,dx' = \delta(x-\xi). \tag{1.126}$$

Consider as a second example a functional of the form

$$F[y] = \int f(y, y', \ldots)\,dx \tag{1.127a}$$

and suppose that we know how to express (cf. below) $F[y+\delta y]$ in the form

$$F[y+\delta y] = \int f(y, y', \ldots)\,dx + \int \mathcal{F}(y, y', \ldots)\,\delta y(x)\,dx,$$

according to formulae (1.123) and (1.124):

$$\frac{\delta F[y]}{\delta y(x)} = \mathcal{F}(y, y', \ldots). \tag{1.127b}$$

The preceding formula may be still further extended: according to the definition of the functional derivative, the expression $\dfrac{\delta F[y]}{\delta y(\xi)}\,\delta y(\xi)\,d\xi$ represents the increase $\delta_\xi F \equiv F[y+\delta y]-F[y]$ in which $\delta y(\xi)$ is non-zero only in the neighbourhood of the point ξ. If we wish to calculate the increase δF when $y(x)$ becomes $y(x)+\delta y(x)$ with $\delta y(x) \neq 0$ in a finite interval a, b, we obtain by an integration

$$\delta F[y, \delta y] = \int_a^b \frac{\delta F[y]}{\delta y(\xi)}\,\delta y(\xi)\,d\xi. \tag{1.128}$$

By iterating the definition of the first derivative, we can define the functional derivatives of higher order

$$\frac{\delta^n F[y]}{\delta y(\xi_1) \ldots \delta y(\xi_n)},$$

the derivative of nth order is a functional of y which is also a function of the n points $\xi_1, \ldots, \xi_n$.

We can prove that, as for ordinary derivatives, we can invert the order of differentiation, and that we can also obtain a "Taylor formula" for the functionals†

$$F[y(x)+\varphi(x)] \tag{1.129}$$
$$= F[y]+\sum\frac{1}{n!}\int\frac{\delta^n F[y]}{\delta y(\xi_1)\dots\delta y(\xi_n)}\varphi(\xi_1)\dots\varphi(\xi_n)\,d\xi_1\dots d\xi_n.$$

In § 2 of this chapter we introduced a special class of functions: the space-like surfaces σ. We shall be led to define functionals of these functions and, in particular, a vector of a linear space: the state vector $\Psi[\sigma]$. We shall therefore now give a few formulae concerning these functionals.

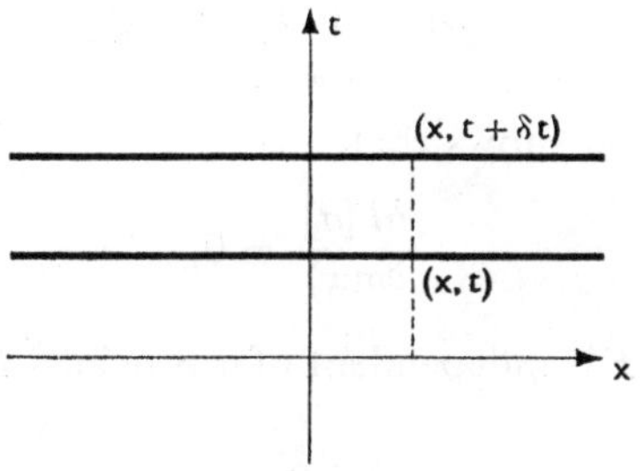

FIG. 1.4.

First of all we shall see what the derivative $\dfrac{\delta\Psi[\sigma]}{\delta\sigma(x)}$ becomes when σ is the equitemporal plane t=constant. According to formula (1.128), we have

$$\delta\Psi[\sigma,\delta t] = \int\frac{\delta\Psi[\sigma]}{\delta\sigma(x)}\,dt\,d^3x$$

from which it follows that

$$\frac{\partial\Psi(t)}{\partial t} = \int\frac{\delta\Psi[\sigma]}{\delta\sigma(x)}\,d^3x. \tag{1.130}$$

Let us now recall Gauss's theorem:

$$\int_{\sigma_{II}}^{\sigma_I}\partial_\mu f(x)\,d^4x = \int_{\sigma_I} f(x)\,d\sigma_\mu - \int_{\sigma_{II}} f(x)\,d\sigma_\mu, \tag{1.131}$$

an interesting application of which is the proof of the following formula:

$$\frac{\partial}{\partial x_\mu}f(x) = \frac{\delta}{\delta\sigma(x)}\int_{\sigma(x)} f(x')\,d\sigma'_\mu, \tag{1.132}$$

where $d\sigma'_\mu$ denotes the element of area at the point x'.

† For these proofs, cf. V. Volterra and J. Pérès, *op. cit.*, p. 85.

Indeed, using the definition of the functional derivative, we may write

$$\frac{\delta}{\delta\sigma(x)}\int_{\sigma(x)} f(x')d\sigma'_\mu = \lim_{\delta\omega=0}\frac{1}{\delta\omega}\left\{\int_\sigma - \int_{\sigma'} f(x')\,d\sigma'_\mu\right\}$$

$$= \lim_{\delta\omega=0}\frac{1}{\delta\omega}\int_{\delta\omega}\frac{\partial f}{\partial x'_\mu}\,d^4x' = \frac{\partial f}{\partial x_\mu};$$

$\delta\omega$ is the element of volume enclosed between the two infinitely close surfaces σ and σ' and the last but one term of the preceding formula results from a direct application of (1.131).

As an application of (1.132) let us consider the functional

$$F[\sigma] = \int_\sigma F_\mu(x')\,d\sigma'_\mu;$$

the preceding theorem shows that

$$\frac{\delta F[\sigma]}{\delta\sigma(x)} = 0, \tag{1.133a}$$

in other words that F is independent of σ provided that F_μ satisfies a continuity equation.

$$\frac{\partial F_\mu(x)}{\partial x_\mu} = 0. \tag{1.133b}$$

The theorem is easily extended when f is a function of the point x and a functional of σ; it then takes the form

$$\frac{\delta}{\delta\sigma(x)}\int_{\sigma(x)} f[\sigma|x']\,d\sigma'_\mu = \frac{\partial}{\partial x_\mu}f[\sigma|x] + \int_{\sigma(x)}\frac{\delta|f[\sigma|x']}{\delta\sigma(x)}\,d\sigma'_\mu. \tag{1.134}$$

Alongside these first applications let us note two other formulae: applying the theorem (1.131) to each of the components of a vector $f_\mu(x)$ and carrying out the summation, we obtain Green's theorem:

$$\int_{\sigma_{II}}^{\sigma_I}\partial_\mu f_\mu(x)\,d^4x = \int_{\sigma_I} f_\mu(x)\,d\sigma_\mu - \int_{\sigma_{II}} f_\mu(x)\,d\sigma_\mu. \tag{1.135}$$

We also note by the way that if $D_{\mu\nu}(\partial)$ is a symmetrical differential operator in μ and ν (or more simply, a numerical coefficient) and if $f_{\mu\nu}(x)$ is a skew-symmetrical tensor, we have

$$D_{\mu\nu}(\partial)f_{\mu\nu}(x) = 0, \tag{1.136}$$

in particular

$$\partial_\mu\partial_\nu f_{\mu\nu}(x) = 0, \tag{1.137}$$

One further notion of functional analysis which is worth introducing, is that of the functional of a functional. Let $u[y(x)|\xi]$ be both a functional of $y(x)$ and a function of the point ξ. If we consider the functional $F[u[y|\xi]]$, we have

$$\delta F = \int \frac{\delta F}{\delta u[y|\xi]} \delta u[y|\xi]\, d\xi = \int \frac{\delta F}{\delta u[y|\xi]} \frac{\delta u[y|\xi]}{\delta y(x)} \delta y(x)\, d\xi\, dx,$$

and consequently,

$$\frac{\delta F}{\delta y(x)} = \int \frac{\delta F}{\delta u[y|\xi]} \frac{\delta u[y|\xi]}{\delta y(x)}\, d\xi. \tag{1.138}$$

The calculus of the variations is an important part of functional analysis. We shall conclude this chapter by expressing a few formulae which will be particularly useful hereafter.

Let $Q_A(x)$ be functions of the space–time point x, the subscript A representing either a single index or a collection of indices. Consider a Lagrangian density

$$\mathcal{L}(Q_A(x), Q_{A,\mu}(x), Q_{A,\mu,\nu}(x), \ldots)$$

function of the Q_A's and of their successive derivatives which we have represented as follows:

$$\frac{\partial Q_A(x)}{\partial x_\mu} = Q_{A,\mu}(x), \qquad \frac{\partial^2 Q_A(x)}{\partial x_\mu \partial x_\nu} = Q_{A,\mu,\nu}(x). \tag{1.139}$$

Consider the action $\mathfrak{A}[Q]$ a functional of the $Q_A(x)$:

$$\mathfrak{A}[Q] = \int \mathcal{L}(Q_A(x), Q_{A,\mu}(x), Q_{A,\mu,\nu}(x), \ldots)\, d^4x, \tag{1.140}$$

where

$$d^4x = d^3\boldsymbol{x}\, dx_0$$

and the integral is extended over the entire space. We have to compute the variation $\delta\mathfrak{A}$ of $\mathfrak{A}$ when the Q_A's undergo infinitesimal variations $\delta Q_A(x)$ which vanish at infinity; we get:

$$\delta\mathfrak{A} = \int \left\{ \frac{\partial\mathcal{L}}{\partial Q_A} \delta Q_A + \frac{\partial\mathcal{L}}{\partial Q_{A,\mu}} \delta Q_{A,\mu} + \frac{\partial\mathcal{L}}{\partial Q_{A,\mu,\nu}} \delta Q_{A,\mu,\nu} + \ldots \right\} d^4x, \tag{1.141}$$

in which we sum over the dummy indices A, μ, ν. By taking into account the boundary conditions satisfied by the δQ_A's we can express the second integral in the form

$$\int \frac{\partial\mathcal{L}}{\partial Q_{A,\mu}} \delta Q_{A,\mu}\, d^4x = \int \left\{ \left(\frac{\partial\mathcal{L}}{\partial Q_{A,\mu}} \delta Q_A \right)_{,\mu} - \left(\frac{\partial\mathcal{L}}{\partial Q_{A,\mu}} \right)_{,\mu} \delta Q_A \right\} d^4x$$
$$= -\int \left(\frac{\partial\mathcal{L}}{\partial Q_{A,\mu}} \right)_{,\mu} \delta Q_A(x)\, d^4x.$$

By treating in the same way the third integral of (1.141), we obtain

$$\int \frac{\partial \mathcal{L}}{\partial Q_{A,\mu,\nu}} \delta Q_{A,\mu,\nu}\, d^4x = \int \left\{ \left(\frac{\partial \mathcal{L}}{\partial Q_{A,\mu,\nu}} \delta Q_A \right)_{,\mu,\nu} - \left(\frac{\partial \mathcal{L}}{\partial Q_{A,\mu,\nu}} \right)_{,\mu,\nu} \delta Q_A \right.$$

$$\left. -2 \left(\frac{\partial \mathcal{L}}{\partial Q_{A,\mu,\nu}} \right)_{,\mu} \partial Q_{A,\nu} \right\} d^4x$$

$$= \int \left(\frac{\partial \mathcal{L}}{\partial Q_{A,\mu,\nu}} \right)_{,\mu,\nu} \delta Q_A(x)\, d^4x,$$

from which it finally follows that

$$\delta \mathfrak{A}[Q] = \int \left[\frac{\partial \mathcal{L}}{\partial Q_A} - \partial_\mu \frac{\partial \mathcal{L}}{\partial Q_{A,\mu}} + \partial_\mu \partial_\nu \frac{\partial \mathcal{L}}{\partial Q_{A,\mu,\nu}} - \dots \right] \delta Q_A(x)\, d^4x \tag{1.142}$$

and

$$\frac{\delta \mathfrak{A}[Q]}{\delta Q_A(x)} = \frac{\partial \mathcal{L}}{\partial Q_A} - \partial_\mu \left\{ \frac{\partial \mathcal{L}}{\partial Q_{A,\mu}} \right\} + \partial_\mu \partial_\nu \left\{ \frac{\partial \mathcal{L}}{\partial Q_{A,\mu,\nu}} \right\} - \dots . \tag{1.143}$$

If $\mathcal{L}$ does not depend on the derivatives of higher order, the equations obtained by writing that the functional derivative $\frac{\delta \mathfrak{A}}{\delta Q_A(x)} = 0$ are the well-known Lagrangian equations

$$\frac{\partial \mathcal{L}}{\partial Q_A} - \frac{\partial}{\partial x_\mu} \frac{\partial \mathcal{L}}{\partial Q_{A,\mu}} = 0. \tag{1.144}$$

We obtain slightly different formulae by considering in the place of the action the total Lagrangian $L[Q_A|t]$:

$$\left. \begin{aligned} L[Q(x)\,|\,t] &= \int \mathcal{L}(Q_A, Q_{A,j}, \dot{Q}_A)\, d^3x, \\ \mathfrak{A}|\dots] &= \int_{t_I}^{t_{II}} L[Q(x)\,|\,t']\, dt', \end{aligned} \right\} \tag{1.145}$$

L is a functional of the Q_A's and a function of the parameter t and we have ascribed to the variable t a specific role in the Lagrangian.

A treatment analogous to the one which led us to (1.143) shows that

$$\frac{\delta L[Q\,|\,t]}{\delta Q_A(x)} = \frac{\partial \mathcal{L}}{\partial Q_A} - \frac{\partial}{\partial x_k} \frac{\partial \mathcal{L}}{\partial Q_{A,k}}, \tag{1.146}$$

from which follows another form of the Lagrangian equations

$$\frac{\delta L}{\delta Q_A(x)} = \frac{\partial}{\partial t} \frac{\partial \mathcal{L}}{\partial \dot{Q}_A(x)}, \tag{1.147}$$

a form which proves useful when one studies the Hamilton–Jacobi system.

One property of the Lagrangian equation should be noted: "One does not change the form of the Lagrangian equations by adding to the Lagrangian density a four-divergence."

We can very easily prove that the actions

$$\mathfrak{A}[Q] = \int \mathcal{L}\, d^4x \qquad \text{et} \qquad \mathfrak{A}'[Q] = \int (\mathcal{L}(Q_A, \ldots) + B_{\mu,\,\mu}(Q_A, \ldots))\, d^4x$$

lead to the same Lagrangian equations by using (1.135) and taking into account that B_μ and its variation δB_μ vanish at infinity.

CHAPTER II

The quantum pictures

WHAT is meant by "quantizing" a field? In the course of this chapter we shall endeavour to answer that question. Broadly speaking, it means associating with the wave aspect of a field, as it appears in the classical representation, a corpuscular aspect arising fundamentally out of a new interpretation of the measurable physical quantities: the physical quantities which are numbers (c-numbers) in the classical theory and, functions of the space–time point x, are replaced in quantum theory by "observables", Hermitian linear operators, functions as before of the point x.

Quantum electrodynamics was the first field to be subjected to the process of quantization: the results obtained were very satisfactory from the outset and have continuously improved over the years, thus confirming the fact that the method of quantum field theory, when combined with certain auxiliary techniques of calculation (renormalization) is undoubtedly valid in this vast area. Most of the examples we shall present in illustration of the general theories will be taken from electrodynamics.

The application of quantum field theory to the nuclear field has been less successful: the results obtained, although qualitatively very interesting, conflict with experimental results as soon as quantitative data are sought. Lastly, the investigation of the properties of the elementary particles forms a still further chapter of quantum field theory: the classification of these particles and the investigation of their interaction are the two fundamental problems which arise. The production by means of accelerators of mesons (and other particles), which until recently could only be observed in cosmic rays, is continually extending the field of this study and multiplying the questions raised by experiments.

Thus a great number of problems still await solution in quantum field theory and although it is impossible to predict whether they are capable of solution within its present framework, this theory has obtained definite and extremely important results.

1. The fields and their quantum description

In our rapid survey of the principles of quantum mechanics (Chap. I, § 5) we confined ourselves to systems with a finite number of degrees of freedom.†

In this section we are going to specify a number of new features which appear when the system under investigation is a field.

As a concrete example, let us therefore consider a field, the electromagnetic field. To describe this field macroscopically, one requires two Euclidean vectors at each space–time point: the electric field $\boldsymbol{E}$ and the magnetic field $\boldsymbol{H}$ which we can infer from the four-vector $A_\mu(x)$. If we assume that the transition from classical to quantum theory occurs in the same manner, whether the system under consideration has a finite or an infinite number of degrees of freedom, we are led to present the problem in the following logical structure:

Suppose a fundamental volume V of Euclidean space is divided into a large number N of cells, the nth being specified by the coordinate $\boldsymbol{X}_n$ of its centre. One may then approximate the distribution of the field at a particular time t by choosing the N four-vectors

$$A_\mu(\boldsymbol{X}_1, t), \ldots, \quad A_\mu(\boldsymbol{X}_N, t) \quad (\mu = 1, \ldots, 4)$$

To bring out the corpuscular properties of this field, we have only to proceed as for systems with a finite number of degrees of freedom and

(a) transform the A_μ's into operators acting on a given Hilbert space in such a way that all observables of the system such as the total current, the energy-momentum of the field and its angular momentum, etc., are Hermitian operators (consequently, with real eigenvalues);

(b) let the number N tend to infinity and therefore consider a system with an infinite number of degrees of freedom, then study the limits of observables such as those referred to in (a).

We must now go into a few details. We must first note that we shall not carry out this programme to the letter: we shall in fact treat quantum field theory directly as a theory of continuous media: fundamental volumes and cells will only be introduced occasionally.

We must also note without going deeply into the difficulties, that by adopting this limit method, we confine ourselves to considering "local" fields which by definition are described by functions of space–time points. It is not, of course, proved that this is the only adequate description of

† The first attempt to quantize the electromagnetic field was carried out by P. A. M. Dirac, *Proc. Roy. Soc.*, *A* **114,** 243 (1928) and P. Jordan and W. Pauli, *Z. Phys.*, **47,** 151 (1928).

the fields actually observed, but since the various attempts to consider fields of another type (but which are not point functions) or "non-local" fields have so far yielded no conclusive results, we shall confine ourselves almost exclusively hereafter to "local fields".†

The "wave functions $A_\mu(x)$" constitute therefore one of the elements of the mathematical framework within which we shall attempt to describe a quantum field. We shall denote them as a rule by the name of "field variables" to recall the fact that they are *linear operators*: $Q_A(x)$, functions of the parameter A which will be added to represent a set of indices (or a single index) each taking numerical values which compose a discrete spectrum (A, for instance, is the index of spin for the spinors) and of the parameter x specifying a point in space–time and therefore taking a continuous series of values. These operators are defined in a particular vector-space $\mathcal{R}$, generally an infinite-dimensional space, which we shall treat as a Hilbert space: this assumption is, however, certainly too limited for the fields we shall be considering. We shall see indeed that the interaction between two different fields introduces elements which do not belong to a Hilbert space and, in order to obtain coherent and definitive results, we shall be obliged to introduce a number of more or less arbitrary and artificial mathematical techniques known as "renormalization", which will give a meaning to matrix elements of the field variables which are otherwise meaningless.

Finally, whenever the quantum system has a classical equivalent, as for instance in the case of the electromagnetic field, its field variables are obtained by replacing the wave functions which describe the classical field by operators: the algebra of these operators is fixed by giving particular commutation (or anticommutation) relations and by assuming the existence of a particular state vector describing the vacuum-state, etc., as will be seen in the course of this chapter. This procedure corresponds to the "quantization of the field under consideration".

To revert to the example of the electromagnetic field, $A_\mu(x)$ will therefore be replaced, after quantization, by a linear operator, a function of the variables μ and x, the former taking the values 1 to 4. Every physical quantity of the system becomes therefore, after quantization, a Hermitian operator (which will also be called an observable), each of whose eigenvalues (all real) may represent the result of a precise measurement of this quantity. It must be noted furthermore that every observable may be either

† A distinction must be made between non-local fields and "non-local interactions" where the interaction of two local fields does not occur at given points, but concerns an elementary domain of space–time.

local or non-local. The various densities (current, energy momentum tensor, etc.) are local quantities, whereas the total charge, energy and momentum of the field, which are functionals of the wave functions, are non-local quantities.

Every eigenvalue of an observable may be obtained by measurement, but the state which corresponds to it (i.e. the corresponding eigenvector) is only uniquely determined if this eigenvalue is not degenerate. We shall consider fields such that we can define "complete sets of commuting observables". Such a set is composed of a set of observables $G_1, G_2, \ldots, G_m$ (generally non-localized) such that (a) they are commuting and are therefore simultaneously measurable; (b) only one eigenvector common to this set corresponds to each of their sets of eigenvalues. In other words, if we suppose that all these operators have been simultaneously diagonalized and if we specify each of the common eigenvectors by their eigenvalues $G'_1, G'_2, \ldots, G'_m$:

$$|G'_1, G'_2, \ldots, G'_m\rangle,$$

these vectors will be uniquely determined, without any degeneracy. We know, moreover, that any observable commuting with the observables of a complete set is a function of these observables.†

A measurement concerning such a set of observables and carried out at the time t_0 defines a "state of the system at t_0" specified by the normalized vector

$$|G'_1, G'_2, \ldots, G'_m; t_0\rangle$$

the expectation value of an arbitrary observable G at this time is the real number

$$\langle t_0; G'_1, \ldots, G'_m | G | G'_m, \ldots, G'_1; t_0\rangle.$$

The time t_0 which may be the time of the initial conditions must be replaced by the surface σ_0 if we replace the equitemporal planes by space-like surfaces σ.

Two observables localized respectively at the points x and y commute with each other when the four-dimensional distance is space-like:

$$(x-y)(x-y) = \sum_{\mu} (x_\mu - y_\mu)(x_\mu - y_\mu) > 0.$$

It is, in fact, impossible in this case for a perturbation produced by the measurement of the observable at x to be propagated as far as y at a speed less than that of light. Such is the form taken by the "causality principle" in quantum field theory.

In our survey of the principles of quantum mechanics (Chap. I, § 4)

† P. A. M. Dirac, *The Principles of Quantum Mechanics*, p. 78.

we chose as our framework the "Heisenberg picture" in which the state vector Ψ is independent of t. This is the case in quantum field theory, when the quantum properties of a field completely described by a classical model are studied. Reverting to our example of the Maxwell field, free photons are completely described by the Maxwell equations and the quantization f these e quations automatically obliges us to adopt the Heisenberg picture. Weocan, however, as will be shown hereafter, describe photons in interaction with matter by the field variables of the free photons, provided we consider the state vector of the field as depending on t; we shall then no longer be in the Heisenberg picture, but in another picture, that of interaction.

It is necessary therefore, in a general formulation of quantum field theory, to assume that the normalized vector Ψ is a functional of σ: $\Psi(\sigma)$, or a function of t: $\Psi(t)$ according as we specify the time t by surfaces σ or by equitemporal surfaces. Ψ is therefore a vector of the space $\mathcal{R}$ which reduces when $t = t_0$ to the vector $| G_1' \ldots, G_m'; t_0)$ (or alternatively which reduces on the initial surface σ_0 to the vector $(G_1', \ldots, G_m'; \sigma_0\rangle)$. The state vector enables us to calculate the expectation values of the observables of the system. We shall suppose, indeed, in the case of the quantum-mechanical systems, that the probability of the result of a precise measurement of the observable G being G' at the time t (or on the surface σ) is

$$|\langle \Psi(t) | G' \rangle|^2$$

and by a calculation similar to that which enabled us to obtain eqn. (1.75), we obtain as the expectation value of G:

$$\langle G \rangle = \langle \Psi(t) | G | \Psi(t) \rangle \tag{2.1a}$$

or

$$\langle G \rangle = \langle \Psi[\sigma] | G | \Psi[\sigma] \rangle. \tag{2.1b}$$

In the course of the above introduction, we have endeavoured to bring out and define in their main lines the basic concepts of quantum field theory. The rest of this chapter will be devoted to defining them more explicitly and in particular to studying the functional equations which govern the evolution of the state vector and of the field variables.

PART 1

THE STATE VECTOR

The foregoing paragraphs may be summarized as follows:

(1) A quantized field is described in a particular quantum picture;

(2) A quantum picture is completed by giving, in a particular linear space $\mathcal{R}$ (and it can be shown that in the case of free fields, this space is restricted to a Hilbert space):

(a) a set of linear operators which we shall refer to as field variables and which, together with their derivatives, will enable us to form the integral density of the observables of the field. These operators $Q_A(x)$ depend on two variables x and A. The variable x is continuous and specifies a space–time point; A is a discontinuous variable; it may symbolize equally well the type of particle (electron, photon, neutron, proton, etc.) or its spin or isobaric spin, etc. We shall frequently drop the subscript A by denoting by $Q(x)$ a column matrix, the elements of which are functions of x, in the space of the index A;

(b) a state vector $\Psi[\sigma]$ (or $\Psi(t)$) by means of which we may calculate the expectation values of these observables with the aid of formulae (2.1).

We intend to investigate the functional relations which define the space–time evolution of the field variables and of the state vector; the investigation of the structure of the observables of the field is deferred to the next chapter.

It will be noted finally, that at least in the course of this general introduction, we shall not attempt to construct the linear space $\mathcal{R}$. In other words, we shall not explicitly give a complete orthonormal basis spanning it. We shall restrict ourselves therefore to a purely formal exposition of quantum field theory, which in order to be complete would require the addition of a theorem formulating the conditions of existence of $\mathcal{R}$. This problem cannot unfortunately be solved in its general aspect, even though in the case of free fields (Chaps. IV and V) we can explicitly construct their definition space $\mathcal{R}$.

2. Evolution of the state vector

Whichever picture is being investigated, the norm of the state vector is independent of t (or of σ). This property combined with a further assumption (the complete determination of the state vector from Cauchy data) suffices to define the form of its evolution equation.

We could not anyway expect, within the limits of the general treatment adopted for this part of the chapter, to define Ψ completely for a given type of field. This determination will be completed in Part 3 of this chapter.

We shall therefore assume that the state vector is completely determined on a space-like surface σ when it is given on an initial space-like surface σ_0 which is arbitrary but has been fixed once for all, namely

$$\Psi[\sigma] = \mathcal{U}(\Psi[\sigma_0]). \tag{2.2a}$$

The determination of the form of $\mathcal{U}$ is easy since, according to the general principle of superposition of the states of a quantum system, $\mathcal{U}$ is required to be a linear operator, a functional of σ and σ_0; we may therefore write

$$\Psi[\sigma] = \mathcal{U}[\sigma, \sigma_0]\Psi[\sigma_0]. \tag{2.2b}$$

On the other hand, by virtue of its very definition, $\Psi[\sigma]$ has a constant norm taken to be equal to I, therefore $\mathcal{U}[\sigma, \sigma_0]$, since it has to be continuous in its two arguments σ and σ_0, is a *unitary operator*. Moreover, the initial condition requires

$$\mathcal{U}[\sigma_0, \sigma_0] = I, \tag{2.3}$$

$\mathcal{U}[\sigma, \sigma_0]$ is the *evolution operator of the state vector.*

What may be said about the choice of the initial state vector $\Psi[\sigma_0]$? None of the assumptions made restricts our choice: $\Psi[\sigma_0]$ could be a particular normalized vector of the linear space $\mathcal{R}$. Nevertheless, Wigner has shown that there are elements of the Hilbert space which might correspond to no physical state. We thus come to formulate superselection rules, as will be seen in Chap. VII (§ 9).

Setting aside for the time being a discussion of these subtle matters, we shall assume that in the system under consideration we can find a set of observables which are constants of motion [cf. formula (2.106)] and which form a complete set of commuting operators. We shall certainly obtain a state vector corresponding to a physically realizable state by choosing as initial vector an eigenvector common to this set. In the case of the Heisenberg picture (cf. end of § 11), this vector will be the state vector of the system at all stages of its evolution.

To state the differential equation obeyed by $\mathcal{U}[\sigma, \sigma_0]$ we shall make use of the following two lemmas:

LEMMA 1. *Let $A(\alpha)$ be an operator, a differentiable function of the parameter α and possessing an inverse which is also a differentiable function of α; we have*

$$\frac{dA^{-1}(\alpha)}{d\alpha} = -A^{-1}(\alpha)\frac{dA(\alpha)}{d\alpha}A^{-1}(\alpha). \tag{2.4}$$

To show this, one need merely express explicitly the following derivative:

$$\frac{d}{d\alpha}\{A(\alpha)A^{-1}(\alpha)\} = 0.$$

LEMMA 2. *If furthermore $A(\alpha)$ is unitary, the expression*

$$i\frac{dA(\alpha)}{d\alpha}A^{-1}(\alpha)$$

is Hermitian.

Taking its Hermitian conjugate

$$\left\{i\frac{dA(\alpha)}{d\alpha}A^{-1}(\alpha)\right\}^{\sim} = -iA(\alpha)\frac{dA^{-1}(\alpha)}{d\alpha} = i\frac{dA(\alpha)}{d\alpha}A^{-1}(\alpha),$$

we verify at once Lemma 2.

Returning to the unitary operator $\mathcal{U}[\sigma, \sigma_0]$, we readily obtain from Lemma 2

$$i\frac{\delta\mathcal{U}[\sigma, \sigma_0]}{\delta\sigma(x)}\mathcal{U}^{-1}[\sigma, \sigma_0] = \Omega[\sigma\,|\,x], \tag{2.5}$$

where the operator Ω is a Hermitian operator, a functional of σ and a function of x. $\mathcal{U}[\sigma, \sigma_0]$ is thus a solution of†

$$i\frac{\delta\mathcal{U}[\sigma, \sigma_0]}{\delta\sigma(x)} = \Omega[\sigma\,|\,x]\mathcal{U}[\sigma, \sigma_0], \tag{2.6a}$$

$$\mathcal{U}[\sigma_0, \sigma_0] = I. \tag{2.6b}$$

Let us now consider the operator $\mathcal{U}[\sigma, \sigma']$ which enables us to go from a state defined on an arbitrary space-like surface σ' to another state on the surface σ:

$$\Psi[\sigma] = \mathcal{U}[\sigma, \sigma']\Psi[\sigma']. \tag{2.7}$$

We clearly have

$$\Psi[\sigma] = \mathcal{U}[\sigma, \sigma_0]\Psi[\sigma_0], \quad \Psi[\sigma'] = \mathcal{U}[\sigma', \sigma_0]\Psi[\sigma_0],$$

such that

$$\Psi[\sigma] = \mathcal{U}[\sigma, \sigma_0]\mathcal{U}^{-1}[\sigma', \sigma_0]\Psi[\sigma'].$$

By definition, we can then set

$$\mathcal{U}[\sigma, \sigma'] \equiv \mathcal{U}[\sigma, \sigma_0]\mathcal{U}^{-1}[\sigma', \sigma_0]. \tag{2.8}$$

It is then easy to see that $\mathcal{U}[\sigma, \sigma']$ possesses the following properties:

(a) $\mathcal{U}[\sigma', \sigma'] = I$;

(b) $\mathcal{U}[\sigma, \sigma'] = \mathcal{U}^{-1}[\sigma', \sigma]$;

(c) $\mathcal{U}[\sigma, \sigma_n]\mathcal{U}[\sigma_n, \sigma_{n-1}] \ldots \mathcal{U}[\sigma_1, \sigma'] = \mathcal{U}[\sigma, \sigma']$;

(d) $\Psi = \mathcal{U}[\sigma, \sigma']\Psi[\sigma']$ is independent of the choice of σ' and depends solely on σ; we therefore have

$$\Psi \equiv \Psi[\sigma];$$

(e) $\mathcal{U}[\sigma, \sigma']$ satisfies

$$i\frac{\delta\mathcal{U}[\sigma, \sigma']}{\delta\sigma(x)} = \Omega[\sigma\,|\,x]\mathcal{U}[\sigma, \sigma'], \tag{2.9a}$$

$$\mathcal{U}[\sigma', \sigma'] = I, \tag{2.9b}$$

the operator $\Omega[\sigma|x]$ being the same as in formula (2.6*a*).

† σ_0 being a surface chosen once for all, we do not write explicitly the dependence of Ω on σ_0.

The same operator Ω obeys a so-called "integrability" condition:

$$\frac{\delta^2 \mathcal{U}[\sigma, \sigma_0]}{\delta\sigma(x')\delta\sigma(x)} = \frac{\delta^2 \mathcal{U}[\sigma, \sigma_0]}{\delta\sigma(x)\delta\sigma(x')}$$

which we can also express in the following form:

$$\frac{\delta\Omega[\sigma \mid x]}{\delta\sigma(x')} - \frac{\delta\Omega[\sigma \mid x']}{\delta\sigma(x)} = i[\Omega[\sigma \mid x], \Omega[\sigma \mid x']]. \tag{2.10}$$

We can easily prove (2.10) by functionally differentiating (2.9*a*), written for the point x, with respect to $\sigma(x')$, then differentiating the same formula (2.9*a*), written for the point x', with respect to $\sigma(x)$ and finally subtracting the results thus obtained.

Note that the adjoint of (2.9*a*) is:

$$-i \frac{\delta \mathcal{U}[\sigma, \sigma']}{\delta\sigma'(x)} = \mathcal{U}[\sigma, \sigma']\Omega[\sigma' \mid x]. \tag{2.11}$$

To pass from (2.9*a*) to an equation bringing out clearly the time derivation, we need only replace σ by the equitemporal surface t (the functionals of σ becoming functions of t), then according to formula (1.30), integrate (2.9*a*) over the entire Euclidean space after multiplying both sides by $d^3\boldsymbol{x}$, by setting

$$\Omega(t) = \int_V \Omega[\sigma \mid x]\, d^3\boldsymbol{x} \quad (\sigma \text{ denoting the surface } t = \text{constant}), \tag{2.12}$$

we thus find the equation obeyed by $\mathcal{U}$:

$$i \frac{\partial \mathcal{U}(t, t_0)}{\partial t} = \Omega(t)\mathcal{U}(t, t_0), \qquad \mathcal{U}(t_0, t_0) = 1 \tag{2.13a}$$

and the equation of the state vector

$$i \frac{\partial \Psi(t)}{\partial t} = \Omega(t)\Psi(t). \tag{2.13b}$$

It is easy to express in condensed form the solution of (2.13*b*), when Ω is independent of t. We set

$$\Psi(t) = \Phi e^{-iEt}, \tag{2.14}$$

Φ and E are respectively a normalized eigenvector and the corresponding eigenvalue of the equation

$$\Omega\Phi = E\Phi. \tag{2.15}$$

By introducing a density $\varrho(E)$ of the spectrum of the eigenvalues E, $\Psi(t)$ can be expressed with the aid of the Stieltjes integral as follows:

$$\Psi(t) = \int C(E)\Phi(E)e^{-iE(t-t_0)}\, d\varrho(E) \tag{2.16}$$

and the function $C(E)$ must be determined such that for $t=t_0$, $\Psi(t)$ reduces to $\Psi(t_0)$;

$$\Psi(t_0) = \int C(E)\, \Phi(E)\, d\varrho(E). \tag{2.17}$$

When in particular the spectrum of Ω is discrete, $C(E)$ takes a series of discrete values $C(E_n)$

$$C(E)_n = \langle \Psi(t_0) \Phi(E_n) \rangle \tag{2.18a}$$

and

$$\varrho(E) = \sum_n \delta(E-E_n). \tag{2.18b}$$

Note. It is interesting to investigate the unitarity of the operator $\mathcal{U}(t, t_0)$. Consider the equation obeyed by $\tilde{\mathcal{U}}(t, t_0)$:

$$-i \frac{\partial \tilde{\mathcal{U}}(t, t_0)}{\partial t} = \tilde{\mathcal{U}}(t, t_0)\Omega. \tag{2.19}$$

Multiplying (2.13*a*) by $\tilde{\mathcal{U}}$ on the left and (2.19) by $\mathcal{U}$ on the right and subtracting, we obtain

$$\frac{\partial}{\partial t} \{\tilde{\mathcal{U}}(t, t_0)\mathcal{U}(t, t_0)\} = 0. \tag{2.20}$$

Let $\Psi(t_0)$ be a vector of the space $\mathcal{R}$; we set†

$$\begin{aligned}\frac{\partial}{\partial t} \| \mathcal{U}(t, t_0)\Psi(t_0) \|^2 &= \frac{\partial}{\partial t} \langle \Psi(t_0) | \tilde{\mathcal{U}}(t, t_0)\mathcal{U}(t, t_0) | \Psi(t_0) \rangle \\ &= \langle \Psi(t_0) | \frac{\partial}{\partial t} \{\tilde{\mathcal{U}}(t, t_0)\mathcal{U}(t, t_0)\} | \Psi(t_0) \rangle, \\ &= 0,\end{aligned}$$

from which follows the well-known theorem:

If $\mathcal{U}(t, t_0)$ is a solution of the evolution equation of the state vector, the norm of the vector $\Psi(t) = \mathcal{U}(t, t_0)\Psi(t_0)$ is conserved in the course of its evolution in time.

The norms of $\Psi(t)$ and $\Psi(t_0)$ are equal when the initial condition $\mathcal{U}(t_0,t_0)=I$ is satisfied. Taking this condition into account, and using (2.20), we verify that we have:

$$\tilde{\mathcal{U}}(t, t_0)\mathcal{U}(t, t_0) = 1 \tag{2.21}$$

but not the relation

$$\mathcal{U}(t, t_0)\tilde{\mathcal{U}}(t, t_0) = 1. \tag{2.22}$$

† $\| \Phi \|$ denotes the norm of the vector Φ: $\| \Phi \| = +\sqrt{\langle \Phi, \Phi \rangle}$.

We can, moreover, easily verify that

$$\frac{\partial}{\partial t}\{\mathcal{U}(t, t_0)\tilde{\mathcal{U}}(t, t_0)\} = i[\mathcal{U}(t, t_0)\tilde{\mathcal{U}}(t, t_0), \Omega]. \tag{2.23}$$

The equation of the state vector allows us therefore to infer (2.21) alone: consequently, $\mathcal{U}$ is only unitary in the domain in which $\mathcal{U}$ and $\tilde{\mathcal{U}}$ are both regular. The full significance of this remark will emerge when we come to discuss the unitarity of the operators $\mathcal{S}$ in a problem involving bound states (Chap. VIII, Part 2).

PART 2
FIELD VARIABLES

Let us now consider the field variables $Q_A(x)$; we propose to discuss their space–time evolution and we shall divide this investigation into two parts:

In A, we shall introduce the class of quantum systems we are going to consider by defining "canonical systems".

In B, we shall prove the existence of systems obeying the conditions stated in A, by showing that this is in fact the case for any quantized field with a "classical equivalent", that is to say for any field whose field equations may be inferred from an action principle and written in Hamiltonian form.

It is extremely important to note, however (and we repeat here what we said in the introduction to this chapter), that the existence of such systems is purely *formal*, in that the investigation of interacting fields (at least as regards such forms of interaction as have been so far discovered) leads us to consider elements which do not belong to Hilbert space (hence the need to renormalize), and that we have so far met with as little success in specifying the space $\mathcal{R}$ which would sufficiently extend the Hilbert space to make it possible to include these elements, as in formulating interactions between fields describing known particles which would introduce only such expressions as are meaningful in Hilbert space.†

† A simple example of the type of difficulty raised by the definition of the space $\mathcal{R}$ was given by L. Van Hove, *Physica*, **18**, 145 (1952).

A. DEFINITION AND PROPERTIES OF CANONICAL SYSTEMS

3. Evolution of field variables

The systems we shall be considering are the "canonical systems" which by definition obey the following properties:

(1) For any quantized field or system of quantized fields in interaction, we can determine a "canonical set" including a number of components (specified by their index) of the field variables, in terms of which we can express all the others. The elements of this set are the dynamically independent "components" of the field variables.†

(2) Let $Q_a(x)$ be any one of these components. We shall complete the definition of canonical systems by assuming that at two points x and x' of space–time, we must have

$$Q_a(x) = V(x, x')Q_a(x')V^{-1}(x, x'), \tag{2.24}$$

where $V(x, x')$, an operator of the space $\mathcal{R}$ depending on the two parameters x and x', is a unitary operator which we shall call "the evolution operator of the field variables".

Within the limits we have set ourselves, we need not justify the choice of such a definition: it is substantiated by all its consequences, particularly by eqn. (2.24) common to all quantum formulation.

We can perhaps clarify it by stressing the fact that wave equations are generally invariant under the space–time translation group. The operator $V(x, x')$ is then a function of $x-x'$ [cf. (2.32)] and (2.24) can be regarded as defining a unitary representation of this group on the space $\mathcal{R}$, a representation induced by the field variables $Q(x)$.‡ We shall see moreover in § 4 that the microcausality (or local commutativity) principle is satisfied

† We are obliged to introduce the canonical set since frequently all the Q_A's are not independent: this is the case when, alongside the wave equations, we are led to introduce auxiliary conditions (for instance, the Lorentz condition in Electrodynamics). It follows that a canonical transformation of the independent components will not induce a canonical transformation of all the Q_A's (some of these field variables are expressed by means of the derivatives of the dynamically independent components).

‡ Denote by ξ the coordinates of the space–time points and consider the translation $\xi \to \xi' = \xi + \varepsilon$; it transforms $Q(\xi)$ into $Q'(\xi') = Q(\xi)$ and leads us to define a unitary operator $\mathcal{V}(\varepsilon)$ on $\mathcal{R}$ such that

$$Q'(\xi') = \mathcal{V}^{-1}(\varepsilon)Q(\xi')\mathcal{V}(\varepsilon)$$

or

$$Q(\xi') = \mathcal{V}(\varepsilon)Q(\xi)\mathcal{V}^{-1}(\varepsilon).$$

To return to the variables x: $\xi = x'$, $\xi' = x$: set

$$Q(x) = \mathcal{V}(x-x')Q(x')\mathcal{V}^{-1}(x-x').$$

from which it follows that the operators $\mathcal{V}$ and V are identical.

when we have a unitary representation of the group of space translations: V is then of the form $V(\boldsymbol{x}-\boldsymbol{x}'; t, t')$.

Thus to any canonical quantum system there corresponds a canonical picture determined by giving simultaneously the operators $\mathcal{U}$ and V: this notion of a canonical picture will be examined in detail in § 3 of this chapter. The operator $V(x, x')$ (like $\mathcal{U}[\sigma, \sigma']$) has the following properties:

$$V(x', x') = 1, \tag{2.25a}$$

$$V(x, x') = V^{-1}(x', x), \tag{2.25b}$$

$$V(x, x_n)V(x_n, x_{n-1}) \dots V(x_1, x') = V(x, x'). \tag{2.25c}$$

It is easy to find the differential equation of $V(x, x')$: by differentiating with respect to x''_α the relation $V(x, x'')V(x'', x') = V(x, x')$; making use of the relations (2.25), we get

$$i\,\frac{\partial V(x'', x)}{\partial x''_\alpha}\,V^{-1}(x'', x) = i\,\frac{\partial V(x'', x')}{\partial x''_\alpha}\,V^{-1}(x'', x'); \tag{2.26}$$

if therefore we set

$$\mathcal{D}_j(x) = i\,\frac{\partial V(x, x')}{\partial x_j}\,V^{-1}(x, x'); \quad \mathcal{D}_0(x) = -i\,\frac{\partial V(x, x')}{\partial x_0}\,V^{-1}(x, x'), \tag{2.27}$$

the relation (2.26) shows that, as we have just written, $\mathcal{D}_j$ and $\mathcal{D}_0$ depend on x alone and not on x' and are, in accordance with Lemma 2 on p. 46, Hermitian operators.

By defining the operator $\mathcal{D}_\mu$ in $\mathcal{R}$ the four components of which are

$$\mathcal{D}_j(x): \quad \mathcal{D}_j(x), \qquad \mathcal{D}_4 = i\mathcal{D}_0, \tag{2.28}$$

we see that $V(x, x')$ satisfies the equation

$$i\,\frac{\partial V(x, x')}{\partial x_\mu} = \mathcal{D}_\mu(x)V(x, x'), \qquad V(x', x') = 1. \tag{2.29}$$

We shall show furthermore that the operators $\mathcal{D}_\mu$ and V obey the following theorems:

THEOREM 1. *$\mathcal{D}_\mu(x)$ satisfies the integrability condition*

$$\partial_\nu\mathcal{D}_\mu(x) - \partial_\mu\mathcal{D}_\nu(x) = i[\mathcal{D}_\mu(x), \mathcal{D}_\nu(x)]. \tag{2.30}$$

To prove this we have only to write (2.29) for $\mathcal{D}_\mu$ and $\mathcal{D}_\nu$ and note that the derivatives $\partial_\mu\partial_\nu V(x, x')$ and $\partial_\nu\partial_\mu V(x, x')$ are equal.†

† We shall see in Chap. III that $\mathcal{D}_j(x)$ and $-i\mathcal{D}_4$ represent the total momentum and the energy of the system under consideration. It follows that for an isolated system ($\mathcal{D}_\mu$ independent of x) we can simultaneously measure all the components of the energy–momentum vector.

THEOREM 2. *$\mathcal{P}_\mu$ is the space–time translation operator of the canonical field variables; in other words, the field equations can be written in the form of canonical equations (2.31).*

Proof: Differentiate (2.24):

$$\frac{\partial Q}{\partial x_\mu} = \frac{\partial V(x, x')}{\partial x_\mu} Q(x')V^{-1}(x, x') + V(x, x')Q(x') \frac{\partial V^{-1}(x, x')}{\partial x_\mu}$$

$$= -i\mathcal{P}_\mu(x)V(x, x')Q(x')V^{-1}(x, x') + iV(x, x')Q(x')V^{-1}(x, x')\mathcal{P}_\mu(x),$$

consequently

$$i\frac{\partial Q}{\partial x_\mu} = [\mathcal{P}_\mu(x), Q(x)]. \tag{2.31}$$

Note that from this relation, it is easy to verify that $Q(x)$ as defined by (2.24) is independent of x'.

THEOREM 3. *In order that (2.24) and consequently (2.31) be invariant under the translation group, it is necessary and sufficient that $\mathcal{P}_\mu$ be independent of x.*

Proof: From eqn. (2.29) we see that in order that $V(x, x')$ be a function of $x-x'$, it is necessary and sufficient that $\mathcal{P}_\mu$ be independent of x. The solution of (2.29) is in that case readily expressed as:

$$V(x-x') = e^{-i(x_\mu - x'_\mu)\mathcal{P}_\mu} \tag{2.32a}$$

and

$$V(\varepsilon) = 1 - i\varepsilon_\mu \mathcal{P}_\mu, \tag{2.32b}$$

where $\varepsilon = x - x'$.

THEOREM 4. *(2.31) is the canonical form of field equations.*

Proof: We shall show that this equation is covariant under any canonical transformation of $Q(x)$:

$$Q'(x) = S(x)Q(x)S^{-1}(x), \tag{2.33}$$

where $S(x)$ is unitary and consequently satisfies an equation of the form (2.29):

$$i\frac{\partial S}{\partial x_\mu} = \Sigma_\mu S. \tag{2.34}$$

Indeed, we differentiate $Q'(x)$:

$$\frac{\partial Q'(x)}{\partial x_\mu} = \frac{\partial}{\partial x_\mu}\{S(x)Q(x)S^{-1}(x)\} = -i[\Sigma_\mu, Q'(x)] + S(x)\frac{\partial Q(x)}{\partial x_\mu}S^{-1}(x);$$

the last term may be written

$$-iS(x)[\mathcal{P}_\mu(x), Q(x)]S^{-1}(x) = -i[S(x)\mathcal{P}_\mu(x)S^{-1}(x), Q(x)],$$

such that finally we have

$$i \frac{\partial Q'(x)}{\partial x_\mu} = [\Sigma_\mu + S(x)\mathcal{P}_\mu(x)S^{-1}(x), Q'(x)]. \qquad (2.35a)$$

The relation (2.31) is therefore seen to be covariant under any canonical transformation.

Conversely, suppose that the operator $\mathcal{P}_\mu$ is the sum of two operators $\mathcal{P}_\mu^{(1)}$ and $\mathcal{P}_\mu^{(2)}$, such that (2.31) may be written

$$i \frac{\partial Q}{\partial x_\mu} = [\mathcal{P}_\mu^{(1)} + \mathcal{P}_\mu^{(2)}, Q(x)],$$

the canonical transformation

$$Q'(x) = S(x)Q(x)S^{-1}(x),$$

where

$$-i \frac{\partial S}{\partial x_\mu} = S\mathcal{P}_\mu^{(2)}S^{-1}S$$

gives

$$i \frac{\partial Q'}{\partial x_\mu} = [S\mathcal{P}_\mu^{(1)}S^{-1}, Q'(x)], \qquad (2.35b)$$

a formula we shall make use of in Part 3 of this chapter.

4. Algebraic and topological considerations in quantum field theory

We have noted more than once that the physical interpretation of quantum field theory requires the introduction of a linear space $\mathcal{R}$ on which the field operators $Q_A(x)$ act and to which the state vector $\Psi(t)$ belongs. If we are interested in specifying the logical structure of this theory, it is convenient to distinguish between the contributions of metric geometry and those of linear algebra.

On the one hand, the necessity of imposing a norm upon the state vector obliges us to define it as a vector of a metric linear space which is generally treated as a Hilbert space.

On the other hand, the mathematical definition of field variables is a more delicate matter: they can in fact be simply defined to start with as elements of a particular algebra before identifying them with operators operating on $\mathcal{R}$†. We can easily prove this by going back over all the definitions and theorems in this section and noting that they can be readily modified in this way, the linear operator $V(x, x')$ of (2.24) being simply

† The albegra thus obtained is an A* algebra, since alongside the operations of multiplication, multiplication by a real or complex scalar, and addition, a further operation denoted by $\sim$ is formally introduced, which possesses, by definition, all the properties of the Hermitian conjugation operation. Cf., for instance, I. E. Segal, *Mathematical Problems of Relativistic Physics*, 1963.

defined as a regular operator. We thus have, at each point x, a particular sub-algebra generated by the operator $Q_A(x)$, where A belongs to a particular discrete set (which for the scalar field is reduced to a single element), a sub-algebra the elements of which are the polynomials in $Q_A(x), Q_B(x), \ldots$. The relation (2.24) specifies the structure of the algebra by stating that the sub-algebras relative to two arbitrary points x and x' are isomorphic (equivalent).

We then have to investigate as thoroughly as possible the properties of this algebra; we shall not do so here, but simply note that it is possible to extend the principles considered in Chap. I (§ 4, pp. 20 *et seq.*). We also note that:

(a) the condition of local commutativity (microcausality) which we shall be investigating in the next section;

(b) the assumption that the field variables between two surfaces σ and σ' form a complete ring, namely that any element which commutes with them is a multiple of the identity operator (cf. Part 4 of this chapter, pp. 80 *et seq.*);

(c) the postulate asserting that the field equations are local (cf. Part 4 of this chapter, pp. 88 *et seq.*)

are so many conditions fixing this structure precisely.

The next step consists in attempting to obtain a representation of this algebra such that its elements act upon vectors of the space $\mathcal{R}$. This is obviously the step which involves the greater number of physical consequences: the space $\mathcal{R}$ treated as a Hilbert space is, in fact, built up from eigenvectors of observables which are Hermitian operators formed by means of field variables; it is this latter point which raises the greatest number of mathematical difficulties, which can only be solved in the simplest cases, those of free fields.

By considering the space $\mathcal{R}$ as a Hilbert space, we can then make the operator $V(x, x')$ of (2.24) a unitary operator. This operator enables us by means of eqs. (2.27) to define four other Hermitian operators $\mathcal{P}_j(x)$, $\mathcal{P}_0(x)$ with which we try to associate physical quantities.

Such a correspondence is easy to state when a classical equivalent of the field under investigation exists (cf. p. 57 *et seq.*). We can see in that case that by identifying $\mathcal{P}_\mu$ with the total momentum-energy of the field it is necessary, in order that eqns. (2.31) should coincide with the differential equations of the field, to state certain commutation relations between the field variable and other operators called conjugate variables. We shall study this point in Part B (p. 57) and Part 4 of this chapter (p. 79).

One final problem, too subtle for treatment here, must be mentioned in conclusion: the question as to how far these commutation relations determine the representation of the field variables in the space $\mathcal{R}$. We know that they uniquely determine this representation when quantum mechanics is involved, but this problem raises considerable difficulties in quantum field theory.†

5. Local commutativity or microcausality

Let us consider two points x and x' belonging to a space-like surface σ: since no perturbation produced at x can be propagated as far as x' (its speed of propagation would be greater than c) two observables defined respectively at x and x' can be simultaneously measured.

This property is generally extended to certain field variables and we write the so-called local commutativity or microcausality:‡

$$[\mathcal{Q}_A(x), \mathcal{Q}_A(x')] = 0 \quad \text{if} \quad x, x' \in \sigma. \tag{2.36}$$

In particular, let us choose as surface σ the equitemporal plane t; formula (2.36) shows that we also have

$$\frac{\partial}{\partial x_4} [\mathcal{Q}_A(\boldsymbol{x}, x_4), \mathcal{Q}_A(\boldsymbol{x}', x_4)] = 0. \tag{2.37}$$

Making use of (2.31), this expression may be developed as follows:

$$[[\mathcal{P}_4(\boldsymbol{x}, x_4), \mathcal{Q}_A(\boldsymbol{x}, x_4)], \mathcal{Q}_A(\boldsymbol{x}', x_4)] + [[\mathcal{Q}_A(\boldsymbol{x}', x_4), \mathcal{P}_4(\boldsymbol{x}', x_4)], \mathcal{Q}_A(\boldsymbol{x}, x_4)] = 0. \tag{2.38}$$

Add the term

$$[[\mathcal{Q}_A(\boldsymbol{x}, x_4), \mathcal{Q}_A(\boldsymbol{x}', x_4)], \mathcal{P}_4(\boldsymbol{x}, x_4)]$$

which is zero in virtue of (2.36).

By applying the Jacobi identity (1.72c) we then obtain

$$[[\mathcal{Q}_A(\boldsymbol{x}, x_4), \mathcal{P}_4(\boldsymbol{x}, x_4) - \mathcal{P}_4(\boldsymbol{x}', x_4)], \mathcal{Q}_A(\boldsymbol{x}', x_4)] = 0. \tag{2.39}$$

This necessary condition is always satisfied for the form of $\mathcal{P}_\mu$ which we shall use; referring to (2.85) we see in fact that $\mathcal{P}_\mu$ is a non-localized observable (it is in fact obtained by integrating a point function over an equitemporal plane or over a surface σ). It follows that $\mathcal{P}_4$ is independent of $\boldsymbol{x}$ and consequently that (2.39) is actually satisfied.

† A. S. Wightman and S. S. Schweber, *Phys. Rev.*, **98**, 812 (1955) have studied the conditions ensuring the uniqueness of this representation.

‡ There is no summation over the index A. We shall also see in Chap. XI the particularly important part played by this assumption in the axiomatic formulation of quantum field theory.

This choice of $\mathcal{P}_\mu$ also shows that the evolution operator $V(x, x')$ is a function of $\boldsymbol{x}-\boldsymbol{x}'$ and that it supplies a unitary representation of the group of translations in Euclidean space.

We noted at the beginning of this section that (2.36) is valid only for "certain" field variables. We shall in fact see later on that the commutator of this expression has to be replaced by an anticommutator, when the particles corresponding to the field under investigation satisfy Fermi–Dirac statistics instead of Bose–Einstein ones.

B. CANONICAL PICTURES WITH CLASSICAL EQUIVALENTS. HEISENBERG–PAULI QUANTIZATION METHOD†

Consider now a field with a classical equivalent, in other words one whose equations derive from an action principle and can be expressed in Hamiltonian form. We shall show that the determination of $\mathcal{P}_\mu$ introduced in (2.27) leads us to postulate commutation relations between field variables and conjugate variables.

This procedure presents both advantages and disadvantages. It brings out clearly the central fact of quantization: the commutation relations between field variables and conjugate variables are defined in such a way that the differential equations governing the evolution of the non-quantized wave functions are the same as eqns. (2.31) which in the last analysis express quite simply the uniqueness of the algebra of the quantized field variables. It conceals however the relativistic invariance of the expressions which are introduced, since it ascribes a quite special role to the equitemporal surfaces; the method is moreover difficult of application, since the identity we have just mentioned between equations only occurs for the dynamically independent components of wave functions.

Finally, the commutation relations thus obtained, together with an additional assumption, namely the existence of an eigenvector $|0\rangle$ of $\mathcal{P}_\mu$ corresponding to the eigenvalue 0, will enable us to give in the simple case of free fields [cf. (5.130) and (5.135)], a uniquely determined representation of the ring generated by the field variables.

The investigation of fields with classical equivalents will occupy the greater part of this book; in the last chapter, however, we shall investigate quantum theories such that the existence of an action is no longer assumed.

† W. Heisenberg and W. Pauli, *Z. Phys.*, **56**, 1 (1929); **59**, 168 (1930); and L. Rosenfeld, *Z. Phys.*, **63**, 574 (1930).

6. Hamiltonian form of field equations

We shall begin by rapidly summarizing the classical Hamiltonian field formalism. The wave functions $Q(x)$ obey a system of partial differential equations which we shall suppose to be obtained from the action†

$$\mathcal{A}[Q] = \int \mathcal{L}(Q(x), Q_{,\mu}(x), x)\, d^4x; \tag{2.40}$$

these equations are therefore written

$$\frac{\partial \mathcal{L}}{\partial Q_A} - \partial_\mu \frac{\partial \mathcal{L}}{\partial Q_{A,\mu}} = 0. \tag{2.41}$$

We propose to replace this Lagrangian system by a Hamiltonian one; for this purpose, let us bring out the time derivative in (2.41) by writing these equations in the form

$$\frac{\partial}{\partial t} \frac{\partial \mathcal{L}}{\partial \dot{Q}_A(x)} = \frac{\partial \mathcal{L}}{\partial Q_A} - \partial_j \frac{\partial \mathcal{L}}{\partial Q_{A,j}}, \tag{2.42}$$

the dot denoting a time derivative. By defining the conjugate variables

$$\Pi_A(x) = \frac{\partial \mathcal{L}}{\partial \dot{Q}_A} = -i \frac{\partial \mathcal{L}}{\partial Q_{A,4}}, \tag{2.43}$$

eqns. (2.42) can be written

$$\dot{\Pi}_A(x) = \frac{\partial \mathcal{L}}{\partial Q_A} - \partial_j \frac{\partial \mathcal{L}}{\partial Q_{A,j}}. \tag{2.44}$$

Suppose now that in eqns. (2.43) the $\dot{Q}_A$'s can be expressed in terms of the conjugate variables Π_A and of the derivatives $Q_{A,j}$, and define a Hamiltonian density

$$\Lambda = \Pi_A(x)\dot{Q}_A(x) - \mathcal{L}, \tag{2.45}$$

where we sum over the repeated indices A and where the Q_A's are expressed in terms of the conjugate variables: Λ is therefore a function of the Q_A's, the Π_A's and the $Q_{A,j}$'s. In fact, we shall suppose it to depend also on the $\Pi_{A,j}$'s in order to include more general cases which we shall investigate later on.

To the Hamiltonian density Λ corresponds the Hamiltonian

$$P_0[Q, \Pi | t] = \int \Lambda(Q_A, Q_{A,j}, \Pi_A, \Pi_{A,j}, x)\, d^3\boldsymbol{x}. \tag{2.46}$$

† In so far as they are functions of x, the field variables are assumed to converge strongly to zero in all space directions when $|\boldsymbol{x}| \to \infty$; integrals such as (2.40) are therefore meaningful.

Finally, from the Lagrangian density we define a Lagrangian, a functional of the Q and a function of t:

$$L[Q\,|\,t] = \int \mathcal{L}(Q_A, Q_{A,\mu}, x)\, d^3\boldsymbol{x}, \tag{2.47}$$

such that the Hamiltonian is written

$$P_0[Q, \Pi\,|\,t] = \int \Pi_A(x)\dot{Q}_A(x)\, d^3\boldsymbol{x} - L, \tag{2.48}$$

and where the $\dot{Q}$'s have been expressed in terms of Q_A, Π_A, $Q_{A,j}$ and $\Pi_{A,j}$. On the one hand, to variations δQ and $\delta\Pi$ corresponds the variation

$$\delta P_0[Q, \Pi\,|\,t] = \int [\delta\Pi_A(x)\dot{Q}_A(x) + \Pi_A(x)\,\delta\dot{Q}_A(x)]\, d^3\boldsymbol{x} - \delta L. \tag{2.49}$$

On the other hand, according to the definition (2.47) of L:

$$\delta L = \int \left[\left(\frac{\partial\mathcal{L}}{\partial Q_A} - \partial_j \frac{\partial\mathcal{L}}{\partial Q_{A,j}}\right)\delta Q_A(x) + \frac{\partial\mathcal{L}}{\partial\dot{Q}_A}\,\delta\dot{Q}_A(x)\right] d^3\boldsymbol{x} \tag{2.50}$$

and with the aid of (2.43) and (2.44) this variation can be expressed in the form

$$\delta L[Q, \Pi\,|\,t] = \int [\dot{\Pi}_A(x)\,\delta Q_A(x) + \Pi_A(x)\,\delta\dot{Q}_A(x)]\, d^3\boldsymbol{x}, \tag{2.51}$$

such that finally we have

$$\delta P_0[Q, \Pi\,|\,t] = \int [\delta\Pi_A(x)\dot{Q}_A(x) - \dot{\Pi}_A(x)\,\delta Q_A(x)]\, d^3\boldsymbol{x}. \tag{2.52}$$

But in accordance with the definition (1.128) of the differential $\delta P_0[\ldots]$, we can write

$$\delta P_0[Q, \Pi\,|\,t] = \int \left[\frac{\delta P_0[Q, \Pi]}{\delta Q(x)}\,\delta Q(x) + \frac{\delta P_0[Q, \Pi]}{\delta\Pi(x)}\,\delta\Pi(x)\right] d^3\boldsymbol{x}. \tag{2.53}$$

Comparing (2.52) with (2.53) and making use of formula (1.147), we obtain the Hamiltonian equations:†

† We remind readers that we are considering dynamically independent wave functions, in other words such that (a) they possess non-zero conjugate variables; (b) wave functions and conjugate variables obey the Hamiltonian equations. The various field components we shall be considering do not all have this property. For instance, it is well known that the Maxwell free field is described by a four-vector $A_\mu(x)$ (with respect to the Lorentz group), the fourth component of which is not dynamically independent. Its conjugate variable is, in fact, zero in accordance with the Lorentz condition. Field theories in which the field variables are not dynamically independent will in all cases be studied separately.

$$\frac{\partial Q_A(x)}{\partial t} = \frac{\delta P_0[Q, \Pi | t]}{\delta \Pi_A(x)} \equiv \frac{\partial \Lambda}{\partial \Pi_A} - \frac{\partial}{\partial x_j} \frac{\partial \Lambda}{\partial \Pi_{A,j}}, \qquad (2.54a)$$

$$\frac{\partial \Pi_A(x)}{\partial t} = -\frac{\delta P_0[Q, \Pi | t]}{\delta Q_A(x)} \equiv -\left(\frac{\partial \Lambda}{\partial Q_A} - \frac{\partial}{\partial x_j} \frac{\partial \Lambda}{\partial Q_{A,j}} \right). \qquad (2.54b)$$

We can also introduce the Poisson bracket of two functionals $A[Q, \Pi]$ and $B[Q, \Pi]$ by setting, by definition,

$$[A, B]_{\text{cl}} = \int d^3\boldsymbol{x}' \left\{ \frac{\delta A[Q, \Pi]}{\delta Q_A(x')} \frac{\delta B[Q, \Pi]}{\delta \Pi_A(x')} - \frac{\delta A[Q, \Pi]}{\delta \Pi_A(x')} \frac{\delta B[Q, \Pi]}{\delta Q_A(x')} \right\}. \qquad (2.55)$$

Any time derivative of a functional $G[Q, \Pi]$ can be expressed by means of the Poisson brackets. Consider indeed the function:

$$G[Q_A, \Pi_A] = \int \mathcal{G}[Q_A, \Pi_A, Q_{A,j}, \boldsymbol{x}]\, d^3\boldsymbol{x},$$

where the Q_A's and Π_A's are functions of t.

The time derivative is readily written

$$\frac{\partial G}{\partial t} = \int \left(\frac{\partial \mathcal{G}}{\partial Q_A} \frac{\partial Q_A}{\partial t} + \frac{\partial \mathcal{G}}{\partial \Pi_A} \frac{\partial \Pi_A}{\partial t} + \frac{\partial \mathcal{G}}{\partial Q_{A,j}} \frac{\partial Q_{A,j}}{\partial t} + \frac{\partial \mathcal{G}}{\partial \Pi_{A,j}} \frac{\partial \Pi_{A,j}}{\partial t} \right) d^3\boldsymbol{x}. \qquad (2.56)$$

Note on the other hand that

$$\int \frac{\partial \mathcal{G}}{\partial Q_{A,j}} \frac{\partial Q_{A,j}}{\partial t} d^3\boldsymbol{x} = -\int \left(\partial_j \frac{\partial \mathcal{G}}{\partial Q_{A,j}} \right) \frac{\partial Q_A}{\partial t} d^3\boldsymbol{x},$$

since all the functions are supposed to converge strongly to zero when $|\boldsymbol{x}| \to \infty$; we have an analogous formula for the fourth term of (2.56). Introducing the expressions $\frac{\partial Q_A}{\partial t}$ and $\frac{\partial \Pi_A}{\partial t}$ taken from the Hamilton equations, $\frac{\partial G}{\partial t}$ takes the form

$$\frac{\partial G}{\partial t} = \int \left[\frac{\delta G[Q, \Pi]}{\delta Q_A(x)} \frac{\delta P_0[Q, \Pi]}{\delta \Pi_A(x)} - \frac{\delta G[Q, \Pi]}{\delta \Pi_A(x)} \frac{\delta P_0[Q, \Pi]}{\delta Q_A(x)} \right] d^3\boldsymbol{x} \equiv [G, P_0]_{\text{cl}}. \qquad (2.57)$$

We can also verify that the Poisson brackets satisfy the same system of algebraic relations (1.72) as the commutators.

Considering finally $Q(x)$ and $\Pi(x)$ as functionals with $\delta(\boldsymbol{x}-\boldsymbol{x}')$ as kernel [cf. (1.126)]

$$Q(x) = \int_{t'=t} \delta(\boldsymbol{x}-\boldsymbol{x}')Q(x')\, d^3\boldsymbol{x}', \quad \Pi(x) = \int_{t'=t} \delta(\boldsymbol{x}-\boldsymbol{x}')\Pi(x')\, d^3\boldsymbol{x}', \tag{2.58}$$

it is easy to write the Hamilton equations by means of the Poisson brackets:

$$\frac{\partial Q}{\partial t} = [Q(\boldsymbol{x}, t), P_0]_{\text{cl}} = \frac{\delta P_0}{\delta \Pi(x)}, \tag{2.59a}$$

$$\frac{\partial \Pi}{\partial t} = [\Pi(\boldsymbol{x}, t), P_0]_{\text{cl}} = -\frac{\delta P_0}{\delta Q(x)}. \tag{2.59b}$$

Note 1. It is sometimes convenient to express the Hamilton equations in a different form. We first replace $\dfrac{\partial Q_A}{\partial t}$ by $D_{(n)}Q_A(x)$ as defined by (1.27). The Lagrangian density $\mathcal{L}$ is then a function of the $Q_A(x)$'s, of the tangential derivatives $D_\mu Q_A(x)$'s [defined by (1.28)] with respect to a particular surface σ (this surface plays exactly the same role as the three-dimensional fundamental volume of the preceding section) and of the normal derivative $D_{(n)}Q_A(x)$. One may define the total Lagrangian:

$$L[Q, \sigma] = \int_\sigma \mathcal{L}(Q_A(x'), D_\mu Q_A(x'), D_{(n)}Q_A(x'))\, d\sigma' \tag{2.60}$$

and the conjugate variables to the wave functions:

$$\Pi_A(x) = \frac{\partial \mathcal{L}}{\partial\{D_{(n)}Q_A(x)\}}. \tag{2.61}$$

Consider a variation δQ_A: by analogy with the property used in eqns. (1.140) *et seq.*: $\delta Q_{A,\mu} = \partial_\mu \delta Q_A$, we suppose that

$$\delta D_\mu Q_A = D_\mu\, \delta Q_A, \qquad \delta D_{(n)}Q_A = D_{(n)}\, \delta Q_A, \tag{2.62}$$

the total variation δL can then be written

$$\delta L = \int_\sigma \left[\left(\frac{\partial \mathcal{L}}{\partial Q_A(x')} - D_\mu \frac{\partial \mathcal{L}}{\partial\{D_\mu Q_A\}} \right) \delta Q_A + D_\mu \left\{ \frac{\partial \mathcal{L}}{\partial\{D_\mu Q_A\}}\, \delta Q_A(x') \right\} + \frac{\partial \mathcal{L}}{\partial\{D_{(n)}Q_A\}}\, \delta D_{(n)}Q_A \right] d\sigma'.$$

The term after the first bracket makes no contribution in accordance with Green's theorem (1.135). We therefore obtain a form of δL which is very similar to (2.50); we can form the Hamiltonian in agreement with (2.45) by eliminating $D_{(n)}Q_A$ from the Lagrangian density (2.60) by means of eqn. (2.50) defining the conjugate variables.

The method of obtaining the Hamilton equations is then entirely analogous to the one used in the preceding section, and it can be easily seen that they are written on a particular surface σ as follows:

$$D_{(n)}Q_A(x) = \frac{\delta P_0[Q, \Pi]}{\delta \Pi_A(x)}, \tag{2.63a}$$

$$D_{(n)}\Pi_A(x) = -\frac{\delta P_0[Q, \Pi]}{\delta Q_A(x)}. \tag{2.63b}$$

Note, on the other hand, that

$$\frac{\partial \mathcal{L}}{\partial\{D_{(n)}Q_A\}} = \frac{\partial \mathcal{L}}{\partial Q_{B,\mu}} \frac{\partial Q_{B,\mu}}{\partial\{D_{(n)}Q_A\}}$$

and that by virtue of formula (1.131):

$$\frac{\partial Q_{B,\mu}}{\partial\{D_{(n)}Q_A\}} = n_\mu\, \delta_{AB},$$

we therefore finally obtain

$$\Pi_A \equiv \frac{\partial \mathcal{L}}{\partial\{D_{(n)}Q_A\}} = n_\mu \frac{\partial \mathcal{L}}{\partial Q_{A,\mu}}. \tag{2.64}$$

Note 2. It is by no means necessary to require the wave functions to be real quantities. We shall see in the next chapter that for charged particle fields it is necessary to make them complex.

Each of the components $Q_A(x)$ is built up with a real part $Q_A^{(1)}(x)$ and an imaginary part $Q_A^{(2)}(x)$, namely

$$\left.\begin{aligned} Q_A^{(1)}(x) &= \frac{Q_A(x)+Q_A^*(x)}{2}, \quad Q_A^{(2)}(x) = \frac{Q_A(x)-Q_A^*(x)}{2i}, \\ Q_A(x) &= Q_A^{(1)}(x)+iQ_A^{(2)}(x). \end{aligned}\right\} \tag{2.65}$$

Since these last two components are independent, the variation of the Lagrangian will be obtained from their independent variations.

We see therefore, finally, that we can equally well consider $Q_A(x)$ and $Q_A^*(x)$ as two independent quantities the variations of which must also be independent. Consequently, a complex field is completely described by twice the number of the components of the wave functions $Q_A(x)$.

In quantized theory, we shall replace the sign* of the complex conjugation by the sign $\sim$ of the Hermitian conjugation in the $\mathcal{R}$ space.

Note 3. As for systems with a finite number of degrees of freedom, the Hamilton equations can be obtained directly from a variation principle.

$$\delta\left\{\int_{t_I}^{t_{II}} dt \left(\int \Pi_A(x)\dot{Q}_A(x)\, d^3\mathbf{x} - P_0[Q, \Pi]\right)\right\} = 0. \tag{2.66}$$

This principle is useful in that it enables us to easily define contact transformations. Let $\hat{Q}(x)$, $\hat{\Pi}(x)$ be another system of field variables and of conjugate variables; the new variables obey the Hamilton equations provided that

$$\int \Pi_A(x)\dot{Q}_A(x)\,d^3\boldsymbol{x} - P_0[Q,\Pi] - \left(\int \hat{\Pi}_A(x)\dot{\hat{Q}}_A(x)\,d_3\boldsymbol{x} - \hat{P}_0[\hat{Q},\hat{\Pi}]\right) \tag{2.67}$$
$$= \frac{dF[\Pi, Q, \hat{\Pi}, \hat{Q}\,|\,t]}{dt}$$

as we can easily see. The functional $F[\ldots]$ is arbitrary; let us therefore choose it as follows:

$$F[\ldots] = \int (Q_A(x)\hat{\Pi}_A(x) - \hat{Q}_A(x)\hat{\Pi}_A(x))\,d^3\boldsymbol{x} - \Delta G[Q,\hat{\Pi}], \tag{2.68}$$

where $\Delta G[\ldots]$ is supposed to be infinitesimal, such that the contact transformation is also infinitesimal.

By expanding $\dfrac{dF[\ldots]}{dt}$, the above relation can be written

$$\int [(\Pi_A(x) - \hat{\Pi}_A(x))\dot{Q}_A(x) + (\hat{Q}_A(x) - Q_A(x))\dot{\hat{\Pi}}_A(x)]\,d^3\boldsymbol{x}$$
$$-P_0[\ldots] - \hat{P}_0[\ldots] = \frac{d\Delta G[\ldots]}{dt}$$
$$= \int \left(\frac{\delta\Delta G[\ldots]}{\delta Q_A(x)}\dot{Q}_A(x) + \frac{\delta\Delta G[\ldots]}{\delta\hat{\Pi}_A(x)}\dot{\hat{\Pi}}_A(x)\right)d^3\boldsymbol{x}.$$

Identifying the coefficients of $\dot{Q}_A$ and $\dot{\hat{\Pi}}_A$, we have

$$\delta\Pi_A(x) = -\frac{\delta\Delta G[Q,\hat{\Pi}]}{\delta Q_A(x)}, \qquad \delta Q_A(x) = \frac{\delta\Delta G[Q,\hat{\Pi}]}{\delta\hat{\Pi}_A(x)},$$
$$P_0[\ldots] = \hat{P}_0[\ldots].$$

We can replace $\hat{\Pi}$ by Π in these relations, since the transformation under consideration is infinitesimal and by using the Poisson brackets, we can write, as may be easily seen

$$\delta\Pi_A(x) = [\Pi_A(x), \Delta G]_{\text{cl}}, \tag{2.69a}$$
$$\delta Q_A(x) = [Q_A(x), \Delta G]_{\text{cl}}. \tag{2.69b}$$

These two formulae will serve as a guideline when we come to state the Schwinger variational principle in § 13 of this chapter.

7. Hamiltonian form of the field operator equations (canonical equations). Quantization

Let us compare the results of Part 1 of this chapter with those we have just obtained. We were led to introduce a displacement operator, the time component of which

$$\mathcal{P}_4 = i\mathcal{P}_0,$$

is such that in accordance with (2.29) and (2.31), we have:

$$i\frac{\partial Q_A(x)}{\partial t} = [Q_A(x), \mathcal{P}_0]. \tag{2.70a}$$

On the other hand we have just obtained for a theory in which the field variables are c-numbers:

$$\frac{\partial Q_A(x)}{\partial t} = \frac{\delta P_0[Q, \Pi]}{\delta \Pi_A(x)} \equiv [Q_A(x), P_0]_{\text{cl}}. \tag{2.70b}$$

We noted in fact, at the end of § 4, Chapter I, the close analogies between commutators and Poisson brackets and we pointed out the extreme importance of these analogies for quantum mechanics.

Bearing these analogies in mind, let us therefore suppose that P_0 and $\mathcal{P}_0$ are expressed by the same functional of the Q's and Π's (in the expression of P_0 they are c-numbers whereas in that of $\mathcal{P}_0$ they will be operators) and let us ask ourselves in what conditions quantum theory and classical theory lead to the same system of field equations.† For this purpose eqn. (2.70a) must be identical with eqns. (2.70b), i.e.

$$-i[Q_A(x), P_0] = \frac{\delta P_0[Q, \Pi]}{\delta \Pi_A(x)} \equiv \frac{\partial \Lambda}{\partial \Pi_A} - \partial_j \frac{\partial \Lambda}{\partial \Pi_{A,j}} \tag{2.71a}$$

or alternatively by introducing into the commutator the density Λ of P_0:‡

$$[Q_A(x), \Lambda(Q(\boldsymbol{x}', t), \Pi(\boldsymbol{x}', t), \ldots)] = i\frac{\delta P_0[Q, \Pi]}{\delta \Pi_A(x)}\,\delta(\boldsymbol{x}-\boldsymbol{x}'). \tag{2.71b}$$

† It is interesting to note, however, that to a P_0 given in classical theory there correspond various P_0 in quantized theory differing from each other in the order of the factors which enter into them. The correspondence principle does not determine the operator P_0 in a unique way; it has to be postulated. It is generally symmetrized, as in quantum mechanics.

‡ We suppose here, as for the theorem on p. 20, that the increments $\delta\Pi_A$, δQ_A commute with all the field variables and conjugate variables; they are c-numbers. The functional derivatives in (2.71) are therefore uniquely determined, even when P_0 is an operator expressed as an analytical functional of the Q_A's and Π_A's provided we pay attention to the order of the factors when we take derivatives. This is no longer the case in the Schwinger–Feynman variation principle (Part 4 of this chapter), where we take advantage of the possibility of defining δQ increments which are operators.

Let us also note that since P_0 represents the total energy of the system, the operator P_0 which corresponds to it is necessarily a Hermitian operator.

The relation (2.71*b*) will lead us to ascribe certain values to the commutators between field variables and conjugate variables (2.74). We shall show recurrently that if the equality (2.71*b*) is realized for the following two kinds of Hamiltonian:

$$P_0^{(1)}[Q] \equiv \int Q_B(\boldsymbol{x}', t)\, d^3\boldsymbol{x}' = \int \delta(\boldsymbol{x}'-\boldsymbol{x}'')Q_B(\boldsymbol{x}'', t)\, d^3\boldsymbol{x}'\, d^3\boldsymbol{x}'', \qquad (2.72a)$$

$$P_0^{(2)}[\Pi] \equiv \int \Pi_B(\boldsymbol{x}', t)\, d^3\boldsymbol{x}' = \int \delta(\boldsymbol{x}'-\boldsymbol{x}'')\Pi_B(\boldsymbol{x}'', t)\, d^3\boldsymbol{x}'\, d^3\boldsymbol{x}'', \qquad (2.72b)$$

it will also be true for all the forms of Hamiltonian that are met with in quantum field theory. On the one hand, one has indeed:

$$\frac{\delta P_0^{(1)}}{\delta \Pi_A(x)} = 0, \qquad (2.73a)$$

$$\frac{\delta P_0^{(2)}}{\delta \Pi_A(x)} = \delta_{AB}. \qquad (2.73b)$$

On the other hand, the Hamiltonians (2.72) lead to the following quantum brackets:

$$[Q_A(\boldsymbol{x}, t), P_0^{(1)}] \equiv \int [Q_A(\boldsymbol{x}, t), Q_B(\boldsymbol{x}', t)]\, d^3\boldsymbol{x}', \qquad (2.73c)$$

$$[Q_A(\boldsymbol{x}, t), P_0^{(2)}] \equiv \int [Q_A(\boldsymbol{x}, t), \Pi_B(\boldsymbol{x}', t)]\, d^3\boldsymbol{x}'. \qquad (2.73d)$$

Then comparing the relations (2.73), we obtain†

† Consider a linear differential equation with constant coefficients:

$$\mathcal{L}\left\{\frac{d}{dx}\right\} u(x) = f(x).$$

The solution of the commutation rule in quantum mechanics

$$[x, p] = i$$

enables us to write the former differential equation in the form:

$$\mathcal{L}(ip)\,|\,u\rangle = |f\rangle$$

the interpretation of which is given by formulae (1.59) *et seq.*

Finally, the inversion of the operator $\mathcal{L}(ip)$ and the calculation of the matrix elements of p in the system of eigenvectors of the operator x:

$$\langle \xi_1 \,|\, p^n \,|\, \xi_2 \rangle = (-i)^n\, \delta^n(\xi_1 - \xi_2)$$

do in fact enable us to obtain the solution.

What can be said of the commutation rules (2.74)? Generally speaking (for the mathematical considerations involved are extremely subtle), a representation of $\Pi(x')$ is

$$[Q_A(\boldsymbol{x}, t), Q_B(\boldsymbol{x}', t)] = 0, \tag{2.74a}$$

$$[Q_A(\boldsymbol{x}, t), \Pi_B(\boldsymbol{x}', t)] = i\delta_{AB}\delta(\boldsymbol{x}-\boldsymbol{x}'). \tag{2.74b}$$

Let us now show that the commutation rules (2.74) assure the equality (2.71) for the other two forms of Hamiltonian:

$$P_0^{(3)}[\Pi] = \int \Pi_{B,j}(\boldsymbol{x}', t)\, d^3\boldsymbol{x}' \tag{2.75a}$$

$$= -\int \frac{\partial\delta(\boldsymbol{x}'-\boldsymbol{x}'')}{\partial x_j''} \Pi_B(\boldsymbol{x}'', t)\, d^3\boldsymbol{x}'\, d^3\boldsymbol{x}'',$$

$$P_0^{(4)}[Q] = \int Q_{B,j}(\boldsymbol{x}', t)\, d^3\boldsymbol{x}' \tag{2.75b}$$

$$= -\int \frac{\partial\delta(\boldsymbol{x}'-\boldsymbol{x}'')}{\partial x_j''} Q_B(\boldsymbol{x}'', t)\, d^3\boldsymbol{x}'\, d^3\boldsymbol{x}''.$$

We verify, for instance, that we have for (2.75*a*):

$$\frac{\delta P_0^{(3)}[Q, \Pi]}{\delta \Pi_A(x)} = -\delta_{AB} \int \frac{\partial\delta(\boldsymbol{x}-\boldsymbol{x}')}{\partial x_l}\, d^3\boldsymbol{x}', \tag{2.76}$$

and that finally

$$[Q_A(\boldsymbol{x}, t), P_0^{(3)}] = \int \left[Q_A(\boldsymbol{x}, t), \frac{\partial}{\partial x_j'} \Pi_B(\boldsymbol{x}', t) \right] d^3\boldsymbol{x}' \tag{2.77}$$

$$= \int \frac{\partial}{\partial x_j'} [Q_A(\boldsymbol{x}, t), \Pi_B(\boldsymbol{x}', t)]\, d^3\boldsymbol{x}'$$

$$= -i \int \frac{\partial\delta(\boldsymbol{x}-\boldsymbol{x}')}{\partial x_j'} \delta_{AB} d^3\boldsymbol{x}' = i \frac{\delta P_0^{(3)}[Q, \Pi]}{\delta \Pi_A(x)}.$$

(in the case of a field with a single component and at the time $t=0$ for the sake of simplicity):

$$\Pi(\boldsymbol{x}') = -i \frac{\delta}{\delta Q(\boldsymbol{x}')}$$

in virtue of (1.126). Consequently, any linear functional differential equation with constant coefficients

$$\mathscr{L}\left\{\frac{\delta}{\delta Q(\boldsymbol{x})}\right\} u[Q] = b[Q]$$

can be transformed into the operator equation:

$$\mathscr{L}(i\Pi(\boldsymbol{x}))u[Q] = b[Q]$$

and in these two equations Q is now considered as a simple function of $\boldsymbol{x}$: $Q(\boldsymbol{x})$.

There is therefore clearly a parallelism between the two cases, only the domain of application changes. Quantum mechanics deals with functions whereas quantum field theory deals with functionals.

Let us next take the Hamiltonian

$$P_0 \equiv \int \Lambda'(Q, \Pi)\Lambda''(Q, \Pi)\, d^3\boldsymbol{x}' \tag{2.78}$$

such that (2.71) is separately obeyed by Λ' and Λ'' which are supposed to be independent of the $Q_{A,j}(x)$ and of $\Pi_{A,j}(x)$.

We first verify that†

$$\frac{\delta P_0\,[Q, \Pi]}{\delta \Pi_A(x)} = \frac{\delta P_0'[Q, \Pi]}{\delta \Pi_A(x)}\Lambda''(Q,\Pi) + \Lambda'(Q,\Pi)\frac{\delta P_0''[Q, \Pi]}{\delta \Pi_A(x)} \tag{2.79}$$

since P_0' and P_0'' do not in fact depend on Q, j and Π, j.

If we now consider the quantum bracket, well-known rules for the calculation of the commutators lead us to write

$$\begin{aligned}
&[Q_A(\boldsymbol{x}, t), P_0] \qquad\qquad (2.80)\\
&= \int [Q_A(\boldsymbol{x}, t), \Lambda'(Q(\boldsymbol{x}', t), \Pi(\boldsymbol{x}', t))\Lambda''(Q(\boldsymbol{x}', t), \Pi(\boldsymbol{x}', t))]\, d^3\boldsymbol{x}'\\
&= \int [Q_A(\boldsymbol{x}, t), \Lambda'(Q(\boldsymbol{x}'\, t), \Pi(\boldsymbol{x}', t))]\Lambda''(Q(\boldsymbol{x}', t), \Pi(\boldsymbol{x}'\, t))\, d^3\boldsymbol{x}'\\
&\quad + \int \Lambda'(Q(\boldsymbol{x}', t)\Pi(\boldsymbol{x}', t))[Q_A(\boldsymbol{x}, t), \Lambda''(Q(\boldsymbol{x}', t), \Pi(\boldsymbol{x}', t))]\, d^3\boldsymbol{x}'
\end{aligned}$$

and by virtue of the assumptions concerning Λ' and Λ'' we do in fact obtain (2.71).

Let us finally consider a general Hamiltonian whose density is supposed to be an analytical function of the variables Q, Π, Q_j, and $\Pi_{,j}$, (space derivatives). This is in fact the kind of Hamiltonian encountered in quantum field theory where the analytical function degenerates into a polynomial of these same variables.

First of all, we can readily see that if the identification expressed by (2.71) is valid for each of the terms of this sum, it remains so for the sum. On the other hand, the introduction of the derivatives of the Dirac function enables us to consider densities depending exclusively on the Q and Π (and on $\boldsymbol{x}$) and not on their derivatives.

For instance, the Hamiltonian

$$P_0[\Pi] = \int \Pi_{B,l}(x)\Pi_{B,l}(x)\, d^3\boldsymbol{x}$$

† Formula (2.79) can be readily proved; indeed, we have

$$\delta P_0[Q, \Pi] = \int\left\{\frac{\delta\Lambda'[Q, \Pi]}{\delta\Pi_A(\boldsymbol{x}'\, t)}\delta\Pi_A(\boldsymbol{x}', t)\Lambda''(Q, \Pi) + \Lambda'(Q, \Pi)\frac{\delta\Lambda''[Q, \Pi]}{\delta\Pi_A(\boldsymbol{x}', t)}\delta\Pi_A(\boldsymbol{x}', t)\right\} d^3\boldsymbol{x}'.$$

can be written in the form

$$P_0[\Pi] = \int \frac{\partial\delta(\boldsymbol{x}-\boldsymbol{x}')}{\partial x_l'} \frac{\partial\delta(\boldsymbol{x}-\boldsymbol{x}'')}{\partial x_l''} \Pi_B(\boldsymbol{x}', t)\Pi_B(\boldsymbol{x}'', t)\, d^3\boldsymbol{x}\, d^3\boldsymbol{x}'\, d^3\boldsymbol{x}'',$$

which no longer contains derivatives of the Π.

Finally, the considerations which immediately follow formula (2.78) prove that this identification also holds for the general type of Hamiltonian we have in mind.

The field quantities are thus equally well given by (2.70*a*) or (2.70*b*) provided that $\mathcal{P}_0$ is given by the same expression as P_0 and that Q and Π verify the commutation rules (2.74). By adding to the commutation rules (2.74) the commutation rule

$$[\Pi_A(\boldsymbol{x}, t), \Pi_B(\boldsymbol{x}', t)] = 0, \tag{2.81}$$

we could prove, as we have just done for the Q_A, that the change in time of the conjugate variables is in classical theory

$$\frac{\partial \Pi_A}{\partial t} = [\Pi_A(\boldsymbol{x}, t), P_0]_{\text{cl}} \tag{2.82}$$

and in quantized theory†

$$\frac{\partial \Pi_A}{\partial t} = -i[\Pi_A(\boldsymbol{x}, t), \mathcal{P}_0]. \tag{2.83}$$

We can also prove, making use of (2.70) and (1.72*c*), that the commutation rules (2.74) and (2.81) continue to be obeyed during the time evolution of the system, in other words that the time derivative of (2.74) and (2.81) is zero.‡

We repeat, in conclusion, that generally not all the components of the field variables and their conjugate variables obey the commutation rules and the canonical equations; some of them can be expressed in terms of others (for instance, in the case of the non-zero mass vector field, three

† Both of eqns. (2.70*a*) and (2.83) are equivalent to the Lagrange equations.
‡ We have

$$\frac{\partial}{\partial t}[Q_A(\boldsymbol{x}, t), \Pi_B(\boldsymbol{x}', t)] = \left[\frac{\partial Q_A(\boldsymbol{x}, t)}{\partial t}, \Pi_B(\boldsymbol{x}', t)\right] + \left[Q_A(\boldsymbol{x}, t), \frac{\partial \Pi_B(\boldsymbol{x}', t)}{\partial t}\right]$$
$$= -i\{[[Q_A(\boldsymbol{x}, t), P_0(t)], \Pi_B(\boldsymbol{x}', t)] - [\Pi_B(\boldsymbol{x}', t), [P_0(t), Q_A(\boldsymbol{x}, t)]] - [P_0(t), [Q_A(\boldsymbol{x}, t), \Pi_B(\boldsymbol{x}', t)]]\}.$$

The first two terms cancel, the last one is zero since P_0 naturally commutes with $i\delta_{AB}\,\delta(\boldsymbol{x}-\boldsymbol{x}')$.

out of four components can be arbitrarily given), others are identically zero (cf. Dirac equation). The set of the Q_A and Π_A which—after elimination of the dependent variables—obeys the canonical relations (commutators and equations) forms the canonical set we introduced in Part 1 of this chapter.

8. The space–time translation operator

Let us return to the canonical form (2.31) of the wave equations

$$i\frac{\partial Q}{\partial x_\mu} = [\mathcal{P}_\mu, Q(x)]. \tag{2.84}$$

In the last two sections we determined the time component of $\mathcal{P}_\mu$: the operator $P_0[Q, \Pi]$.

What can be said of the space components of this four-vector?

Write down the Hamiltonian density Λ given by (2.45) in covariant notation:

$$\Lambda = \frac{\partial \mathcal{L}}{\partial Q_{A,4}} Q_{A,4} - \mathcal{L}(x)\,\delta_{44},$$

and, by a natural extension, form the tensor $T_{\mu\nu}$ (which, as we shall see in the next chapter, is the energy–momentum tensor) and the four-vector P_μ (or energy–momentum vector):

$$T_{\mu\nu} = -\frac{\partial \mathcal{L}}{\partial Q_{A,\nu}} Q_{A,\mu} + \mathcal{L}\,\delta_{\mu\nu}, \tag{2.85a}$$

$$P_\mu[\sigma] = \int_{\sigma(x)} T_{\mu\nu}(x')\,d\sigma'_\nu, \tag{2.85b}$$

$$\Lambda = -T_{44}; \tag{2.85c}$$

then, comparing with the preceding formula, one sees:

$$P_4 = i\int \Lambda(x)\,d^3\boldsymbol{x} = i\int \Lambda\,d\sigma = -\int \Lambda\,d\sigma_4.$$

Let us show that P_j can be made identical with $\mathcal{P}_j$:

$$P_j(t) = -i\int T_{j4}(\boldsymbol{x}', t)\,d^3\boldsymbol{x}' = -\int \Pi_A(\boldsymbol{x}', t) Q_{A,j}(\boldsymbol{x}', t)\,d^3\boldsymbol{x}'. \tag{2.85d}$$

We compute

$$[P_j[Q, \Pi\,|\,t, Q(x)]$$

which, by virtue of commutation rules (2.74), may also be written:

$$[P_j[Q, \Pi \mid t], Q_A(x)] = \int [\Pi_B(\boldsymbol{x}', t)Q_{B,j}(\boldsymbol{x}', t), Q_A(\boldsymbol{x}, t)]\, d^3\boldsymbol{x}'$$

$$= \int \{\Pi_B(\boldsymbol{x}', t)[Q_{B,j}(\boldsymbol{x}', t), Q_A(\boldsymbol{x}, t)]$$

$$+[\Pi_B(\boldsymbol{x}' t), Q_A(\boldsymbol{x}, t)]Q_{B,j}(\boldsymbol{x}', t)\}\, d^3\boldsymbol{x}'$$

$$= i\,\frac{\partial Q_A}{\partial x_j}(\boldsymbol{x}, t).$$

The equations concerning space components of $\mathcal{P}_\mu$ are therefore satisfied with the choice we have just made. We may verify in similar fashion

$$i\,\frac{\partial \Pi(x)}{\partial x_\mu} = [P_\mu, \Pi(x)]. \tag{2.86}$$

9. Commutation rules for different times

The time evolution of canonical variables $Q(x)$ and $\Pi(x)$ is described by the system of Hamiltonian differential equations (2.54) which is of the first order with respect to time.† This system may therefore be solved for Cauchy data: if $Q(x)$ and $\Pi(x)$ are given at the time t_0, we are able to determine its solution at any particular time; the solution of the differential system (2.54) may therefore be expressed by means of two differential operators $W^{(Q)}(t, \nabla)$ and $W^{(\Pi)}(t, \nabla)$ as follows:

$$Q(\boldsymbol{x}, t) = W^{(Q)}(t, t_0;\ \nabla)Q(\boldsymbol{x}, t_0) + W^{(\Pi)}(t, t_0);\ \nabla)\Pi(\boldsymbol{x}, t_0). \tag{2.87}$$

Since these two operators must be independent of the initial data $Q(\boldsymbol{x}, t_0)$ and $\Pi(\boldsymbol{x}, t_0)$, we have

$$W^{(Q)}(t_0, t_0;\ \nabla) = 1, \qquad W^{(\Pi)}(t_0, t_0;\ \nabla) = 0. \tag{2.88}$$

Let us therefore calculate the expression:

$$[Q_A(x), Q_B(x')] = W^{(Q)}_{AA'}(t, t';\ \nabla)[Q_{A'}(\boldsymbol{x}, t')\, Q_B(\boldsymbol{x}', t')]$$
$$+ W^{(\Pi)}_{AA'}(t, t';\ \nabla)[\Pi_{A'}(\boldsymbol{x}, t'), Q_B(\boldsymbol{x}', t')],$$

the first commutator is zero, the second has the value

$$-i\delta_{A'B}\, \delta(\boldsymbol{x}-\boldsymbol{x}').$$

Consequently,

$$[Q_A(x), Q_B(x')] = -iW^{(\Pi)}_{AB}(t, t';\ \nabla)\delta(\boldsymbol{x}-\boldsymbol{x}'). \tag{2.89}$$

† We are still considering the dynamically independent components. In this section we are generalizing a method initiated by G. Wentzel, *Quantum Theory of Fields*, Interscience Publishers, 1949, p. 19 *et seq.*

The problem of determining the right side of (2.89) is simplified when each of the components $Q_A(x)$ verifies the equation (free fields)

$$(\Box - m^2)Q_A(x) = 0. \tag{2.90}$$

The operator W, a function of $t-t'$, must then obey

$$(\Box - m^2)W^{(\Pi)}(t, \nabla) = 0, \tag{2.91a}$$

with

$$W^{(\Pi)}(0, \nabla) = 0, \qquad \left.\frac{\partial W^{(\Pi)}}{\partial t}\right]_{t=0} = W(\nabla), \tag{2.91b}$$

as its initial conditions and the differential operator with constant coefficients $W(\nabla)$ depends on the kind of field which is being considered or more precisely (as will be shown later on by another method) on its spin.

It is easy to obtain a solution of (2.91) which will satisfy these conditions; we shall express it by Taylor expansion about the point $t=0$. This equation may be written down in the form:

$$\frac{\partial^2 W^{(\Pi)}(t, \nabla)}{\partial t^2} = \{\nabla^2 - m^2\} W^{(\Pi)}(t, \nabla) \tag{2.92}$$

and the initial conditions (2.91) require that

$$\left.\frac{\partial^{2n} W^{(\Pi)}}{\partial t^{2n}}\right]_{t=0} = 0, \qquad \left.\frac{\partial^{2n+1} W^{(\Pi)}}{\partial t^{2n+1}}\right]_{t=0} = \{\nabla^2 - m^2\}^n W(\nabla), \tag{2.93}$$

such that $W^{(\Pi)}$ which is a function of $t-t'$ is expanded as follows:

$$W^{(\Pi)}(t, \nabla) = \sum_{n=0} \frac{t^{2n+1}}{(2n+1)!} \{\nabla^2 - m^2\}^n W(\nabla). \tag{2.94}$$

In this way we obtain

$$W^{(\Pi)}(t-t, \nabla)\delta(\boldsymbol{x}-\boldsymbol{x}') = \sum \frac{(t-t')^{2n+1}}{(2n+1)!} \{\nabla^2 - m^2\}^n W(\nabla)\delta(\boldsymbol{x}-\boldsymbol{x}'). \tag{2.95}$$

Using in the above expression the exponential representation of $\delta(\boldsymbol{x})$, its right-hand side becomes

$$\left(\frac{1}{2\pi}\right)^3 \int d^3\boldsymbol{k} e^{i\boldsymbol{k}(\boldsymbol{x}-\boldsymbol{x}')} W(i\boldsymbol{k}) \sum_n \frac{(t-t')^{2n+1}}{(2n+1)!} (-1)^n \frac{(\sqrt{\boldsymbol{k}^2+m^2})^{2n+1}}{\sqrt{\boldsymbol{k}^2+m^2}}$$

$$= -\left(\frac{1}{2\pi}\right)^3 \int d^3\boldsymbol{k} e^{i\boldsymbol{k}(\boldsymbol{x}-\boldsymbol{x}')} W(i\boldsymbol{k}) \frac{\sin\{(t-t')\sqrt{\boldsymbol{k}^2+m^2}\}}{\sqrt{\boldsymbol{k}^2+m^2}}.$$

Denoting by $\Delta(\boldsymbol{x}, t)$ the function

$$\Delta(\boldsymbol{x}, t) = \frac{-1}{(2\pi)^3} \int e^{i\boldsymbol{k}.\boldsymbol{x}} \frac{\sin\{t\sqrt{\boldsymbol{k}^2+m^2}\}}{\sqrt{\boldsymbol{k}^2+m^2}} d^3\boldsymbol{k}, \tag{2.96}$$

one sees that, in accordance with the well-known rules† concerning a product of two Fourier transforms, (2.95) may be written in the form

$$W^{(\Pi)}(t-t', \nabla)\delta(\boldsymbol{x}-\boldsymbol{x}') = -W(\nabla)\Delta(\boldsymbol{x}-\boldsymbol{x}', t-t').$$

We shall prove further on that $\Delta(\boldsymbol{x}, t)$ is a function of the four-dimensional argument x invariant under Lorentz transformations. The commutator (2.89) may thus be finally expressed in the form:

$$[Q_A(x), Q_B(x')] = +iW_{AB}(\nabla)\Delta(x-x'). \tag{2.97}$$

The relativistic invariance of the commutation rules depends therefore exclusively on the behaviour of the differential operator W and will have to be studied separately for each type of field. Examples of applications of this method will be found in Chaps. V [formula (5.26) *et seq.*] and VI [formula (6.1.80) *et seq.*].

10. Non-Hermitian fields

Note 2 in § 6 enables us to apply the above formalism to the field variables which are non-Hermitian operators corresponding as we have said to complex wave functions.

The commutation rules then take the following form:‡

$$\begin{aligned} &[Q_A(\boldsymbol{x}, t), Q_B(\boldsymbol{x}', t)] \\ &= [\tilde{Q}_A(\boldsymbol{x}, t), \tilde{Q}_B(\boldsymbol{x}', t)] = [\Pi_A(\boldsymbol{x}, t), \Pi_B(\boldsymbol{x}', t)] \\ &\qquad = [\tilde{\Pi}_A(\boldsymbol{x}, t), \tilde{\Pi}_B(\boldsymbol{x}', t)] = 0, \\ &[Q_A(\boldsymbol{x}, t), \Pi_B(\boldsymbol{x}', t)] = [\tilde{Q}_A(\boldsymbol{x}, t), \tilde{\Pi}_B(\boldsymbol{x}', t)] = i\delta_{AB}(\boldsymbol{x}-\boldsymbol{x}'). \end{aligned} \tag{2.98}$$

The independence of Q and $\tilde{Q}$ leads us to modify formula (2.87) as follows:

$$\begin{aligned} Q(\boldsymbol{x}, t) &= W^{(Q)}(t, t_0; \nabla)Q(\boldsymbol{x}, t_0) + W^{(\tilde{Q})}(t, t_0; \nabla)\tilde{Q}(\boldsymbol{x}, t_0) \\ &\quad + W^{(\Pi)}(t, t_0; \nabla)\tilde{Q}(\boldsymbol{x}, t_0) + W^{(\tilde{\Pi})}(t, t_0; \nabla)(\tilde{\Pi}\boldsymbol{x}, t_0), \end{aligned} \tag{2.99}$$

and commutation rules have to be computed as in (2.89); for instance, we have

$$[Q_A(x), \tilde{Q}_B(x')] = -iW^{(\tilde{\Pi})}_{AB}(t, t'; \nabla)\delta(\boldsymbol{x}-\boldsymbol{x}'). \tag{2.100}$$

† We have in fact

$$\int d^3\boldsymbol{x} e^{-i\boldsymbol{k}.\boldsymbol{x}} \mathcal{F}(\nabla)g(\boldsymbol{x})$$

$$= \int d^3\boldsymbol{x} e^{-i\boldsymbol{k}.\boldsymbol{x}} \mathcal{F}(\nabla)\left\{\frac{1}{(2\pi)^3}\int e^{i\boldsymbol{k}'.\boldsymbol{x}}(G\boldsymbol{k}')d^3\boldsymbol{k}'\right\} = \int \mathcal{F}(i\boldsymbol{k})G(\boldsymbol{k})\, d^3\boldsymbol{k}.$$

‡ We denote here by $\tilde{\Pi}$ the conjugate variable of $\tilde{Q}$, which may be different from the Hermitian conjugate of the conjugate variable of Q.

In the event that the field Q obeys eqn. (2.90) we obtain, setting

$$\left.\frac{\partial W^{(\tilde{\Pi})}(t, t'; \nabla)}{\partial t}\right]_{t=t'} = W(\nabla) \tag{2.101a}$$

an expression similar to (2.89)

$$Q_A(x), Q_B(x')] = -iW_{AB}(\nabla)\Delta(x-x'). \tag{2.101b}$$

PART 3

CANONICAL TRANSFORMATIONS AND PICTURES

11. General principles. Heisenberg picture

We have shown in Parts 1 and 2 that the evolution of the state vector and of the field variables was described by two unitary operators $\mathcal{U}[\sigma, \sigma_0]$ and $V(x, x')$ verifying

$$\left.\begin{aligned} i\frac{\delta\mathcal{U}[\sigma, \sigma_0]}{\delta\sigma(x)} &= \Omega[\sigma\,|\,x]\mathcal{U}[\sigma, \sigma_0], \\ i\frac{\partial V(x, x')}{\partial x_\mu} &= P_\mu V(x, x'), \end{aligned}\right\} \tag{2.102}$$

where

$$P_\mu = \int_{\sigma(x)} \left\{-\frac{\partial \mathcal{L}}{\partial Q_{A,\nu}(x')} Q_{A,\mu}(x') + \mathcal{L}\delta_{\mu\nu}\right\} d\sigma'_\nu \tag{2.103}$$

and that, furthermore,

$$\Psi[\sigma] = \mathcal{U}[\sigma, \sigma_0]\Psi[\sigma_0], \quad Q(x) = V(x, x')Q(x')V^{-1}(x, x'). \tag{2.104}$$

The quantum picture to be chosen is uniquely determined if we give both the operator $\mathcal{U}$ and V. From this we can compute the expection value of any local or non-local variable G according to the well-known formula

$$\langle G(t)\rangle \equiv \langle\Psi(t)\,|\,G(t)\,|\,\Psi(t)\rangle. \tag{2.105}$$

The derivative $\dfrac{d\langle G(t)\rangle}{dt}$ may also be calculated very easily:†

† The observable G may depend on t, either explicitly or through the time dependen field variables and their conjugate variables. The parameter t of $G(t)$ expresses only the former dependence.

$$\frac{d\langle G(t)\rangle}{dt} = \left\langle \frac{\partial \Psi(t)}{\partial t} \,|\, G(t) \,|\, \Psi(t) \right\rangle \tag{2.106}$$

$$+\left\langle \Psi(t) \,\middle|\, \frac{\partial G(t)}{\partial t} \,\middle|\, \Psi(t) \right\rangle + \left\langle \Psi(t) \,\middle|\, G(t) \,\middle|\, \frac{\partial \Psi(t)}{\partial t} \right\rangle$$

$$= \left\langle \Psi(t) \,\middle|\, \frac{\partial G(t)}{\partial t} + i[P_0(t) - \Omega(t), G(t)] \,\middle|\, \Psi(t) \right\rangle,$$

and we infer from this formula the following important theorem:

THEOREM. *The expectation value of any observable $G(t)$ is independent of t, provided that $G(t)$ is explicitly independent of t and that it commutes with $P_0 - \Omega$.*

We then say that G is a "constant of motion" in the chosen picture.

The description of a quantum field requires therefore a knowledge of the operators Ω and $\mathcal{D}_\mu$; an essential difference must however be noted between these operators: the latter of the two is determined as soon as we know the Lagrangian density, whereas the former is not. A further description is then necessary to determine Ω, namely:

Whenever a quantum system is completely *described by the quantization of a classical system, the state of this system does not change in time.*

This is, moreover, a consequence of the correspondence principle, since the classical physical quantity $G(x)$ and the expectation value $\langle \Psi | G(x) | \Psi \rangle$ of the observable which corresponds to it obey the same formal differential equation.

We thus define a particular quantum picture as "the Heisenberg picture", characterized by

$$\mathcal{U}[\sigma_0, \sigma_0] = 1, \quad V(x, x') \equiv V^{(H)}(x, x'), \tag{2.107}$$

From this it follows that

$$\Psi[\sigma] = \Psi[\sigma_0]. \tag{2.108}$$

We shall denote the Hamiltonian P_0 in this picture by H so that the canonical form of the field equations is

$$\frac{\partial Q^{(H)}}{\partial t} = i[H(t), Q^{(H)}(\boldsymbol{x}, t)] \tag{2.109}$$

the solution of which may be expressed in explicit form (writing V in place of $V^{(H)}$):

$$Q^{(H)}(\boldsymbol{x}, t) = V(t, t')Q^{(H)}(\boldsymbol{x}, t')V^{-1}(t, t'), \tag{2.110}$$

with

$$-i\frac{\partial V(t, t')}{\partial t} = H(t)V(t, t'), \quad V(t', t') = 1. \tag{2.111}$$

If H is independent of t, V is a function of $t-t'$ and can be expressed in exponential form:

$$V(t-t') = e^{+i(t-t')H}. \tag{2.112}$$

We note lastly that formula (2.106) proves that any observable $G^{(H)}$ which does not explicitly depend on t (but could depend on it through the time-dependent field and conjugate variables) and which commutes with H is a constant of motion.

Its eigenvalues are then "good quantum numbers", that is to say that they do not change in time and can help determine the constant state vector Ψ, an eigenvector of a complete set of commuting observables.

12. The various quantum pictures

It is well known that a canonical transformation generated by a unitary operator $\mathcal{U}$ does not modify the matrix elements and in particular the expectation value of the observables: it therefore modifies none of the results of measurements concerning the quantum system under consideration.†

From the Heisenberg picture ($\mathcal{U}=I$, $V^{(H)}$), we shall obtain any one of the canonical pictures f characterized by the evolution operator $\mathcal{U}^{(f)}$ of the state vector‡

$$\Psi^{(f)}(t) = \mathcal{U}^{(f)}(t, t_0)\,\Psi^{(H)}(t_0), \tag{2.113a}$$

$$Q^{(f)}(\boldsymbol{x}, t) = \mathcal{U}^{(f)}(t, t_0)Q^{(H)}(\boldsymbol{x}, t)\mathcal{U}^{(f)-1}(t, t_0), \tag{2.113b}$$

$\mathcal{U}^{(f)}(t, t_0)$ which is a unitary operator, consequently obeys the equation

$$i\frac{\partial\mathcal{U}^{(f)}(t, t_0)}{\partial t} = \Omega^{(f)}(t)\mathcal{U}^{(f)}(t, t_0), \quad \mathcal{U}^{(f)}(t_0, t_0) = 1, \tag{2.114}$$

where $\Omega^{(f)}$ is a Hermitian operator expressed with the help of the variablee $Q^{(f)}$.

It can easily be proved [cf. (2.26) for an analogous argument] that the operator $\Omega^{(f)}(t)$ in (2.114) depends only on t and not on t_0. Let us in fact suppose that $\Omega^{(f)}$ is of the form $\Omega^{(f)}(t, t_0)$ and let us differentiate with respect to t' the expression

$$\mathcal{U}^{(f)}(t, t')\mathcal{U}^{(f)}(t', t_0) = \mathcal{U}^{(f)}(t, t_0).$$

† The notion of a quantum picture first appeared in quantum mechanics in the form of Dirac's "Heisenberg and Schrödinger pictures", *loc. cit.*, p. 108 *et seq.* The interaction picture was introduced by S. Tomonaga, *Progr. Theor. Phys.*, **1**, 27 (1946); J. Schwinger, *Phys. Rev.*, **74**, 1439 (1948).

‡ We suppose that $Q^{(H)}$ and $Q^{(f)}$ represent the dynamically independent components of the field variables.

Making use of the property (c) of $\mathcal{U}^{(f)}(t, t')$ on p. 47, we easily obtain the following equality:

$$\Omega^{(f)}(t', t) = \Omega^{(f)}(t', t_0),$$

$\Omega^{(f)}(t, t_0)$ is therefore indeed independent of t_0.

When the density of $\Omega^{(f)}(t)$ is an analytical function of the variables $Q^{(f)}$ and $\Pi^{(f)}$ and of their space derivatives, we can easily express $\Omega^{(f)}$ in terms of Heisenberg operators. Indeed, we obtain

$$\Omega^{(f)}(t) = \int \Omega^{(f)}(Q^{(f)}(\boldsymbol{x}, t), \Pi^{(f)}, (\boldsymbol{x}, t) \ldots ; t) d^3\boldsymbol{x} \tag{2.115}$$

$$= \mathcal{U}^{(f)}(t, t_0) \int \Omega^{(f)}(Q^{(H)}, \Pi^{(H)}, \ldots ; t)\, d^3\boldsymbol{x}\, \mathcal{U}^{(f)-1}(t, t_0).$$

We see, therefore, that $\Omega^{(f)}$ is the canonically transformed operator of the integral $\int d^3\boldsymbol{x}\Omega^{(f)}(\ldots)$ where we have replaced the variables $Q^{(f)}, \Pi^{(f)}$. of the analytical function $\Omega^{(f)}(\ldots)$ by the Heisenberg variables $Q^{(H)}, \Pi^{(H)}$

Let us now differentiate (2.113*b*): by treatments analogous to those used to prove (2.35), we obtain

$$\frac{\partial Q^{(f)}(\boldsymbol{x}, t)}{\partial t} = i[Q^{(f)}(\boldsymbol{x}, t), \Omega^{(f)}(t)] \tag{2.116}$$

$$+ i\mathcal{U}^{(f)}(t, t_0)[H(t), Q^{(f)}(\boldsymbol{x}, t)]\mathcal{U}^{(f)-1}(t, t_0)$$

$$= i[\mathcal{U}^{(f)}(t, t_0)H(t)\mathcal{U}^{(f)-1}(t, t_0) - \Omega^{(f)}(t), Q^{(f)}(\boldsymbol{x}, t)].$$

Setting

$$\mathcal{U}^{(f)}(t, t_0)H(t)\mathcal{U}^{(f)-1}(t, t_0) \equiv H^{(f)}(t), \tag{2.117}$$

eqn. (2.116) takes the form

$$\frac{\partial Q^{(f)}(\boldsymbol{x}, t)}{\partial t} = i[H^{(f)}(t) - \Omega^{(f)}(t), Q^{(f)}(\boldsymbol{x}, t)] \equiv i[P_0^{(f)}(t), Q^{(f)}(\boldsymbol{x}, t)] \tag{2.118}$$

and this relation defines $P_0^{(f)}$; we have an analogous formula for the conjugate variables $\Pi(x)$.

Formula (2.118) may be interpreted like formula (2.115), when the Hamiltonian density $\mathcal{H}$ in the Heisenberg picture is an analytical functional of the Q's and Π's and their space derivatives

$$H^{(f)}(t) = \mathcal{U}^{(f)}(t, t_0) \int \mathcal{H}(Q^{(H)}(\boldsymbol{x}, t);\, \Pi^{(H)}(\boldsymbol{x}, t), \ldots ; t)\, d^3\boldsymbol{x}\, \mathcal{U}^{(f)-1}(t, t_0)$$

$$= \int \mathcal{H}(Q^{(f)}(\boldsymbol{x}, t), \Pi^{(f)}(\boldsymbol{x}, t), \ldots ; t)\, d^3\boldsymbol{x}, \tag{2.119}$$

in other words, to obtain $H^{(f)}(t)$, we must integrate over the whole three-dimensional space, the Hamiltonian density in the Heisenberg picture in which we have replaced the variables $Q^{(H)}$ and $\Pi^{(H)}$ by the variables $Q^{(f)}$and $\Pi^{(f)}$.

Consider in particular the Schrödinger picture, which we shall denote by the superscript S. It is characterized by the following choice of $\Omega^{(f)}(t)$:

$$\Omega^{(S)}(t) = H^{(S)}(t); \tag{2.120a}$$

formula (2.118) then proves that

$$\frac{\partial Q^{(S)}(\boldsymbol{x}, t)}{\partial t} = 0. \tag{2.120b}$$

Suppose, furthermore, that the field under consideration is conservative in the Heisenberg picture, in other words that the density $\mathcal{H}$ depends on t only through the $Q^{(S)}$'s and $\Pi^{(S)}$'s. Formulae (2.119) and (2.120*b*) prove that $H^{(S)}$ is independent of t and by virtue of (2.113) and (2.114) we obtain:

$$\Psi^{(S)}(t) = e^{-i(t-t_0)H^{(S)}}\Psi^{(H)}(t_0). \tag{2.121}$$

In all cases we infer from the above formulae that the state vectors and field variables in the Heisenberg or Schrödinger pictures coincide at the time $t = t_0$. Note, however (and we shall have occasion to return to this point), that the Heisenberg picture in which the field variables are functions of the space–time point is better suited to the formalism of quantum field theory than the Schrödinger picture in which they are only functions of the three-space point and not of the time; in particular, the relativistic variance of the formulae is more apparent in the Heisenberg picture.

Assume now that in the Heisenberg picture, the Hamiltonian is the sum of two partial Hamiltonians† (2.122)

$$H(t) = H_0(t)+H_1(t), \tag{2.122}$$

to which $H_0^{(f)}(t)$ and $H_1^{(f)}(t)$ correspond according to formula (2.117). We then obtain, by virtue of (2.118),

$$P_0^{(f)}(t) = H_0^{(f)}(t)+H_1^{(f)}(t)-\Omega^{(f)}(t).$$

We shall define the "interaction picture" (described by the superscript I) by choosing

$$\Omega^{(I)}(t) = H_1^{(I)}(t).$$

† Note that if, in a conservative system, H is independent of t, the partial Hamiltonians depend on it in such a way that $\frac{\partial H_0}{\partial t}+\frac{\partial H_1}{\partial t} = 0.$

The equations governing the evolution of the state vector and of the field variables [equations corresponding to (2.113*a*) and (2.113*b*)] are

$$i\frac{\partial \mathcal{U}^{(I)}(t, t_0)}{\partial t} = H_1^{(I)}(t)\mathcal{U}^{(I)}(t, t_0), \tag{2.123a}$$

$$\frac{\partial Q^{(I)}(\boldsymbol{x}, t)}{\partial t} = i[H_0^{(I)}(t), Q^{(I)}(\boldsymbol{x}, t)]. \tag{2.123b}$$

Turning now to the field variables, let us assume that $\mathcal{H}_0$ is an analytical function of the $Q^{(I)}$'s $\Pi^{(I)}$'s and their space derivatives; we then obtain

$$H_0^{(I)}(t) = \int \mathcal{H}_0(Q^{(I)}(\boldsymbol{x}, t), \Pi^{(I)}(\boldsymbol{x}, t), \ldots; t)\, d^3\boldsymbol{x}$$

and a similar relation for $H_1^{(I)}$.

On the other hand, in a plane t = constant, the commutation rules are independent of the picture adopted [(2.74), (2.81)], consequently the equation (2.123*b*) expresses the fact that the differential equation obeyed by the $Q^{(I)}$'s is the same as the one which determines the evolution of the field variables describing the field defined by the Hamiltonian density $\mathcal{H}_0$.

This picture is used whenever the field equations defined by $H_0(t)$ have a simple form: for instance, H_0 represents two non-interacting free fields, whereas H_1 represents their interaction. The interaction picture then treats the field variables of interacting systems as those of a free field, while the evolution of the state vector of the entire system is described by (2.123*a*). This picture is of the greatest importance in the calculation of scattering cross-sections.

Finally, let us examine the relations between the Schrödinger picture and the interaction picture; let $U^{(SI)}(t, t_0)$ be the transformation operator expressed with the help of the Schrödinger variables:

$$\Psi^{(S)}(t) = \mathcal{U}^{(SI)}(t, t_0)\Psi^{(I)}(t). \tag{2.124}$$

By introducing into the above formula the evolution operators $\mathcal{U}^{(S)}(t, t_0)$ of $\Psi^{(S)}$ and $\mathcal{U}^{(I)}(t, t_0)$ of $\Psi^{(I)}$, we obtain

$$\mathcal{U}^{(S)}(t, t_0) = \mathcal{U}^{(SI)}(t, t_0)\mathcal{U}^{(I)}(t, t_0). \tag{2.125}$$

By differentiating this formula with respect to t, and using the evolution equations of $\mathcal{U}^{(S)}$ and $\mathcal{U}^{(I)}$, we get

$$i\frac{\partial \mathcal{U}^{(S)}}{\partial t} \equiv (H_0^{(S)} + H_1^{(S)})\mathcal{U}^{(S)} = i\frac{\partial \mathcal{U}^{(SI)}}{\partial t}\mathcal{U}^{(I)} + \mathcal{U}^{(SI)}H_1^{(I)}\mathcal{U}^{(I)}.$$

Let us then express $\mathcal{U}^{(S)}$ by means of (2.125) and $H_1^{(I)}(t)$ with the aid of $H_1^{(S)}$ [cf. (2.124)]; we obtain as an equation for $\mathcal{U}^{(SI)}$:

$$i\frac{\partial \mathcal{U}^{(SI)}(t, t_0)}{\partial t} = H_0^{(S)}\mathcal{U}^{(SI)}(t, t_0), \quad \mathcal{U}^{(SI)}(t_0, t_0) = 1. \tag{2.126}$$

When in particular $H_0^{(S)}$ is independent of t:

$$\mathcal{U}^{(SI)}(t, t_0) = e^{-i(t-t_0)H_0{}^{(S)}} \tag{2.127}$$

Note finally that the time t_0, the time at which all the pictures coincide, is relegated in the applications to $t_0 = -\infty$. This assumption involves some highly delicate problems when taking the limit. We shall examine it in Chap. VIII, § 9.

PART 4

VARIATIONAL FORMULATION

13. Schwinger–Feynman variation principle

In § 6 of this chapter, we used an action principle which yields the Lagrangian and Hamiltonian equations. The Lagrangian density we thus introduced was classical, and therefore a c-number of the space $\mathcal{R}$.

Our purpose is now to formulate, following Schwinger, a variational principle concerning a Lagrangian density which is an operator on the space $\mathcal{R}$. This principle contains, as we shall be seeing, a large part of quantum field theory; it possesses great strength and elegance, but its application involves somewhat "tricky" considerations.†

Let the Hermitian action be

$$\mathcal{A}[\sigma(x), \sigma_0, Q] = \int_{\sigma_0}^{\sigma(x)} \mathcal{L}(Q(\xi), Q_{,\mu}(\xi), \xi)\, d^4\xi, \tag{2.128}$$

where $\sigma(x)$ represents a space-like surface σ passing through the point x and σ_0 is the surface of the very remote past where all the field variations vanish. We shall state our variational principle as follows:

Any Hermitian infinitesimal variation $\delta\mathcal{A}[\ldots]$ of the action induces a canonical transformation of the vector space in which the quantum system is defined, and the generator of this transformation is this same operator $\delta\mathcal{A}[\ldots]$.

We shall immediately apply this general proposition to the special case of the dynamically independent components of the field variables:

† J. Schwinger, *Phys. Rev.*, **82**, 914 (1951); **91**, 713 (1953). R. P. Feynman, *Rev. Mod. Phys.*, **20**, 367 (1948).

Let the Hermitian infinitesimal variation $\delta\mathcal{A}[\ldots]$ *be induced by the transformation of* $Q_A(x)$ *into* $Q'_A(x')$*; let us define*

$$\delta Q_A(x) = Q'_A(x) - Q_A(x), \tag{2.129}$$

where the two field variables Q'_A *and* Q_A *are taken at the same space–time point* x. *The Schwinger–Feynman principle may be written*

$$\delta Q_A(x) = i[\delta\mathcal{A}[\sigma(x), \sigma_0, Q, \ldots], Q_A(x)]. \tag{2.130}$$

Let us specify the variations of the field variables $Q_A(x)$: assume first that a variation of the Q_A's is induced by a change of the coordinates of an arbitrary space–time point,

$$x_\mu \to x'_\mu = x_\mu + \delta x_\mu(x), \tag{2.131a}$$

then,

$$Q(x) \to Q'(x') = Q(x) + \Delta Q(x), \tag{2.131b}$$

the points x and x' being infinitely close. We then have

$$\Delta Q(x) = Q'(x') - Q(x) \tag{2.132}$$

$$= Q'(x') - Q'(x) + Q'(x) - Q(x) = \delta x_\mu \frac{\partial Q}{\partial x_\mu} + \delta Q(x)$$

apart from an infinitesimal quantity of the second order, since in the first term of the right side we have replaced $\dfrac{\partial Q'}{\partial x_\mu}$ by $\dfrac{\partial Q}{\partial x_\mu}$.

In particular, if the variation of the field variables is arbitrary and not connected with a variation of the coordinates, we have:

$$\Delta Q(x) = \delta Q(x). \tag{2.133}$$

If on the other hand the variation of the coordinates is a translation δx_μ (and therefore independent of (x))† $Q'(x') = Q(x)$ and consequently,

$$\Delta Q(x) = 0,$$

therefore

$$\delta Q(x) = -\delta x_\mu \frac{\partial Q}{\partial x_\mu}. \tag{2.134}$$

We shall also assume that the field variables $Q_A(x)$ in which x is between two infinitely close surfaces σ form a complete ring of operators in that

† It must be noted that the δx_μ's are independent of x at a finite distance but that they are zero on σ_0 in conformity with the special role played by the latter surface specifying the remote past (cf. page 104, footnote †). It is moreover possible to avoid introducing this particular surface σ_0. One shows that for the variations usually considered, $\delta\mathcal{A}[\sigma_{II}, \sigma_I, \ldots]$ can be expressed as the difference: $F[\sigma_{II}, \ldots] - F[\sigma_I, \ldots]$; the generating operator of the transformation studied is postulated to be the operator $F[\sigma, \ldots]$.

any operator commuting with that set [and therefore with the Q_A's and their derivatives] is a multiple of the identity.†

We shall now make use of formula (2.130) in a case where the variations $\delta^{(1)}Q$, $\delta^{(2)}Q$, ... are independent, in other words such that if we consider the general variation $\delta^{(1)}Q+\delta^{(2)}Q+\ldots$ which corresponds to the variation of the action $\delta^{(1)}\mathcal{A}[\ldots]+\delta^{(2)}\mathcal{A}[\ldots]+\ldots$, we can cancel all the variations $\delta^{(j)}Q$ except one; it is to this one that we shall specifically apply formula (2.130). Note finally, that from a strictly logical standpoint, since we are unable to prove the existence of the operators δQ satisfying (2.130), the problem of the compatibility of the relations we shall infer by the above method remains to be solved. We need hardly add that if we cannot solve it as a whole, the investigation of free fields (Chap. VI) is an example in which this compatibility is achieved.

Let us first consider an arbitrary variation $\delta^{(1)}Q_A(x)$ of $Q_A(x)$ such that it vanishes on $\sigma(x)$:

$$\delta^{(1)}Q(x) = 0, \quad \text{for} \quad x \in \sigma.$$

By a treatment similar to that of (1.142), we easily see that

$$\delta^{(1)}\mathcal{A}[\sigma, \sigma_0, \ldots] = \int_{\sigma_0}^{\sigma} \mathcal{F}_A(\xi)\delta^{(1)}Q_A(\xi)\, d^4\xi, \tag{2.135a}$$

where $\mathcal{F}_A(x)$ is the left-hand side of the Lagrangian equations:

$$\mathcal{F}_A(x) = \frac{\partial \mathcal{L}}{\partial Q_A} - \partial_\mu \frac{\partial \mathcal{L}}{\partial Q_{A,\mu}}. \tag{2.135b}$$

The principle (2.130) thus leads to

$$0 = \left[\int_{\sigma_0}^{\sigma} \mathcal{F}_B(\xi)\delta^{(1)}Q_B(\xi)\, d^4\xi, Q_A(x)\right]. \tag{2.136}$$

Take $\delta^{(1)}Q(x)$ depending on x as a δ-function about the point X enclosed between the surfaces σ_0 and σ; the above formula is reduced to

$$[\mathcal{F}_B(X)\delta^{(1)}Q_B(X), Q_A(x)] = 0 \tag{2.137a}$$

and since the Q's form a complete ring of operators $\mathcal{F}_B(X)_B\delta^{(1)}Q_B(X)$ is a c-number for the space $\mathcal{R}$. On the other hand, let us choose $\delta^{(1)}Q_B(x)$ itself as a c-number; the above formula becomes

$$[\mathcal{F}_B(X), Q_A(x)] = 0. \tag{2.137b}$$

$\mathcal{F}_B(X)$is consequently itself a c-number. If therefore we choose in (2.137a) the variation $\delta^{(1)}Q_B(X)$ as an arbitrary operator, it is necessary that

$$\mathcal{F}_B(X) = 0. \tag{2.138}$$

† Not to be confused with the complete set of commuting observables (cf. p. 43).

We obtain in this way the Lagrangian equations.

An important remark must be made here in the interests of strict reasoning. Since $\mathcal{L}(Q,...)$ and the field variables themselves $Q_A(x)$ are operators on the space $\mathcal{R}$, such derivatives as $\dfrac{\partial\mathcal{L}}{\partial Q_A}$ or $\dfrac{\partial\mathcal{L}}{\partial Q_{A,\mu}}$ are no longer defined in a unique way.† Suppose indeed that this density contains a term like $[Q_A(x)]^3$; the derivative $\dfrac{\partial\mathcal{L}}{\partial Q_A}$ could only be inferred from the increase

$$\delta\mathcal{L} = \delta Q_A(Q_A(x))^2+Q_A(x)\delta Q_A(x)Q_A(x)+(Q_A(x))^2\delta Q_A(x)$$

if we knew by virtue of what rules of commutation between the δQ's and the Q's we can displace the quantities $\delta Q(x)$ on the extreme right (or on the extreme left), defining in this way a right-derivative (or left-derivative). Without these rules, which could not be stated *a priori*, the only conclusion we can reach with any certainty is that if $\delta^{(1)}\mathcal{A}[\ldots]$ can be expressed in the form (2.135*a*) with $\delta^{(1)}Q$ on the right or left of $\mathcal{F}(x)$, then the field variables satisfy (2.138) and in all the cases we shall be considering hereafter these equations will be identical with the Lagrangian ones, as may be easily verified. This remark is of a general nature and applies to the discussions which follow, in particular to the analytical definition of the $\Pi_A(x)$ [for which we further postulate in (2.142) the commutation rules of the δQ's and the field variables], and to the definition of $T_{\mu\nu}$ and P_μ.

Having made these general observations, let us now consider a second variation $\delta^{(2)}Q$ such that it is not necessarily zero on σ and let us return to the calculation of $\delta^{(2)}\mathcal{A}[\ldots]$:

$$\delta^{(2)}\mathcal{A}[\ldots] = \int_{\sigma_0}^{\sigma}\left[\frac{\partial\mathcal{L}}{\partial Q_A}\,\delta^{(2)}Q_A+\frac{\partial\mathcal{L}}{\partial Q_{A,\mu}}\,\delta^{(2)}Q_{A,\mu}\right]d^4\xi \tag{2.139}$$

$$= \int_{\sigma_0}^{\sigma}\left[\frac{\partial\mathcal{L}}{\partial Q_A}-\partial\mu\,\frac{\partial\mathcal{L}}{\partial Q_{A,\mu}}\right]\delta^{(2)}Q_A(\xi)\,d^4\xi$$

$$+\int_{\sigma_0}^{\sigma}\frac{\partial}{\partial\xi_\mu}\left\{\frac{\partial\mathcal{L}}{\partial Q_{A,\mu}}\,\delta^{(2)}Q_A(\xi)\right\}d^4\xi$$

† This difficulty did not arise in Part 3 of this chapter in which the possibility of defining the variations of the field variables as operators was not exploited. If in fact we have a function F of the operator X, the derivative

$$\frac{F(X+\delta X)-F(X)}{\delta X}$$

has a precise meaning when δX is a *c*-number of the form εI where ε is infinitesimal. This was in fact the point of view adopted in this part.

since

$$\partial_\mu \delta^{(2)} Q_A = \delta^{(2)} \partial_\mu Q_A.$$

The second integral is transformed by applying Gauss's theorem, into

$$\int_{\sigma(x)} \frac{\partial \mathcal{L}}{\partial Q_{A,\mu}} \delta^{(2)} Q_A \, d\sigma_\mu = \int_{\sigma(x)} \Pi_A(\xi) \delta^{(2)} Q_A(\xi) \, d\sigma(\xi),$$

where we have set $\Pi_A(x) = \dfrac{\delta \mathcal{L}}{\partial Q_{A,\mu}} n_\mu$. Making use of the Lagrangian equations, the variation of the action takes the form

$$\delta^{(2)} \mathcal{A}[\sigma, \sigma_0, Q, \delta^{(2)} Q] = \int_{\sigma(x)} \Pi_A(\xi) \delta^{(2)} Q_A(\xi) \, d\sigma(\xi) \tag{2.140}$$

and the Schwinger–Feynman principle leads to

$$\delta^{(2)} Q_A(x) = i \left[\int_{\sigma(x)} \Pi_B(\xi) \delta^{(2)} Q_B(\xi) \, d\sigma(\xi), Q_A(x) \right]. \tag{2.141}$$

A certain arbitrariness still exists at this stage as regards the definition of the variations $\delta^{(2)} Q(x)$: we suppose that they commute or anticommute on a space-like surface $\sigma(x)$ with the field variables†

$$\left.\begin{aligned} [\delta^{(2)} Q_A(x), Q_B(x')]_\mp \equiv \delta^{(2)} Q_A(x) Q_B(x') \mp Q_B(x') \delta^{(2)} Q_A(x) = 0 \\ \text{for } x, x' \in \sigma. \end{aligned}\right\} \tag{2.142}$$

The choice of the symbols $[\ldots]_-$ or $[\ldots]_+$ is not arbitrary; it is necessitated by physical considerations concerning the fields under investigation [cf. the remarks following formula (2.145)] and is conditioned by the structure of the Lagrangian density, as will be seen in an example at the end of this section. On the other hand, by virtue of the well-known rule

$$[A_1, A_2 A_3] = [A_1, A_2]_\mp A_3 - A_2 [A_3, A_1]_\mp, \tag{2.143}$$

the right-hand side of (2.141) can also be written

$$\begin{aligned} \delta^{(2)} Q_A(x) = & -i \int_{\sigma(x)} [Q_A(x), \Pi_B(\xi)]_\mp \delta^{(2)} Q_B(\xi) \, d\sigma(\xi) \\ & + i \int_{\sigma(x)} \Pi_B(\xi) [\delta^{(2)} Q_B(\xi), Q_A(x)]_\mp \, d\sigma(\xi) \end{aligned} \tag{2.144}$$

† The existence of operators $\delta^{(2)} Q$ commuting with all the field variables on a surface σ presents no difficulties since we may choose them as c-numbers; the partial derivatives of the Lagrangian density and therefore the conjugate variables are then perfectly defined. The same cannot be said for variations anticommuting with the field variables: their existence seriously restricts the choice of the $Q(x)$.

and according to (2.142) the second integral is zero and we obtain

$$[Q_A(x), \Pi_B(x')]_{\mp} = i\delta_{AB}\delta^{(\sigma)}(x-x') \quad \text{for} \quad x, x' \in \sigma, \tag{2.145}$$

where the definition of $\delta^{(\sigma)}(x)$ can be readily formulated [cf. also (5.12)].

Thus the anticommutation relations appear quite naturally in the course of the calculations. We shall see further on that the anticommuting fields or "fermion fields" obey Fermi statistics in which it is impossible to create more than one particle in a particular state; the commuting fields or "boson fields" obey Bose statistics in which an indeterminate number of particles can be created in the same state. Electrons and nucleons are fermions; photons and Π mesons are bosons.

Consider finally a displacement of space–time points accompanied by a variation of the field variables in accordance with formulae (2.131).

As a result of the variation of the coordinates the surface σ comes into σ' and the variation of the action is

$$\delta\mathcal{A}[\ldots] = \int_{\sigma_0}^{\sigma'} \mathcal{L}(Q'(\xi'), \ldots)\, d^4\xi' - \int_{\sigma_0}^{\sigma} \mathcal{L}(Q(\xi), \ldots)\, d^4\xi, \tag{2.146}$$

with

$$d^4\xi' = \frac{D(\xi_1', \xi_2', \xi_3', \xi_0')}{D(\xi_1, \xi_2, \xi_3, \xi_0)}\, d^4\xi = \left(1 + \frac{\partial\, \delta\xi_\nu}{\partial \xi_\nu}\right) d^4\xi.$$

We thus obtain

$$\int_{\sigma_0}^{\sigma'} \mathcal{L}(Q'(\xi'), \ldots)\, d^4\xi' = \int_{\sigma_0}^{\sigma} \mathcal{L}(Q'(\xi+\delta\xi), \ldots)\left(1 + \frac{\partial\delta\xi_\nu}{\partial\xi_\nu}\right) d^4\xi$$
$$= \int_{\sigma_0}^{\sigma} \left\{ \mathcal{L}(Q'(\xi), \ldots) + \frac{\partial\mathcal{L}}{\partial Q_A'} \frac{\partial Q_A'}{\partial \xi_\mu}\, \delta\xi_\mu \right.$$
$$\left. + \frac{\partial\mathcal{L}}{\partial Q'_{A,\nu}} \frac{\partial Q'_{A,\nu}}{\partial \xi_\mu}\, \delta\xi_\mu + \frac{\partial\mathcal{L}}{\partial \xi_\mu}\, \delta\xi_\mu \right\} \left\{ 1 + \frac{\partial\delta\xi_\lambda}{\partial\xi_\lambda} \right\} d^4\xi,$$

and by neglecting the infinitesimal quantities of higher order and carrying out the products, one gets

$$\int_{\sigma_0}^{\sigma'} \mathcal{L}(Q'(\xi), \ldots)\, d^4\xi'$$
$$= \int_{\sigma_0}^{\sigma} \left[\mathcal{L}(Q'(\xi), \ldots) + \frac{\partial}{\partial \xi_\mu} \{\mathcal{L}(Q'(\xi), Q_\lambda'(\xi), \xi)\, \delta\xi_\mu\} \right] d^4\xi.$$

By applying Green's theorem to the second integral of the right side, the above expression becomes

$$\int_{\sigma_0}^{\sigma} \mathcal{L}(Q'(\xi), \ldots)\, d^4\xi + \int_{\sigma(x)} \mathcal{L}(Q'(\xi), \ldots)\, \delta\xi_\mu\, d\sigma_\mu(\xi)$$

and noting that apart from infinitesimal quantities of higher order, we can replace in the surface integral $Q'(\xi)$ by $Q(\xi)$, the variation of the action is finally expressed in the form

$$\delta\mathcal{A}[\ldots] = \int_{\sigma_0}^{\sigma} [\mathcal{L}(Q'(\xi), \ldots)-\mathcal{L}(Q(\xi), \ldots)]\, d^4\xi + \int_{\sigma} \mathcal{L}(Q(\xi), \ldots)\delta\xi_\mu\, d\sigma_\mu. \tag{2.147}$$

In the above volume integral it will be noted that the two Lagrangian densities are taken at the same point ξ. Noting that the symbol δ introduced in (2.129) commutes with ∂_μ, since†

$$\left.\begin{aligned} \partial_\mu \delta Q &= \delta_\mu\{Q'(x)-Q(x)\} = \partial_\mu Q'(x)-\partial_\mu Q(x), \\ \delta\partial_\mu Q &= \partial_\mu Q'(x)-\partial_\mu Q(x), \end{aligned}\right\} \tag{2.148}$$

we can use the same method which led to (2.140) and write

$$\delta\mathcal{A}[\ldots] = \int_{\sigma(x)} \Pi_A(\xi)\,\delta Q_A(\xi)\, d\sigma(\xi) + \int_{\sigma(x)} \mathcal{L}(Q(\xi), \ldots)\,\delta\xi_\mu\, d\sigma_\mu. \tag{2.149}$$

Returning to the independent variations ΔQ and δx_μ, we can express the variation of the action in the form

$$\delta\mathcal{A}[\ldots] = \int_{\sigma(x)} \Pi_A(\xi)\,\Delta Q_A(\xi)\, d\sigma(\xi) + \int_{\sigma(x)} T_{\nu\mu}(\xi)\,\delta\xi_\nu\, d\sigma_\mu \tag{2.150}$$

where $T_{\nu\mu}$ is expressed as above:

$$T_{\nu\mu} = -\frac{\partial\mathcal{L}}{\partial Q_{A,\mu}}\, Q_{A,\nu} + \mathcal{L}\delta_{\mu\nu}. \tag{2.151}$$

Formula (2.150) will not be used for the more general variations until the next chapter; for the time being we shall content ourselves with assum-

† It should be noted that this is no longer the case for the variation Δ, since from $\Delta Q = Q'(x')-Q(x)$ we infer

$$\begin{aligned} \partial_\mu \Delta Q = \partial_\mu \{Q'(x+\delta x)-Q(x)\} &= \partial_\mu\left\{Q'(x)+\delta x_\lambda \frac{\partial Q'(x)}{\partial x_\lambda} - Q(x)\right\} \\ &= \partial_\mu Q'(x)-\partial_\mu Q(x)+\partial_\mu\{\delta x_\varrho \partial_\varrho Q'\}. \end{aligned}$$

On the other hand

$$\Delta\partial_\mu Q'(x) = \frac{\partial Q'(x+\delta x)}{\partial x'_\mu} - \frac{\partial Q}{\partial x_\mu} = \partial_\mu \delta Q + \partial_\mu\{\delta x_\varrho \partial_\varrho Q(x')\} - \frac{\partial \delta x_\nu}{\partial x'_\lambda}\partial_\nu Q'(x')$$

so that, to the first order, we have

$$\partial_\mu \Delta Q - \Delta\partial_\mu Q = \frac{\partial\, \delta x_\lambda}{\partial x_\mu}\, \frac{\partial Q}{\partial x_\lambda}.$$

ing that $\delta\xi_\mu$, independent of ξ, represents a translation, in other words a transformation such that

$$\Delta Q(x) = 0. \tag{2.152}$$

Making use of (2.134) the variation of the action takes the simplified form:

$$\delta\mathcal{A}[\ldots] = \delta\xi_\nu \int_{\sigma(x)} T_{\nu\mu}(\xi)\, d\sigma_\mu(\xi) = \delta\xi_\nu P_\nu[\sigma] \tag{2.153}$$

and the Schwinger principle enables us to write the canonical equations

$$i\frac{\partial Q_A}{\partial x_\mu} = [P_\mu[\sigma], Q_A(x)]. \tag{2.154}$$

We have thus inferred from the same variation principle, Lagrangian equations, commutation relations and the existence of a displacement operator; this principle contains, therefore, in a particularly elegant and compact form, the entire canonical formalism of quantum field theory.

Let us now return to the possibility we mentioned earlier, of choosing between commutation and anticommutation rules, in other words of fixing the type of commutation relations. We pointed out that this freedom of choice was restricted by the form of Lagrangian density used. To illustrate this very important point, let us consider the case of a neutral scalar field $\Phi(x)$ [$\Phi(x)$ Hermitian]. The action may be written [cf. (5.1) and (5.21)]:

$$\mathcal{A}[\sigma, \sigma_0, \Phi] = -\frac{1}{2}\int_{\sigma_0}^{\sigma(x)} \left[\frac{\partial\Phi}{\partial\xi_\mu}\frac{\partial\Phi}{\partial\xi_\mu} + m^2\Phi^2\right] d^4\xi. \tag{2.155}$$

By the techniques used above, it is easy to state that to a variation $\delta\Phi$ there corresponds a variation of the action

$$\delta\mathcal{A}[\ldots] = +\frac{1}{2}\int_{\sigma_0}^{\sigma} [\delta\Phi(\xi)\{\Box - m^2\}\Phi(\xi) + \{\Box - m^2\}\Phi(\xi)\delta\Phi(\xi)]\, d^4\xi$$
$$-\frac{1}{2}\int_{\sigma(x)} \left(\delta\Phi(\xi)\frac{\partial\Phi(\xi)}{\partial\xi_\mu}n_\mu + \frac{\partial\Phi}{\partial\xi_\mu}\delta\Phi(\xi)n_\mu\right) d\sigma(\xi). \tag{2.156}$$

Suppose $\delta\Phi$ to be of the type $\delta^{(1)}\Phi$; the surface integral vanishes and we can conclude by virtue of (2.137*a*) that the expression

$$\delta^{(1)}\Phi\{\Box - m^2\}\Phi(x) + \{\Box - m^2\}\Phi(x)\delta^{(1)}\Phi(x)$$

is a *c*-number.

On the other hand, choosing $\delta^{(1)}\Phi$ itself as a *c*-number, we conclude that $\{\Box - m^2\}\Phi(x)$ is a *c*-number. Now even for a $\delta^{(1)}\Phi$ operator, we have just shown that $\{\Box - m^2\}\Phi(x)\delta^{(1)}\Phi(x)$ is also a c-number; so that making use of (2.138) we can write:

$$\{\Box - m^2\}\Phi(x) = 0.$$

Let us now suppose that $\delta\Phi$ is of the type $\delta^{(2)}\Phi$; the variation of the action is of the type (2.140) and according to (2.156):

$$\delta^{(2)}\mathcal{A}[\ldots] = -\frac{1}{2}\int_{\sigma(x)} \left[\delta\Phi(\xi), \frac{\partial\Phi}{\partial\xi_\mu} n_\mu \right]_+ d\sigma(\xi). \tag{2.157a}$$

If $\delta\Phi(\xi)$ commutes with $\Phi(\xi)$ and its derivatives, we then obtain

$$\delta^{(2)}\mathcal{A}[\ldots] = \int_{\sigma(x)} -\frac{\partial\Phi}{\partial\xi_\mu} n_\mu(\xi)\delta^{(2)}\Phi(\xi)\, d\sigma(\xi) = \int_{\sigma(x)} \Pi(\xi)\ \delta^{(2)}\Phi(\xi)\, d\sigma(\xi) \tag{2.157b}$$

from which the conjugate variable $\Pi(\xi)$ and the commutation rules between Φ and Π can be readily obtained.

If on the contrary, $\delta^{(2)}\Phi(\xi)$ anticommutes, then $\delta^{(2)}\mathcal{A}[\ldots]$ is zero and we can no longer define a conjugate variable satisfying the anticommutation relations (2.145), and are therefore no longer able to carry out a canonical quantization.

It is interesting, finally, to interpret (2.130) with the aid of the concept of group representation. Let us assume for this purpose that the variation $\delta\mathcal{A}[\ldots]$ is due to a transformation operating on the field variables and belonging to a particular group of transformations T. Let us further assume that the action remains invariant under this group; it can then be easily proved that $\delta\mathcal{A}[\sigma, \sigma_0, \ldots]$ is independent of σ [cf. text following (3.35)]. Thus (2.130) can be written

$$\begin{aligned}\delta Q_A(x) &= (1+i\delta\mathcal{A}[\ldots])Q_A(x)\,(1+i\delta\mathcal{A}[\ldots])^{-1}\\ &= U^{-1}[\delta Q, \ldots]Q_A(x)U[\delta Q, \ldots]\end{aligned}$$

which enables us to define by iteration the unitary operator $U[Q, \ldots]$ such that

$$Q'(x) = U^{-1}Q(x)U \tag{2.158}$$

and consequently to obtain a unitary representation of the group T on the space $\mathcal{R}$.

Note. By introducing the derivatives of the Dirac function $\delta^{(\sigma)}(x)$, formula (2.154) enables us to write down commutation relations between $T_{\mu\nu}(x')$ and $Q_A(x)$; this formula is indeed expressed in the form

$$i\frac{\partial Q_A}{\partial x_\mu} = \int_\sigma d\sigma'_\nu\,[T_{\mu\nu}(x'), Q_A(x')] = \int_\sigma d\sigma'\, n_\nu(x')\,[T_{\mu\nu}(x'), Q_A(x)],$$

consequently

$$n_\nu(x')\,[T_{\mu\nu}(x'), Q_A(x)] = i\,\frac{\partial}{\partial x_\mu}\,\delta^{(\sigma)}(x-x')\,Q_A(x'). \tag{2.159}$$

14. Local character of the field equations and structure of the Hamiltonian. Applications

Let us consider a Hamiltonian corresponding to anticommuting fields. It is easy to verify that if it is composed as in (2.72), (2.75) of a single anticommuting field, the relation (2.71) is no longer valid and consequently the field equations can no longer be expressed in the canonical form (2.70*a*).

This remark holds if the Hamiltonian is composed of three anticommuting fields; we shall even prove that it is impossible to express the field equations in canonical form when the Hamiltonian is composed of an odd number of anticommuting fields. Some interesting results may be inferred from this property.†

As a direct application, consider the case of Bremsstrahlung: in the presence of a classical external field

$$e \to e+\gamma,$$

the Hamiltonian density expressing this interaction must be expressed by means of the fields appearing in this interaction; it can therefore contain only two anticommuting field variables (those relative to the electron). The field variable of the photon must commute with that of the electrons.

Let us return to the theorem we are interested in, namely *that the Hamiltonian density must be composed of an even number of fields anticommuting with a given field.* To this end, we shall base our argument on an assumption we have not explicitly made use of hitherto: that of the local character of the canonical equations (2.70*a*). It consists in postulating that by writing them down in the form

$$i\,\frac{\partial Q(X,t)}{\partial t} = \int_t [Q(\boldsymbol{X},t), \Lambda(\boldsymbol{x},t)]\,d^3\boldsymbol{x},$$

the commutator of the right side must depend solely on the point $\boldsymbol{X}$ and not on both $\boldsymbol{X}$ and $\boldsymbol{x}$. We shall discuss this point more thoroughly later on.

We shall confine ourselves to three fields, leaving it to the reader to expand the theorem to a larger number of fields.

We assume, therefore, that

$$P_0 = \int Q^{(1)}(\boldsymbol{x},t)Q^{(2)}(\boldsymbol{x},t)Q^{(3)}(\boldsymbol{x},t)\,d^3\boldsymbol{x},$$

† T. Kinoshita, *Phys. Rev.*, **96**, 199 (1954); H. Umezawa, J. Podolanski and S. Oneda, *Proc. Phys. Soc.*, A **68** (1955).

where $Q^{(j)}(x)$ $(j = 1, 2, 3)$ corresponds to three given fields. We propose to calculate the commutator

$$[Q^{(1)}(x)Q^{(2)}(x)Q^{(3)}(x),\ Q(X)]^{(t)}. \tag{2.160}$$

Here $Q(X)$ is another field variable and the superscript (t) means that the points x and X are taken at the same time t.

We shall write the commutation and anticommutation rules of the field Q with the fields $Q^{(j)}$ in the following form:

$$Q^{(j)}(\boldsymbol{x}, t)Q(\boldsymbol{X}, t)-\varepsilon_j Q(\boldsymbol{X}, t)Q^{(j)}(\boldsymbol{x}, t) = \alpha_j \delta(\boldsymbol{x}-\boldsymbol{X}), \tag{2.161}$$

where ε_j is a number equal to $+1$ or -1 according as the fields commute or anticommute and where α_j is another number (which may be zero) but in all cases independent of $\boldsymbol{x}, \boldsymbol{X}$ and t. All the commutation and anticommutation relations considered above belong in fact to this type.

On the other hand, (2.160) can also be written (as may be easily verified) in the form

$$\begin{aligned} &Q^{(1)}(x)Q^{(2)}(x)\,(Q^{(3)}(x)Q(X)-\varepsilon_3 Q(X)Q^{(3)}(x)) \\ &+\varepsilon_3 Q^{(1)}(x)\,(Q^{(2)}(x)Q(X)-\varepsilon_2 Q(X)Q^{(2)}(x))Q^{(3)}(x) \\ &+\ \varepsilon_3\varepsilon_2(Q)^{(1)}(x)Q(X)\ \ -\varepsilon_1 Q(X)Q^{(1)}(x))Q^{(2)}(x)Q^{(3)}(x) \\ &+(\varepsilon_3\varepsilon_2\varepsilon_1-1)Q(X)Q^{(1)}(x)Q^{(2)}(x)Q^{(3)}(x), \end{aligned} \tag{2.162}$$

the times for the points X and x being the same.

All the terms of the preceding expression depend finally on X [through the medium of the $\delta(\boldsymbol{x}-\boldsymbol{X})$ of (2.161)], except the last one, which depends simultaneously on $\boldsymbol{X}$, $\boldsymbol{x}$ and t.

In the case of a local field, we have to cancel this term, in other words

$$\varepsilon_3\varepsilon_2\varepsilon_1-1 = 0.$$

This expression shows that only an even number of fields $Q^{(j)}$ can anticommute with $Q(x)$. The extension to a Hamiltonian density composed of n field variables can be easily obtained.†

Let us discuss as a second application the reaction‡

$$P+P \rightarrow P+N+\Pi^{+}.$$

Consider the field P (proton) as a given field. The Hamiltonian describing this reaction contains a proton field, therefore a field anticommuting with the given field P; it is therefore necessary

† In fact, the latter term gives the following contribution to the bracket expressing $\dfrac{\partial Q(X)}{\partial t}$:

$$(\varepsilon_3\varepsilon_2\varepsilon_1-I)Q(\boldsymbol{X}, t)\int_t Q^{(1)}(\boldsymbol{x}, t)Q^{(2)}(\boldsymbol{x}, t)Q^{(3)}(\boldsymbol{x}, t)\, d^3\boldsymbol{x}.$$

This derivative would therefore depend on all values of field variables $Q^j(x)$ at the time t; and the denial of this dependence is in fact the very essence of localizability.

‡ This reaction is actually a secondary reaction, the primary reaction being $P \rightarrow N+\pi^{+}$.

(a) either that the neutron field N should anticommute with P and that the positive meson field Π^+ should commute with P;

(b) or that the field N should commute with P while Π^+ anticommutes with P.

On the other hand, we have: $P+P \rightarrow P+P+\Pi^0$, which proves by a similar argument that the neutral meson field (Π^0) commutes with P.

If therefore we assume that charged and neutral mesons have the same type of commutation relations (they belong to the same family), we see that the assumption (a) above must be chosen and (b) excluded: the neutron field anticommutes with the proton one.

15. Peierls relation

If G is an operator depending on the field variables $Q(x)$, its variation δG at a point x is clearly given by a formula analogous to (2.130):

$$\delta G(x) = i[\delta \mathcal{A}[\sigma(x), \ldots], G(x)].$$

Let $J_1(x)$ and $J_2(x)$ be two infinitesimal operators depending on the field variables; let us choose the following variations of the Lagrangian density

$$\delta_{J_1}\mathcal{L} = \delta(x-\xi)J_1(x),$$
$$\delta_{J_2}\mathcal{L} = \delta(x-\eta)J_2(x),$$

to which correspond the following variations of the action

$$\delta_{J_1}\mathcal{A} = J_1(\xi), \qquad \delta_{J_2}\mathcal{A} = J_2(\eta)$$

provided that ξ_0 and η_0 are smaller than x_0 [in other words, that the points ξ and η are earlier in time than $\sigma(x)$], otherwise they are zero. As a result of the variation $\delta_{J_1}\mathcal{A}$, the operator J_2 considered as a function of the Q's suffers an increase

$$\delta_{J_2}J_1(x) = i[J_2(\eta), J_1(x)].$$

Similarly,

$$\delta_{J_1}J_2(x) = i[J_1(\xi), J_2(x)]. \tag{2.163}$$

Replace in the first of these equations η by x and x by ξ; it takes the form

$$\delta_{J_2}J_1(\xi) = i[J_2(x), J_1(\xi)]. \tag{2.164}$$

The variations $\delta_{J_2}J_1(\xi)$ and $\delta_{J_1}J_2(\xi)$ are zero when respectively $\xi_0 > x_0$ and $\xi_0 < x_0$; we can therefore group together (2.164) and (2.163) in the Peierls relation:

$$[J_1(\xi), J_2(x)] = i\,\delta_{J_2}J_1(\xi) - i\,\delta_{J_1}J_2(x). \tag{2.165}$$

This formula may be used to obtain the covariant form of the commutation rules (that is to say, the commutator of two field variables taken at two different space–time points x and x').†

† R. Peierls, *Proc. Roy. Soc.*, **A 214**, 143 (1957).

CHAPTER III

Fundamental field observables

WE PROPOSE to study in this chapter a number of non-local observables which characterize a field (or interacting fields): observables which are independent of the surface on which they are defined when the action describing the field under consideration (or the fields and their interaction) does not depend explicitly on time. In the Heisenberg picture, these observables are constants of motion (cf. Chap. II, end of § 11).

The definition of these observables will be based essentially on the invariance of the action under particular groups of transformations and on the Schwinger–Feynman variation principle: we shall establish in this way that the generating operators of these groups are independent of t and we shall give the commutation rules which characterize them.

The inhomogeneous Lorentz group (the continuous group of proper transformations, called also the Poincaré group) will lead us to define the energy–momentum tensor and the spin and orbital angular momentum tensors. The invariance of the action under the space reflections will lead us to define parity. Finally, the invariance of the Lagrangian under gauge transformations of the first kind (a particular case of gauge invariance which will be described more thoroughly in Chap. VII) will enable us to define the charge–current vector and the charge of a field.

We shall also study the charge conjugation (a transformation under which wave equations remain invariant,† but the sign of the current changes), but we shall reserve to a later chapter the study of the time reversal (cf. Chap. V, § 20). It is worth noting that the charge conjugation enables us to discriminate to some extent between the commutator and anticommutator methods of quantization.

† It would be more exact to speak of the "covariance" than the "invariance" of the equations. The vast majority of physicists use the latter term; as a rule we shall follow their example.

The applications of the ideas contained in this chapter are fundamental in the theory of elementary particles: in order to study these applications in a more intensive and concise manner, we have reserved them for Chap. IV, § 2, and Chap. VII, §§ 7, 8 and 9.

1. Invariance of a quantum field theory

The Lorentz invariance or the invariance under any other group of a quantum field theory is a concept which might be liable to a certain ambiguity if we gave no further information about it.

A quantum field theory includes, we have just seen, differential equations, commutation rules and a state vector by means of which expectation values are expressed.

It is possible, when speaking of invariance, to confine oneself to specifying that of the differential equations. One can study, for example, the invariance of the Maxwell equations or of the Dirac equation under the Lorentz group. One can define in this way invariants and covariants for each given field. Consider the Maxwell field as defined by its vector-potential $A_\mu(x)$; we can define as the covariant of this field under the Lorentz group the skew-symmetrical tensor $F_{\mu\nu}$ representing electric and magnetic fields. Similarly, for the Dirac field, we can define the well-known sixteen bilinear covariants: scalar, pseudo-scalar, vector, etc. This is a procedure which can be carried out entirely within the framework of classical physics, the field variables being *c*-numbers of the space $\mathcal{R}$, that is to say wave functions.

We can also consider simultaneously the invariance of the field equations and that of the commutation relations: we then operate within the framework of quantum mechanics. Since differential equations and commutation relations determine in a unique way the algebra of the field variables, we may infer (but for systems with an infinite number of degrees of freedom no conclusive proof has been given) that the transformed variables are the variables canonically transformed (1.68) in the space $\mathcal{R}$.†

These considerations hold within the framework of any picture: in accordance with the distinction made in § 4 of the preceding chapter, they concern the algebra of the field variables.

But when we wish to discuss the variance of the expectation values of observables, it immediately becomes apparent that the choice of a parti-

† An example illustrating this distinction is provided by the study of the time reversal of the Dirac field. If we consider the solution of the Dirac equation simply as a spinor $\psi(x)$, function of x, such a reflection may be perfectly defined by a unitary transformation. Difficulties only arise when the field is quantized (Chap. V, § 20).

cular picture, namely that of Heisenberg, can considerably simplify the exposition of the theory. The correspondence principle next introduces important restrictions; it requires the expectation value of an observable to have the same variance as its classical equivalent. Let us consider, for instance, the expectation value $\langle G_\mu \rangle = \langle \Psi \mid G_\mu \mid \Psi \rangle$ of an observable whose classical equivalent is a four-vector with respect to the Lorentz group:

$$\langle G_\mu \rangle \to \langle G_\mu \rangle' = L_{\mu\nu} \langle G_\nu \rangle \equiv L_{\mu\nu} \langle \Psi | G_\nu | \Psi \rangle.$$

On the other hand, in view of what we said at the beginning of this section:

$$G'_\mu = L_{\mu\nu} G_\nu = D^{-1}(L) G_\mu D(L),$$

where $D(L)$ is a unitary operator of the space $\mathcal{R}$, representing L; let us compare these equalities; we obtain

$$\langle G_\mu \rangle' = \langle \Psi | G'_\mu | \Psi \rangle = \langle \Psi | D^{-1}(L) G_\mu D(L) | \Psi \rangle. \tag{3.1a}$$

The above formula therefore proves that the observables can be supposed to vary alone and that the state vector remains constant, or alternatively that the state vector alone varies and that the observables remain constant. To prove this, we have merely to consider, in the latter case, the new state vector†

$$\Psi' = D(L) \Psi.$$

Conversely, we can define the Lorentz invariance of a theory by placing ourselves from the outset in the space $\mathcal{R}$: a quantum field theory will be said to be invariant under a transformation represented by the operator U of the space $\mathcal{R}$ if this operation leaves the absolute value of the scalar products, that is to say transition probabilities, constant in time. We then infer, in accordance with a theorem of Wigner, that we can choose the operator U either unitary or antiunitary (the importance of the antiunitary character will be brought out when we study the time reversal, Chap. V, § 20). The transformed expectation values of an observable G are given, according to the above remarks, by

$$\langle \Psi' | G | \Psi' \rangle, \quad \text{with} \quad \Psi' = U \Psi.$$

This, in the terminology of Wigner and his school, is "the active viewpoint".

† If the variable G_μ under consideration depended on x, we should have

$$\langle G_\mu(x') \rangle = L_{\mu\nu} \langle \Psi | G_\nu(x) | \Psi \rangle = \langle \Psi | G'_\nu(x') | \Psi \rangle$$
$$= \langle \Psi | D^{-1}(L) G_\nu(x') D(L) | \Psi \rangle. \tag{3.1b}$$

The ideas we have just briefly indicated will be expanded and clarified with the aid of numerous examples drawn from the study of free fields (Chaps. IV, V and VI).†

2. Infinitesimal inhomogeneous Lorentz transformations

Let us consider the inhomogeneous linear transformation

$$x_\mu \to x'_\mu = a_\mu + L_{\mu\nu} x_\nu. \tag{3.2a}$$

It will be called a Lorentz transformation if it conserves the linear element $ds^2 = dx_\mu\, dx_\mu$. We readily verify that we obtain for the $L_{\mu\nu}$ a condition identical to (1.6), $L^{-1} = L^T$, namely

$$\left.\begin{aligned} (L^{-1})_{\mu\nu} &= L_{\nu\mu} \\ L_{\lambda\mu}L_{\lambda\nu} &= L_{\mu\lambda}L_{\nu\lambda} = \delta_{\mu\nu} \end{aligned}\right\} (L^{-1} = L^T). \tag{3.2b}$$

An infinitesimal inhomogeneous Lorentz transformation is defined by the infinitesimal parameters ε_μ and $\varepsilon_{\mu\nu}$ as follows:

$$a_\mu \equiv \varepsilon_\mu, \tag{3.3a}$$

$$L_{\mu\nu} = \delta_{\mu\nu} + \varepsilon_{\mu\nu}, \tag{3.3b}$$

such that the transformation (3.2*a*) is written

$$x_\mu \to x'_\mu = x_\mu + \varepsilon_\mu + \varepsilon_{\mu\nu} x_\nu \tag{3.4a}$$

or

$$\delta x_\mu \equiv x'_\mu - x_\mu = \varepsilon_\mu + \varepsilon_{\mu\nu} x\ . \tag{3.4b}$$

The condition (3.2*b*) concerning the $L_{\mu\nu}$'s shows that the $\varepsilon_{\mu\nu}$'s are antisymmetric:

$$\varepsilon_{\mu\nu} + \varepsilon_{\nu\mu} = 0. \tag{3.5a}$$

Taking account of the $L_{\mu\nu}$'s, we also obtain

$$\varepsilon_{ij} = \varepsilon^*_{ij}, \quad \varepsilon^*_{j4} = -\varepsilon_{j4}. \tag{3.5b}$$

It is very easy to prove that the transformations (3.1) form a group and to give its multiplication law; let us consider the two successive transformations

$$x'_\mu = a_\mu + L_{\mu\nu} x_\nu, \qquad x''_\mu = a'_\mu + L'_{\mu\nu} x'_\nu, \tag{3.6a}$$

we verify that we have

$$x'_\mu = a''_\mu + L''_{\mu\nu} x_\nu, \tag{3.6b}$$

with

$$a''_\mu = a'_\mu + L'_{\mu\nu} a_\nu, \quad L''_{\mu\nu} = L'_{\mu\lambda} L_{\lambda\nu}, \quad L''_{\mu\lambda} L''_{\mu\tau} = \delta_{\lambda\tau}. \tag{3.6c}$$

† E. Wigner, V. Bargman and A. S. Wightman, *Relativistic Invariance in Quantum Theory*, Princeton University Press, 1958. A proof of the Wigner theorem will be found in A. Messiah, *Mécanique quantique*, Dunod, Paris, 1960, Vol. II, p. 540 (English edition).

On the other hand, we shall assume that the field variables have a determined variance under a homogeneous Lorentz transformation, in other words that there exists a matrix $\mathcal{J}(L)$ (defining a representation of the Lorentz group), a function of the Lorentz transformation L under consideration, such that

$$Q(x) \to Q'(x') = \mathcal{J}(L)Q(x) \tag{3.7a}$$

or by exhibiting the indices

$$Q'_A(x') = \mathcal{J}_{AB}(L)Q_B(x). \tag{3.7b}$$

Let us also note that a transformation of this kind does not affect all the indices A. For instance, if $\psi(x)$ describes a nucleon the Lorentz transformation concerns the spin indices and not those of the isospin.

We may also remark that if the field $Q(x)$ under investigation is a four-vector (Maxwell field, etc.) the matrix $\mathcal{J}$ is identical to L; if the field is a tensor, the matrix $\mathcal{J}$ is obtained by tensor products of the matrices L. In the case of the spinor field, $\mathcal{J}$ has a special structure which will be studied later on.

When L is infinitesimal, $\mathcal{J}$ depends on the parameters $\varepsilon_{\mu\nu}$ and apart from infinitesimal quantities of the second order, we can express it in the form

$$\mathcal{J} = I + \frac{1}{2}\,\varepsilon_{\mu\nu}J_{\mu\nu}, \tag{3.8a}$$

with

$$Q'_A(x') = \left\{\delta_{AB} + \frac{1}{2}\,\varepsilon_{\mu\nu}J_{\mu\nu AB}\right\}Q_B(x). \tag{3.8b}$$

On the other hand, since the $\varepsilon_{\mu\nu}$'s are antisymmetric, only the antisymmetric part (with respect to the indices μ and ν) of $J_{\mu\nu}$ contributes in a non-vanishing manner to the preceding formula [cf. (1.136)]; we shall therefore set

$$J_{\mu\nu} = -J_{\nu\mu}. \tag{3.8c}$$

Let us now consider an infinitesimal *inhomogeneous* Lorentz transformation; it splits into a homogeneous Lorentz transformation:

$$x_\mu \to x''_\mu = x_\mu + \varepsilon_{\mu\nu}x_\nu \tag{3.9a}$$

and a translation

$$x''_\mu \to x'_\mu = x''_\mu + \varepsilon_\mu. \tag{3.9b}$$

Under the first transformation, the wave function varies according to formula (3.8):

$$Q(x) \to Q''(x'') = Q(x) + \frac{1}{2}\,\varepsilon_{\mu\nu}J_{\mu\nu}Q(x), \tag{3.10a}$$

whereas under the second:

$$Q''(x'') \to Q'(x') = Q''(x''). \tag{3.10b}$$

We therefore have

$$Q'(x') = Q(x) + \frac{1}{2}\varepsilon_{\mu\nu}J_{\mu\nu}Q(x) \tag{3.11a}$$

and

$$Q'(x') = Q'(x_\mu + \varepsilon_\mu + \varepsilon_{\mu\nu}x_\nu) = Q'(x) + \varepsilon_\mu \frac{\partial Q'}{\partial x_\mu} + \varepsilon_{\mu\nu}x_\nu \frac{\partial Q'}{\partial x_\mu} \tag{3.11b}$$

and by neglecting second order infinitesimal quantities, we can replace Q' by Q in the derivatives $\frac{\partial Q'}{\partial x_\mu}$.

Let us finally compare (3.11*a*) with (3.11*b*); we get

$$Q'_A(x) = Q_A(x) - \varepsilon_\mu \frac{\partial Q_A}{\partial x_\mu} - \frac{1}{2}\varepsilon_{\mu\nu}\{(x_\nu\partial_\mu - x_\mu\partial_\nu)\delta_{AB} - J_{\mu\nu AB}\}Q_B(x), \tag{3.12}$$

and inserting into the above expression the operators†

$$p_\mu = i\frac{\partial}{\partial x_\mu}, \tag{3.13a}$$

$$m_{\mu\nu} = i\{x_\nu\partial_\mu - x_\mu\partial_\nu - J_{\mu\nu}\} \equiv l_{\mu\nu} + s_{\mu\nu}, \tag{3.13b}$$

with $s_{\mu\nu} = -iJ_{\mu\nu}$, we obtain the formula

$$Q'(x) = \left\{I + i\varepsilon_\mu p_\mu + \frac{i}{2}\varepsilon_{\mu\nu}m_{\mu\nu}\right\}Q(x) \tag{3.14}$$

which expresses Q' in terms of Q, Q and Q' concerning the same point x of space–time.

Note that the operators p_μ and $m_{\mu\nu}$ can be easily made explicit in momentum space. Setting

$$Q(k) = \int e^{-ikx}Q(x)\, d^4x,$$

formulae (3.13) can be expressed in another form,

$$p_\mu Q(k) = -k_\mu Q(k), \tag{3.13c}$$

$$m_{\mu\nu}Q(k) = -i\left\{k_\nu\frac{\partial}{\partial k_\mu} - k_\mu\frac{\partial}{dk_\nu} + J_{\mu\nu}\right\}Q(k). \tag{3.13d}$$

Here k_μ is a vector, whereas p_μ was an operator (vector).

† Note that the operators p_μ and $m_{\mu\nu}$ seem to be the momentum and angular momentum operators of wave mechanics, They have not, however, the required dimensions; by multiplying these operators by $\hbar$ we get the known operators of the first quantization.

It is also easy to verify that the operators p_μ and $m_{\mu\nu}$ obey the following relations:

$$[p_\mu, p_\nu] = 0, \tag{3.15a}$$

$$[m_{\mu\nu}, p_\lambda] = i(\delta_{\mu\lambda}p_\nu - \delta_{\nu\lambda}p_\mu); \tag{3.15b}$$

(3.15*b*) can also be written

$$\frac{1}{2}\varepsilon_{\mu\nu}[m_{\mu\nu}, p_\lambda] = i\varepsilon_{\lambda\nu}p_\nu. \tag{3.15c}$$

Assuming $J_{\mu\nu} = 0$, we can easily see that $l_{\mu\nu}$ satisfies the characteristic relation of the angular momenta.

$$[l_{\mu\nu}, l_{\varrho\sigma}] = i(\delta_{\mu\varrho}l_{\nu\sigma} - \delta_{\mu\sigma}l_{\nu\varrho} + \delta_{\nu\sigma}l_{\mu\varrho} - \delta_{\nu\varrho}l_{\mu\sigma}). \tag{3.15d}$$

Various important remarks may be made concerning the results we have just obtained. In Chap. IV, part 3, we shall endeavour to define a scalar product for the wave functions $Q(x)$,† we shall then be able to speak of the Hermitian character of the differential operators p_μ and $m_{\mu\nu}$ and formula (3.14) will enable us to define unitary representations of the Lorentz group.

Finally, formulae (3.15) are of a very general type since they are, in fact, a particular case of the Lie formulae which depend only on the structure of the group under investigation and not on its representation space.

We shall not study these formulae in their general aspect, but we shall apply a less general method of investigation to the particular case of the Lorentz group.‡

It is convenient first of all to express the coefficients $L_{\mu\nu}$ of formula (3.2) in the form

$$L_{\mu\nu} = \delta_{\mu\nu} + \eta_{\mu\nu} \tag{3.16a}$$

such that for $\eta_{\mu\nu} = 0$, the homogeneous Lorentz transformation is reduced to unity. In order to make the notation more homogeneous, we shall set

$$a_\mu = \eta_\mu. \tag{3.16b}$$

† This point must be clarified: for a non-relativistic field $\varphi(x)$, the length of the vector φ is

$$\|\varphi\|^2 \equiv \langle \varphi, \varphi \rangle = \int \varphi^*(x)\varphi(x)\, d^3\boldsymbol{x};$$

φ is therefore an element of the space L^2. For a field with a more complex variance, $\|\varphi\|^2$ is obtained from the definition of the scalar product $\langle \varphi, \psi \rangle$ [cf. formulae (4.75) and (4.76)].

‡ B. L. Van der Waerden, *Die Gruppentheoretische Methode in Quantenmechanik*, Edwards Brothers Inc., 1944. H. Weyl, *The Theory of Groups and Quantum Mechanics*, Dover. E. P. Wigner, *Group Theory and its Application to the Quantum Mechanics of Atomic Spectra*, Academic Press, 1959. See also references in footnote, p. 1.

The η's are not assumed to be infinitesimal; they satisfy in accordance with (3.2) the relation

$$\eta_{\mu\nu}+\eta_{\nu\mu}+\eta_{\mu\lambda}\eta_{\nu\lambda}=0,$$

and formulae (3.6) relative to a product of two Lorentz transformations may then be written

$$\eta''_{\varrho}=\eta'_{\varrho}+\eta_{\varrho}+\eta'_{\varrho\sigma}\eta_{\sigma}, \tag{3.17a}$$
$$\eta''_{\lambda\varrho}=\eta'_{\lambda\varrho}+\eta_{\lambda\varrho}+\eta'_{\lambda\sigma}\eta_{\sigma\varrho}. \tag{3.17b}$$

Let us now consider a representation $U(\eta'_{\mu},\eta'_{\lambda\varrho})$ of the Lorentz transformation $\eta'_{\mu},\eta'_{\lambda\varrho}$: let this operator act on a vector v' of the representation space; we have

$$v'\to v''(\eta'_{\mu},\eta'_{\lambda\varrho})=U(\eta'_{\mu},\eta'_{\lambda\varrho})v'\equiv\left\{I+i\eta'_{\mu}\mathfrak{P}_{\mu}+\frac{i}{2}\eta'_{\mu\nu}\mathfrak{M}_{\mu\nu}+\ldots\right\}v', \tag{3.18a}$$

where $\mathfrak{P}_{\mu}$ and $\mathfrak{M}_{\mu}$ are defined by the relations:

$$i\mathfrak{P}_{\mu}v'=\left(\frac{\partial v''}{\partial\eta'_{\mu}}\right)_{\eta'=0}, \tag{3.18b}$$

$$\frac{i}{2}\mathfrak{M}_{\mu\nu}v'=\left(\frac{\partial v''}{\partial\eta'_{\mu\nu}}\right)_{\eta'=0}. \tag{3.18c}$$

Consider the sequence of transformations $v\to v'\to v''$; we require that the operators U which represent them constitute a group

$$\begin{aligned}v''&=\left\{I+i\eta'_{\mu}\mathfrak{P}_{\mu}+\frac{i}{2}\eta'_{\mu\nu}\mathfrak{M}_{\mu\nu}+\ldots\right\}v' \qquad (3.19)\\
&=\left\{I+i\eta'_{\mu}\mathfrak{P}_{\mu}+\frac{i}{2}\eta'_{\mu\nu}\mathfrak{M}_{\mu\nu}+\ldots\right\}\\
&\quad\times\left\{I+i\eta_{\mu}\mathfrak{P}_{\mu}+\frac{i}{2}\eta_{\mu\nu}\mathfrak{M}_{\mu\nu}+\ldots\right\}v\\
&=\left\{I+i\eta''_{\mu}\mathfrak{P}_{\mu}+\frac{i}{2}\eta''_{\mu\nu}\mathfrak{M}_{\mu\nu}+\ldots\right\}v.\end{aligned}$$

Express the derivatives in formulae (3.18) by introducing the η'''s and making use of (3.17):

$$\begin{aligned}\left(\frac{\partial v''}{\partial\eta'_{\mu}}\right)_{\eta'=0}&=\left(\frac{\partial v''}{\partial\eta''_{\varrho}}\frac{\partial\eta''_{\varrho}}{\partial\eta'_{\mu}}\right)_{\eta'=0}+\left(\frac{\partial v''}{\partial\eta''_{\lambda\varrho}}\frac{\partial\eta''_{\lambda\varrho}}{\partial\eta'_{\mu}}\right)_{\eta'=0}\\
&=\left(\frac{\partial v''}{\partial\eta''_{\varrho}}\right)_{\eta'=0}\delta_{\varrho\mu}=\frac{\partial v'}{\partial\eta_{\mu}}.\end{aligned}$$

It follows that, returning to formula (3.18*a*), we obtain the first set of Lie equations:

$$\frac{\partial v'}{\partial \eta_\mu} = i\mathfrak{P}_\mu v'. \tag{3.20a}$$

Similarly,

$$\left(\frac{\partial v''}{\partial \eta'_{\mu\nu}}\right)_{\eta'=0} = \frac{\partial v'}{\partial \eta_{\mu\nu}} + \eta_{\nu\varrho}\frac{\partial v'}{\partial \eta_{\mu\varrho}} + \eta_\nu \frac{\partial v'}{\partial \eta_\mu};$$

by inserting this derivative into eqn. (3.18*b*) and making use of (3.2), we obtain the second set of Lie equations:

$$\frac{\partial v'}{\partial \eta_{\mu\nu}} = -L_{\lambda\nu}\left(\eta_\lambda \frac{\partial v'}{\partial \eta_\mu} + \frac{i}{2}\mathfrak{M}_{\mu\lambda} v'\right)$$

or alternatively, by making use of (3.20*a*) and of the expression of $L_{\mu\nu}$:

$$\frac{\partial v'}{\partial \eta_{\mu\nu}} = -(\delta_{\lambda\nu} + \eta_{\lambda\nu})\left\{-i\eta_\lambda \mathfrak{P}_\mu + \frac{i}{2}\mathfrak{M}_{\mu\lambda}\right\} v'. \tag{3.20b}$$

The operators $\mathfrak{P}_\mu$, $\mathfrak{M}_{\lambda\varrho}$ cannot be arbitrarily chosen, since we must in fact have the three conditions

$$\frac{\partial^2 v'}{\partial \eta_\mu \partial \eta_\nu} = \frac{\partial^2 v'}{\partial \eta_\nu \partial \eta_\mu}, \qquad \frac{\partial^2 v'}{\partial \eta_\varrho \partial \eta_{\mu\nu}} = \frac{\partial^2 v'}{\partial \eta_{\mu\nu} \partial \eta_\varrho},$$

$$\frac{\partial^2 v'}{\partial \eta_{\mu\nu} \partial \eta_{\varrho\sigma}} = \frac{\partial^2 v'}{\partial \eta_{\varrho\sigma} \partial \eta_{\mu\nu}}.$$

The first of the above relations readily gives (3.15*a*); the second gives

$$[\mathfrak{M}_{\mu\nu}, \mathfrak{P}_\varrho] = -2i\,\delta_{\nu\varrho}\mathfrak{P}_\mu.$$

Writing $\mathfrak{M}_{\mu\nu}$ as a sum of a symmetric operator and an antisymmetric operator with respect to μ and ν, we verify that the antisymmetric part [which is alone affected by the infinitesimal transformation (3.16)] satisfies (3.17*b*). Finally, from the last of the equalities it can be easily seen that the antisymmetric part of the operator $\mathfrak{M}_{\mu\nu}$ verifies:

$$[\mathfrak{M}^{(a,s)}_{\mu\nu}, \mathfrak{M}^{(a,s)}_{\varrho\sigma}] = i(\delta_{\mu\varrho}\mathfrak{M}^{(a,s)}_{\nu\sigma} - \delta_{\mu\sigma}\mathfrak{M}^{(a,s)}_{\nu\varrho} + \delta_{\nu\sigma}\mathfrak{M}^{(a,s)}_{\mu\varrho} - \delta_{\nu\varrho}\mathfrak{M}^{(a,s)}_{\mu\sigma}). \tag{3.21}$$

Now, if we write that the above operator is the sum of an operator $l_{\mu\nu}$ verifying (3.15*d*) and an operator $s_{\mu\nu} = -iJ_{\mu\nu}$ commuting with $l_{\mu\nu}$ [the operators $l_{\mu\nu}$ and $s_{\mu\nu}$ in (3.15*b*) do not operate in the same space], we find that $J_{\mu\nu}$ obeys the following relation:

$$[J_{\mu\nu}, J_{\varrho\sigma}] = -\delta_{\mu\varrho}J_{\nu\sigma} + \delta_{\mu\sigma}J_{\nu\varrho} - \delta_{\nu\sigma}J_{\mu\varrho} + \delta_{\nu\varrho}J_{\mu\sigma}, \tag{3.22}$$

a relation we should have obtained directly by writing the Lie formula (3.21) for the representation $\mathcal{J}(L)$ (3.8*b*) of the Lorentz group. We shall have, moreover, in the following section, a verification of the theorem under consideration: we shall obtain, in fact, a representation of the Lorentz group on the space $\mathcal{R}$ and we shall verify [formula (3.47)] that the generating operators of the representation also satisfy the relations (3.15) and (3.21).

Note also that we have written in (3.19):

$$U(\eta'_{\mu}, \eta'_{\lambda\varrho})U(\eta_{\mu}, \eta_{\lambda\varrho}) = U(\eta''_{\mu}, \eta''_{\lambda\varrho}),$$

further investigation would show that the equality sign of the above formula ought properly to be replaced by the equivalence sign (a state being defined by a "ray" of the space $\mathcal{R}$ rather than by a vector). The equality sign can be justified up to a point by the fact that we can arbitrarily choose a particular phase factor (factor of magnitude 1) in the definition of a state (cf. nevertheless superselection rules, Chap. VII, § 11).

We may finally note that formulae (3.15*b*) are merely a particular case of formulae (2.30).

Note. The most direct illustration of the above theorem is obtained by studying the "natural" representation of the homogeneous Lorentz group, namely that which is induced in the vector space of the space–time vectors X. X is therefore a column matrix with four components and a Lorentz transformation is specified by the 4×4 matrix with elements $\varepsilon_{\mu\nu}$. We have

$$X_{\lambda} \to X'_{\lambda} = X_{\lambda} + \varepsilon_{\lambda\varrho} X_{\varrho},$$

a formula we shall write with the help of the six generating matrices $\mathfrak{M}_{\mu\nu}$ (with elements $\mathfrak{M}_{\mu\nu\lambda\tau}$) in the form

$$X'_{\lambda} = X_{\lambda} + \frac{i}{2}\, \varepsilon_{\mu\nu} \mathfrak{M}_{\mu\nu\lambda\tau} X_{\tau};$$

by comparing these two equations and taking into account the fact that $\mathfrak{M}_{\mu\nu}$ is skew-symmetrical with respect to μ and ν, we obtain†

$$\mathfrak{M}_{\mu\nu\lambda\tau} = -i(\delta_{\mu\lambda}\,\delta_{\nu\tau} - \delta_{\nu\lambda}\,\delta_{\mu\tau}),$$

consequently the λth component of the vector $\mathfrak{M}_{\mu\nu} X$ is

$$\{\mathfrak{M}_{\mu\nu} X\}_{\lambda} = -i(\delta_{\mu\lambda} X_{\nu} - \delta_{\nu\lambda} X_{\mu}).$$

† One may also note that one gets in this way the expression of $J_{\mu\nu}$. Since in this case: $J_{\mu\nu} = -i\mathfrak{M}_{\mu\nu}$, one has:

$$J_{\mu\nu\lambda\tau} = \delta_{\mu\lambda}\delta_{\nu\tau} - \delta_{\nu\lambda}\delta_{\mu\tau}.$$

Furthermore, the above formula has been obtained without any reference to the real or complex character of $X\rangle$; this formula is therefore valid for any group of transformations which leaves $X_{\mu} X_{\mu}$ invariant, where the components X_{μ} are real or complex numbers. Hence, the matrices $J_{\mu\nu}$ are the generators of any infinitesimal transformation of O_4 acting on the four-vector.

Calculate first the αth component of the vector $\{\mathfrak{M}_{\varrho\sigma}Y\}_\alpha$ where Y is a given space–time vector in accordance with the above formula; we next set $Y = \mathfrak{M}_{\mu\nu}X$. Consider then $\{\mathfrak{M}_{\mu\nu}Y\}_\alpha$ and set $Y = \mathfrak{M}_{\varrho\sigma}X$. One obtains in this way

$$\{[\mathfrak{M}_{\mu\nu}, \mathfrak{M}_{\varrho\sigma}]X\}_\alpha$$

and verifies (3.21).

3. Classical energy–momentum tensor

It is worth recalling first of all how the energy–momentum tensor is defined in classical theory. Consider for this purpose the action

$$\mathfrak{A}[Q, \omega] = \int_\omega \mathscr{L}(Q, Q_\mu, \ldots)\, d^4x, \tag{3.23}$$

the fundamental volume ω being limited by the three-dimensional surface S. We are interested in the variation of $\mathfrak{A}$ when each point of the volume ω, including those of its surface S, suffers the translation δx_μ (independent of x):

$$x_\mu \to x'_\mu = x_\mu + \delta x_\mu;$$

the wave functions then vary in such a way that

$$Q(x) \to Q'(x') = Q(x), \tag{3.24}$$

consequently,

$$\delta Q \equiv Q'(x) - Q(x) = -\delta x_\mu \frac{\partial Q}{\partial x_\mu}. \tag{3.25}$$

By calculations similar to those carried out in Chap. II, we obtain

$$\delta\mathfrak{A}[Q, \omega] = \int_\omega -\frac{\partial}{\partial x_\nu}\left\{\frac{\partial \mathscr{L}}{\partial Q_{A,\nu}} Q_{A,\mu}\right\} d^4x\, \delta x_\mu + \oint_S \mathscr{L}(Q, Q_\mu)\, d\sigma_\mu\, \delta x_\mu,$$

where we have already made use of the Lagrange equations.

By transforming the surface integral into a volume integral [cf. (1.131)]:

$$\oint_S \mathscr{L}(Q, Q_\mu)\, d\sigma_\mu\, \delta x_\mu = \int_\omega \frac{\partial \mathscr{L}}{dx_\nu}\, \delta x_\nu\, d^4x,$$

the variation of the action can be expressed in the form

$$\delta\mathfrak{A}[Q, \omega] = \int_\omega \frac{\partial}{\partial x_\nu}\left\{-\frac{\partial \mathscr{L}}{\partial Q_{A,\nu}} Q_{A,\mu} + \mathscr{L}(Q, Q_\varrho)\delta_{\mu\nu}\right\} \delta x_\mu\, d^4x \tag{3.26a}$$

$$\equiv \int_\omega T_{\mu\nu,\nu}(x)\, \delta x_\mu\, d^4x,$$

with

$$T_{\mu\nu}(x) = -\frac{\partial \mathcal{L}}{\partial Q_{A,\nu}} Q_{A,\mu} + \mathcal{L}\,\delta_{\mu\nu}; \tag{3.26b}$$

$T_{\mu\nu}$ is the energy–momentum tensor which we find in classical field theory. The invariance of the action under any space–time translation leads to the law of conservation of $T_{\mu\nu}$:

$$T_{\mu\nu,\,\nu} = 0, \tag{3.27}$$

summation being made over ν. Making use of formula (1.133), we see that by setting

$$P_\mu[\sigma] = \int_{\sigma(x)} T_{\mu\nu}(x')\, d\sigma'_\nu, \tag{3.28a}$$

the relation (3.27) expresses the fact that P_μ is independent of σ:

$$\frac{\delta P_\mu[\sigma]}{\delta\sigma(x)} = 0. \tag{3.28b}$$

Conversely, let us define $T_{\mu\nu}$ by (3.26*b*) and assume that the field equations are satisfied; then we can easily see that

$$\frac{\partial}{\partial x_\nu} T_{\mu\nu} = \frac{\partial \mathcal{L}}{\partial x_\mu}. \tag{3.29}$$

Consequently, $T_{\mu\nu,\,\nu}$ is indeed zero if $\mathcal{L}$ does not explicitly depend on x.†
On the other hand, let us define the tensor $L_{\mu\nu}$ as follows:

$$L_{\mu\nu\lambda} = T_{\mu\lambda}x_\nu - T_{\nu\lambda}x_\mu, \tag{3.30a}$$

$$L_{\mu\nu}[\sigma] = \int_{\sigma(x)} L_{\mu\nu\lambda}(x')\, d\sigma'_\lambda. \tag{3.30b}$$

The space components of this tensor evaluated on an equitemporal plane t = constant:

$$L_{jk} = \int (T_{j4}x_k - T_{k4}x_j)\, d\sigma_4 = -i \int (T_{j4}x_k - T_{k4}x_j)\, d^3\boldsymbol{x}, \tag{3.31}$$

represent the components of the orbital angular momentum, since $-iT_{i4}$ is the density of the momentum vector: L_{jk} is therefore the total angular momentum tensor of the field.

† We have, indeed:

$$T_{\mu\nu,\,\nu} = \partial_\nu \left\{ -\frac{\partial \mathcal{L}}{\partial Q_{A,\nu}} Q_{A,\mu} \right\} + \partial_\mu \mathcal{L}(Q_A, Q_{,\mu}\, x)$$

$$= -\frac{\partial \mathcal{L}}{\partial Q_A} Q_{A,\mu} - \frac{\partial \mathcal{L}}{\partial Q_{A,\nu}} Q_{A,\mu,\nu} + \frac{\partial \mathcal{L}}{\partial Q_A} Q_{A,\mu} + \frac{\partial \mathcal{L}}{\partial Q_{A,\nu}} Q_{A,\nu,\mu} + \frac{\partial \mathcal{L}}{\partial x_\mu}.$$

By computing, on the other hand, the divergence of $L_{\mu\nu\lambda}$, we obtain

$$\frac{\partial}{\partial x_\lambda} L_{\mu\nu\lambda} \equiv \frac{\delta}{\delta\sigma(x)} L_{\mu\nu}[\sigma] = T_{\mu\nu} - T_{\nu\mu}. \tag{3.32}$$

We find again the well-known theorem of classical fields, namely that the total orbital angular momentum of the field is constant if the energy–momentum tensor is symmetrical.

Having recalled these well-known results concerning classical fields, we now propose to consider quantized fields.

4. The quantum energy–momentum and angular momentum tensors. Vacuum state

Let us therefore consider a quantized field with invariant action under the inhomogeneous group of proper Lorentz transformations (3.4).

Let us suppose that its action

$$\mathfrak{A}[\sigma, \sigma_0, Q] = \int_{\sigma_0}^{\sigma} \mathcal{L}(Q, Q_\mu, \ldots)\, d^4x \qquad (\sigma_0, \text{ initial surface}) \tag{3.33}$$

suffers an infinitesimal transformation given by formula (2.150).† We have

$$\begin{aligned} \delta\mathfrak{A}[\sigma, \sigma_0, Q, \Delta Q, \delta x] &= \int_\sigma T_{\nu\mu}(x)\, \delta x_\nu(x)\, d\sigma_\mu + \int_\sigma \frac{\partial \mathcal{L}}{\partial Q_{A,\mu}} \Delta Q_A(x)\, d\sigma_\mu \\ &= \int_\sigma T_{\nu\mu}(x)\, \delta x_\nu(x)\, d\sigma_\mu + \int_\sigma \Pi_A(x)\, \Delta Q_A(x)\, d\sigma, \end{aligned} \tag{3.34}$$

taking the field equations into account. Let us prove that $\delta\mathfrak{A}[\sigma, \sigma_0, \ldots]$ is independent of σ; the assumed invariance of the action:

$$\mathfrak{A}[\sigma_{\rm I}, \sigma_{\rm II}, Q] = \int_{\sigma_{\rm I}}^{\sigma_{\rm II}} \mathcal{L}(Q, Q_\mu, \ldots)\, d^4x. \tag{3.35}$$

under inhomogeneous proper Lorentz transformations means that

$$\delta\mathfrak{A}[\sigma_{\rm I}, \sigma_{\rm II}, Q, \Delta Q, \delta x] = 0$$

provided that the points of $\sigma_{\rm II}$ as well as the points of $\sigma_{\rm I}$ participate in this transformation. Introducing the initial surface σ_0 [on which any variation is zero, cf. footnote, p. 80], the integral (3.35) is written

$$\mathfrak{A}[\sigma_{\rm I}, \sigma_{\rm II}, Q] = \left\{ \int_{\sigma_0}^{\sigma_{\rm II}} - \int_{\sigma_0}^{\sigma_{\rm I}} \mathcal{L}(Q, Q_\mu \ldots)\, d^4x \right\}.$$

† Cf. comments on p. 82 on the positions of the operators δQ, ΔQ, ..., in formulae (3.34) *et seq.*

By considering the variations, we obtain

$$0 = \delta\mathfrak{A}[\sigma_I, \sigma_{II}, Q, \ldots] = \delta\mathfrak{A}[\sigma_{II}, \sigma_0, \ldots] - \delta\mathfrak{A}[\sigma_I, \sigma_0, \ldots]$$

consequently $\delta\mathfrak{A}[\sigma, \sigma_0, \ldots]$ is indeed independent of σ.†

On the other hand, by comparing the definition (2.132) of $\Delta Q(x)$ with formula (3.8*b*), we obtain

$$\Delta Q_A(x) = Q'_A(x') - Q_A(x) \equiv \frac{1}{2}\varepsilon_{\mu\nu} J_{\mu\nu AB} Q_B(x). \tag{3.36}$$

Let us then insert into (3.34) the above expression of $\Delta Q(x)$ and the expression (3.4*b*) of δx; we obtain‡

$$\begin{aligned} \delta\mathfrak{A}[Q, \delta x] &= \int_{\sigma(x)} T_{\nu\mu}(x)(\varepsilon_\nu + \varepsilon_{\nu\lambda} x_\lambda)\, d\sigma_\mu \\ &\quad + \frac{1}{2}\varepsilon_{\nu\lambda} \int_\sigma \frac{\partial \mathcal{L}}{\partial Q_{A,\mu}} J_{\nu\lambda AB} Q_B(x)\, d\sigma_\mu \\ &= \varepsilon_\nu P_\nu + \frac{1}{2}\varepsilon_{\mu\nu} \int_\sigma \left\{ T_{\mu\lambda} x_\nu - T_{\nu\lambda} x_\mu + \frac{\partial \mathcal{L}}{\partial Q_{A,\lambda}} J_{\mu\nu AB} Q_B(x) \right\} d\sigma_\lambda. \end{aligned} \tag{3.37}$$

Note finally that $\delta Q_A(x)$ is given by formula (3.12):

$$\begin{aligned} \delta Q(x) &\equiv Q'(x) - Q(x) \\ &= -\varepsilon_\mu \frac{\partial Q}{\partial x_\mu} - \frac{1}{2}\{\varepsilon_{\mu\nu}(x_\nu \partial_\mu - x_\mu \partial_\nu)\,\delta_{AB} - J_{\mu\nu AB}\} Q_B(x). \end{aligned} \tag{3.38}$$

By using the Schwinger–Feynman variational principle in its form (2.130) and identifying the factors of ε_μ and $\varepsilon_{\mu\nu}$ in the two sides, we obtain the following commutation relations:

$$\frac{\partial Q_A}{\partial x_\mu} = -i[P_\mu, Q_A(x)], \tag{3.39}$$

$$\begin{aligned} &\{(x_\nu \partial_\mu - x_\mu \partial_\nu)\,\delta_{AB} - J_{\mu\nu AB}\} Q_B(x) \\ &\quad = -i\left[\int_\sigma \left\{ x'_\nu T_{\mu\lambda} - x'_\mu T_{\nu\lambda} + \frac{\partial \mathcal{L}}{\partial Q_{A,\lambda}} J_{\mu\nu AB} Q_B(x') \right\} d\sigma'_\lambda, Q_A(x) \right]. \end{aligned} \tag{3.40}$$

† We admit therefore that $\delta\mathfrak{A}[\sigma_{II}, \sigma_I, \ldots] = 0$, but that $\delta\mathfrak{A}[\sigma, \sigma_0, \ldots] \neq 0$. This difference in the behaviour of two variations is due to the assumption that any variation vanishes on σ_0, which is equivalent to saying that for any domain at a finite distance, between the surfaces σ_I and σ_{II} for instance, the ε_μ's and $\varepsilon_{\mu\nu}$'s are constants, but that they converge strongly to zero when one approaches σ_0 (adiabatic hypothesis). We shall meet with a similar difficulty concerning the gauge transformations of the first kind (purely formal difficulties, as we pointed out in the footnote on p. 80).

‡ When the calculations leading to (3.37) are made for two space-like surfaces σ_I and σ_{II} one finds:

$$\delta\mathfrak{A}[\ldots] = \varepsilon_\nu \int_{\sigma_I}^{\sigma_{II}} \partial_\mu T_{\nu\mu}\, d^4x + \frac{1}{2}\varepsilon_{\mu\nu} \int_{\sigma_I}^{\sigma_{II}} \partial_\lambda \left\{ T_{\mu\lambda} x_\nu - T_{\nu\lambda} x_\mu + \frac{\partial \mathcal{L}}{\partial Q_{A,\lambda}} J_{\mu\nu AB} Q_B \right\} d^4x.$$

By writing that $\delta\mathfrak{A}[\ldots] = 0$, we obtain the energy–momentum tensor (cf. previous section) and the angular momentum density tensor, both of which are tensors with zero divergence. This is a particular case of a theorem stated by E. Noether, *Nach. Kgl. Ges. Wiss.*, Göttingen, **235** (1918); cf. also J. Winogradski, *Cahiers de Phys.*, **67** (1956); E. L. Hille, *Rev. Mod. Phys.*, **23**, 253 (1957).

By defining the total momentum operator P_μ:

$$P_\mu = \int_\sigma T_{\mu\nu}\, d\sigma_\nu = \int_\sigma \left\{ -\frac{\partial \mathcal{L}}{\partial Q_{A,\nu}} Q_{A,\mu} + \mathcal{L}(\ldots)\, \delta_{\mu\nu} \right\} d\sigma_\nu, \tag{3.41}$$

concurrently with the total angular momentum operator

$$M_{\mu\nu} = \int_\sigma x_\nu T_{\mu\lambda} - x_\mu T_{\nu\lambda} + \frac{\partial \mathcal{L}}{\partial Q_{A,\lambda}} J_{\mu\nu A B} Q_B(x) \Big\} \, d\sigma_\lambda, \tag{3.42}$$

the relations (3.39) and (3.40) may be written in matrix form:

$$\frac{\partial Q(x)}{\partial x_u} = -i[P_\mu, Q(x)], \tag{3.43a}$$

$$\{(x_\nu \partial_\mu - x_\mu \partial_\nu) - J_{u\nu}\}\, Q(x) = -i[M_{u\nu}, Q(x)]. \tag{3.43b}$$

We thus obtain a representation of the inhomogeneous Lorentz group, but the $Q(x)$'s are now considered as operators of the vector space $\mathcal{R}$.

This representation is unitary, for its generators are the ten operators P_μ and $M_{\lambda\nu}$ which are Hermitian or anti-Hermitian. By iterating (by exponentiation) the unitary infinitesimal operator:

$$I - i\, \delta\mathfrak{A}[Q, \delta x] = I - i\left(\varepsilon_\mu P_\mu + \frac{1}{2}\, \varepsilon_{\mu\nu} M_{\mu\nu} \right) \tag{3.44a}$$

we obtain the unitary operator $D(L)$

$$Q'(x) = D^{-1}(L) Q(x) D(L). \tag{3.44b}$$

Thus the Lorentz transformation

$$x \to x' = Lx$$

induces on a particular subspace of the indices A of the field functions a representation whose matrix $\mathfrak{J}(L)$ satisfies

$$Q(x) \to Q'(x) = \mathfrak{J}(L) Q(L^{-1}x) \tag{3.45}$$

according to formula (3.7).

A second representation of this transformation L is given by the unitary operator $D^{-1}(L)$: the space of this representation is the space $\mathcal{R}$.

A third representation, this time on the space L^2, is obtained by iterating the infinitesimal operator $I + \frac{i}{2}\, \varepsilon_{\mu\nu} m_{\mu\nu}$ in accordance with (3.14). This third representation is unitary and equivalent to the second one so long as $\mathcal{R}$ can be considered as a Hilbert space.

By comparing (3.44) and (3.45), we obtain the important formula

$$D^{-1}(L) Q(x) D(L) = \mathfrak{J}(L) Q(L^{-1}x). \tag{3.46}$$

Finally it is convenient to verify the group property: considering two Lorentz transformations L_1 and L_2, we have

$$Q(x) \to Q'(x) = \mathcal{J}(L_1)Q(L_1^{-1}x) \to Q''(x)$$
$$= \mathcal{J}(L_2)Q'(L_2^{-1}x) = \mathcal{J}(L_2)\mathcal{J}(L_1)Q((L_2L_1)^{-1}x).$$

On the other hand, under L_1:

$$Q(x) \to Q'(x) = D^{-1}(L_1)Q(x)D(L_1),$$

under L_2:

$$Q'(x) \to Q''(x) = D'^{-1}(L_2)Q'(x)D'(L_2);$$

the prime $'$ of the operator D, in the latter formula, indicates that this operator is a functional of the operators $Q'(x)$; we therefore have

$$D'(L_2) = D^{-1}(L_1)D(L_2)D(L_1),$$

from which it finally follows that

$$Q''(x) = D^{-1}(L_1)D^{-1}(L_2)Q(x)D(L_2)D(L_1)$$
$$= D^{-1}(L_2L_1)Q(x)D(L_2L_1);$$

the group property is therefore verified.

Let us now return to the properties of the generating operators: starting from the relations (3.43), we can show that P_μ and $M_{\mu\nu}$ obey the same set of relations (3.15) as p_μ and $m_{\mu\nu}$, namely

$$[P_\mu, P_\nu] = 0, \tag{3.47a}$$

$$[M_{\mu\nu}, P_\lambda] = i(\delta_{\mu\lambda}P_\nu - \delta_{\nu\lambda}P_\mu), \tag{3.47b}$$

$$\frac{1}{2}\,\varepsilon_{\mu\nu}[M_{\mu\nu}, P_\lambda] = i\varepsilon_{\lambda\nu}P_\nu, \tag{3.47c}$$

$$[M_{\mu\nu}, M_{\varrho\sigma}] = i(\delta_{\mu\varrho}M_{\nu\sigma} - \delta_{\mu\sigma}M_{\nu\varrho} + \delta_{\nu\sigma}M_{\mu\varrho} - \delta_{\nu\varrho}M_{\mu\sigma}). \tag{3.47d}$$

These formulae are easily verified: differentiating (3.43*b*) with respect to x_λ and appropriately grouping the terms thus obtained, we find (3.47*d*). Easy calculations enable us to find (3.47*a*, *b*, *c*). Finally, considering the field at a given point x, formula (3.43*b*) is written

$$J_{\mu\nu}Q = i[M_{\mu\nu}, Q] \tag{3.47e}$$

and this relation enables us to calculate $[J_{\mu\nu}, J_{\varrho\sigma}]Q$ (noting that $J_{\alpha\beta}$ commutes with all the $M_{\mu\nu}$'s) and to find (3.22) again from (3.47*d*).

On the other hand, formula (3.37) proves that, since $\delta\mathfrak{A}\,[\sigma, ...]$ is independent of σ and that ε_μ and $\varepsilon_{\mu\nu}$ are arbitrary, the operators P_μ and $M_{\mu\nu}$ are also independent of σ. By applying formulae (1.133) to these two operators we shall obtain interesting relations: from

$$\frac{\delta P_\mu[\sigma]}{\delta\sigma(x)} = 0$$

we obtain the energy–momentum conservation

$$T_{\mu\nu,\,\nu} = 0. \tag{3.48}$$

Let us then set

$$\mathfrak{M}_{\mu\nu\lambda} = \frac{\partial \mathcal{L}}{\partial Q_{A,\,\lambda}} J_{\mu\nu AB} Q_B(x), \qquad \mathfrak{M}_{\mu\nu\lambda} = -\mathfrak{M}_{\nu\mu\lambda}, \tag{3.49}$$

such that, again by virtue of (1.133), the relation

$$\frac{\delta M_{\mu\nu}[\sigma]}{\delta\sigma(x)} = \frac{\delta}{\delta\sigma(x)} \int_{\sigma(x)} \{x_\nu T_{\mu\lambda} - x_\mu T_{\nu\lambda} + \mathfrak{M}_{\mu\nu\lambda}\}\, d\sigma_\lambda = 0 \tag{3.50}$$

implies

$$\frac{\partial}{\partial x_\lambda} \{x_\nu T_{\mu\lambda} - x_\mu T_{\nu\lambda} + \mathfrak{M}_{\mu\nu\lambda}\} = 0,$$

in other words, finally

$$T_{\mu\nu} - T_{\nu\mu} + \mathfrak{M}_{\mu\nu\lambda,\,\lambda} = 0. \tag{3.51}$$

Formulae (3.47*a*) and (3.47*b*) show that the observables P_μ and $M_{\mu\nu}$ commute with P_4: they are therefore constants of motion in the Heisenberg picture where $P_4 = iH$ (Chap. II, § 12). Their eigenvalues are independent of t and a certain number of components of these operators can belong to the complete set of observables, from which one deduces the constant state vector of this picture.

Let us return to formula (3.42): it exhibits in a natural manner the decomposition of the total angular momentum tensor $M_{\mu\nu}$ into an orbital angular momentum tensor $L_{\mu\nu}$ and a spin tensor $\mathfrak{M}_{\mu\nu}$:

$$M_{\mu\nu} = L_{\mu\nu} + \mathfrak{M}_{\mu\nu}, \tag{3.52a}$$

$$L_{\mu\nu}[\sigma] = \int_\sigma (x_\nu T_{\mu\lambda} - x_\mu T_{\nu\lambda})\, d\sigma_\lambda, \tag{3.52b}$$

$$\mathfrak{M}_{\mu\nu}[\sigma] = \int_\sigma \frac{\partial \mathcal{L}}{\partial Q_{A,\,\lambda}} J_{\mu\nu AB} Q_B(x)\, d\sigma_\lambda = \int_\sigma \mathfrak{M}_{\mu\nu\lambda}\, d\sigma_\lambda. \tag{3.52c}$$

Using the relation (2.159), we can obtain the commutation relations of these two operators with the field functions $Q(x)$ and justify the names which have been given to them. Let us in fact calculate the commutators:

$$\begin{aligned} -i[L_{\mu\nu}[\sigma], Q_A(x)] &= -i \int_\sigma d\sigma' n_\nu(x') x'_\lambda [T_{\mu\lambda}(x'), Q_A(x)] \\ &\quad + i \int_\sigma d\sigma' n_\lambda(x') x'_\mu [T_{\nu\lambda}(x'), Q_A(x)] \\ &= \partial_\mu \{x_\nu Q_A(x)\} - \partial_\nu \{x_\mu Q_A(x)\}, \end{aligned}$$

from which it finally follows that

$$\{x_\nu\partial_\mu - x_\mu\partial_\nu\}Q(x) = -i[L_{\mu\nu}[\sigma], Q(x)], \qquad (x\in\sigma). \tag{3.53}$$

We therefore also obtain

$$J_{\mu\nu AB}Q_B(x) = i[\mathfrak{M}_{\mu\nu}[\sigma], Q_B(x)] \qquad (x\in\sigma). \tag{3.54}$$

It must be noted, finally, that in the majority of the cases considered in quantum field theory, the tensor $T_{\mu\nu}$ is not symmetrical. But we can always form a symmetrical tensor which supplies the total momentum and the total angular momentum of the field; let us in fact consider the symmetrical tensor†

$$\Theta_{\mu\nu} = \frac{1}{2}(T_{\mu\nu}+T_{\nu\mu}) - \frac{1}{2}\frac{\partial}{\partial x_\lambda}\{\mathfrak{M}_{\mu\lambda\nu}+\mathfrak{M}_{\nu\lambda\mu}\}, \tag{3.55}$$

which we shall call "symmetrical energy–momentum tensor". In order to derive its conservation law, let us differentiate it with respect to x:

$$\Theta_{\mu\nu,\,\nu} = \frac{1}{2}T_{\mu\nu,\nu} - \frac{1}{2}\mathfrak{M}_{\mu\lambda\nu,\,\lambda,\,\nu} - \frac{1}{2}\mathfrak{M}_{\nu\lambda\mu,\,\nu,\,\lambda}.$$

Now, $\mathfrak{M}_{\mu\nu\lambda}$ being skew-symmetrical with respect to its first two indices, we obtain by virtue of formula (1.136)

$$\partial_\nu\partial_\lambda\mathfrak{M}_{\nu\lambda\mu} = 0,$$

hence

$$\Theta_{\mu\nu,\,\nu} = \frac{1}{2}(T_{\mu\nu,\nu} - \mathfrak{M}_{\mu\lambda\nu,\,\lambda,\,\nu}) = 0, \tag{3.56}$$

and this is an expression which vanishes by virtue of (3.51).

On the other hand, by virtue of the same formula (3.51), $\Theta_{\mu\nu}$ can be written in the form

$$\Theta_{\mu\nu} = T_{\mu\nu} + \frac{1}{2}\partial_\lambda\{\mathfrak{M}_{\mu\nu\lambda}+\mathfrak{M}_{\lambda\mu\nu}+\mathfrak{M}_{\lambda\nu\mu}\}. \tag{3.57}$$

In calculating $\int_\sigma \Theta_{\mu\nu}(x)d\sigma_\nu$ we have to integrate the second term over σ; let σ_0 on the other hand be the surface in the remote past on which all the field variables vanish; in accordance with Gauss's theorem (1.135), this integral can be expressed in the form

$$\frac{1}{2}\int_{\sigma_0}^{\sigma} \partial_\lambda\partial_\nu\{\mathfrak{M}_{\mu\nu\lambda}+\mathfrak{M}_{\lambda\mu\nu}+\mathfrak{M}_{\lambda\nu\mu}\}\, d^4x.$$

Now, if we take into account the symmetry of the operator $\partial_\lambda\,\partial_\nu$ with respect to the indices λ and ν and the skew-symmetrical character of the

† This tensor was first obtained by F. J. Belinfante, *Physica*, **6**, 887 (1939) and L. Rosenfeld, *Mém. Acad. Roy. Belgique*, **18**, 6 (1940).

brackets with respect to the same indices, we conclude, making use once more of (1.136) that this integral is actually zero. We therefore have

$$P_\mu = \int_\sigma T_{\mu\nu}\, d\sigma_\nu = \int \Theta_{\mu\nu}\, d\sigma_\nu. \tag{3.58}$$

We can show by the same method that the total angular momentum $M_{\mu\nu}$ of the field is given in terms of the tensor $\Theta_{\mu\nu}$ by the relation

$$M_{\mu\nu} = \int_\sigma \{x_\nu \Theta_{\mu\lambda} - x_\mu \Theta_{\nu\lambda}\}\, d\sigma_\lambda. \tag{3.59}$$

By calculations analogous to those which led to formula (3.32), one may prove that the conservation of the total angular momentum is a consequence of the symmetry of the tensor $\Theta_{\mu\nu}$.

We now introduce the "vacuum state" by a statement of a property of the operator $D(L)$.

We shall assume—and this is one of the fundamental assumptions of any quantum field theory—that there is one state and one only which is invariant under the Lorentz transformations. Using the notation used in (3.44) and specifying this sate by $|0\rangle$, we may write by definition

$$D(L)|0\rangle = |0\rangle, \tag{3.60}$$

from which it follows that

$$P_\mu |0\rangle = 0, \qquad M_{\lambda\nu}|0\rangle = 0. \tag{3.61}$$

In other words, *there exists therefore a state with zero total momentum and angular momentum.* In a particle interpretation, it would be a state in which there were no particles present. The existence of the vacuum state may also be inferred from the following two assumptions:

(a) the total energy of the system P_0 is positive definite;

(b) the "vacuum state", a state of minimal energy, is an invariant state.

We infer from (a) that the system does really possess a positive or zero minimal eigenvalue. Furthermore, since the "vacuum state" corresponding to this minimal eigenvalue appears as identical to all observers (as is assumed in (b)), this eigenvalue is necessarily 0.

Note. It must be noted, as we have already said, that all the preceding formulae are expressed in symbolic form: in quantum field theory one must take into consideration for each type of Lagrangian density the order of the factors of its terms.

5. Spin of a field

The operator $\mathfrak{M}_{\mu\nu}$ is obviously not a constant of motion, but the operator (in the space of the indices A) $J_{\mu\nu}$ may be used to characterize a field by giving its spin. We shall in fact see in the next chapter that each field may be characterized by two numbers: the mass of the particles which it represents and their spin (or spins).

Let us consider the space components of $J_{\mu\nu}$; they will enable us to define the operator $\boldsymbol{S}$ with components S_1, S_2, S_3 as follows:

$$S_k = -iJ_{lm} \equiv s_{lm}, \tag{3.62}$$

where k, l, m is one of the cyclic permutations of the numbers 1, 2, 3.

With this notation, formula (3.22) becomes

$$[S_k, S_l] = iS_m, \tag{3.63}$$

k, l, m representing a cyclic permutation of the numbers 1, 2, 3.†

The components of the three-vector $\boldsymbol{S}$ therefore obey the same commutation relations as those of an angular momentum. By choosing an orthogonal basis simultaneously diagonalizing $\boldsymbol{S}^2$ and S_3, we can show that the matrix $\boldsymbol{S}^2 = S_1^2+S_2^2+S_3^2$ has as its eigenvalues $S(S+1)$, where S is one of the numbers of the series $0, \frac{1}{2}, 1, \frac{3}{2}, \ldots$. For a given S, the eigenvalues of S_3 are the $2S+1$ numbers of the series

$$-S, \quad -S+1, \quad \ldots, \quad S-1, \quad S. \tag{3.64}$$

S is called the spin of the field.

For a given S, the relation (3.64) proves that $J_{\mu\nu}$ is represented by a $(2S+1)\times(2S+1)$ matrix. We can therefore say that a particle with a spin

† Otherwise: $[S_k, S_l] = i\varepsilon_{klr}S_r$, where ε_{klr} is equal to zero if two of its indices are equal and equal to $+1$ or -1 according as k, l, r are obtained from 1, 2, 3 by an odd or even permutation.

It is, furthermore, clear that the matrices J_{lm} of formula (3.62), when the infinitesimal rotation acts on a three-vector ($\mathcal{D}_1$ representation), may be obtained from the matrices $J_{\mu\nu}$ of the note on p. 100 by the suppression of their fourth row and column. The connection between these matrices and the matrices to which the text refers (S_3 and $\boldsymbol{S}^2$ diagonal) is very simple: the former ones have real elements and they act on three-vectors X with real components X_1, X_2, X_3, while the latter can be inferred from the transformation which transforms the vector $\boldsymbol{X}$ into the vector of components:

$$\xi = \frac{X_1+iX_2}{\sqrt{2}}, \quad \eta = \frac{X_1-iX_2}{\sqrt{2}}, \quad \zeta = X_3.$$

In other words, they may be obtained by a canonical transformation $AJ_{lm}A^{-1}$, where the unitary matrix A is given by:

$$A = \begin{pmatrix} 1/\sqrt{2} & i/\sqrt{2} & 0 \\ 1/\sqrt{2} & -i/\sqrt{2} & 0 \\ 0 & 0 & 1 \end{pmatrix}.$$

S has $2S+1$ independent components and that it corresponds to a unitary representation of order S of the group of the Euclidean space rotations.

Conversely, let us consider a given $Q(x)$: the transformation of $Q(x)$ for a given Lorentz transformation being assumed to be known, the angular momentum $\mathbf{S}$ gives a representation of the rotations of three-dimensional space. If this representation is irreducible and of order S, $Q(x)$ will be said to represent particles with a spin S. If, on the other hand, this representation is reducible into representations $\mathcal{D}_S, \mathcal{D}_{S-1}, \ldots$, we shall say that the field variable represents a mixture of particles with spin $S, S-1, \ldots$, or alternatively, adopting an expression from de Broglie's "fusion theory" (cf. Chap. IV, Part 3) that it describes a particle with a maximal spin S.

6. Orbital angular momentum of a field

Like the spin tensor, the orbital angular momentum tensor $L_{\mu\nu}$ is not a constant of motion, except for fields such that

$$J_{\mu\nu} = 0. \tag{3.65}$$

The scalar and pseudo-scalar fields which are distinguished from one another by their intrinsic parity (cf. § 7) obey this condition. The second of the two represents, as we know, the pions, the quanta of the nuclear field. We shall therefore discuss the properties of the orbital angular momentum.

Consider the space components of formula (3.53); we have

$$[L_{jk}, Q_A(x)] = -i(x_j\partial_k - x_k\partial_j)Q_A(x). \tag{3.66}$$

Since L_{jk} is a skew-symmetrical tensor, it is specified by its three components

$$L_1 = L_{23}, \quad L_2 = L_{31}, \quad L_3 = L_{12}, \tag{3.67}$$

and (3.66) splits into three formulae:

$$[L_1, Q_A(x)] = -i(x_2\partial_3 - x_3\partial_2)Q_A(x) \equiv l_1(\partial)Q_A(x), \tag{3.68a}$$
$$[L_2, Q_A(x)] = -i(x_3\partial_1 - x_1\partial_3)Q_A(x) \equiv l_2(\partial)Q_A(x), \tag{3.68b}$$
$$[L_3, Q_A(x)] = -i(x_1\partial_2 - x_2\partial_1)Q_A(x) \equiv l_3(\partial)Q_A(x), \tag{3.68c}$$

the definition of the differential operators $l_j(\partial)$ being evident.

From the above relations we easily obtain the commutator

$$[L_j^2, Q_A(x)] = l_j^2(\partial)Q_A(x),$$

from which it finally follows that

$$\left[\sum_1^3 L_j^2, Q_A(x)\right] \equiv [L^2, Q_A(x)] = \sum_1^3 l_j^2(\partial)Q_A(x) \equiv l^2(\partial)Q_A(x), \tag{3.69}$$
$$= -\left\{\frac{1}{\sin\theta}\frac{\partial}{\partial\theta}\left\{\sin\theta\frac{\partial}{\partial\theta}\right\} + \frac{1}{\sin^2\theta}\frac{\partial^2}{\partial\varphi^2}\right\}Q_A(x).$$

The right side of (3.69) is the expression in polar coordinates of $l^2(\partial)$. Moreover, the operators $l_3(\partial)$ and $l^2(\partial)$ have, as is well known, the following eigenfunctions and eigenvalues:

$$l_3(\partial)Y_{lm}(\theta, \varphi) = mY_{lm}(\theta, \varphi), \tag{3.70a}$$

$$l^2(\partial)Y_{lm}(\theta, \varphi) = l(l+1)Y_{lm}(\theta, \varphi); \quad |m| \leqslant l \quad (l = 0, 1, 2, 3, \ldots). \tag{3.70b}$$

$Y_{lm}(\Theta, \varphi)$ are the spherical harmonics which constitute a complete orthogonal basis on the sphere of radius 1. Let us split the wave functions in this basis:

$$Q_A(x) = \sum_{l, m} q_A^{(l, m)}(r, t)Y_{lm}(\theta, \varphi), \tag{3.71}$$

where $r = |\boldsymbol{x}|$ and the q_A's are operators. Let us substitute this expansion into (3.68c) and take the orthogonality of the spherical harmonics into account; we obtain the following commutation relation:

$$[L_3, q_A^{(l, m)}(r, t)] = mq_A^{(l, m)}(r, t). \tag{3.72}$$

The existence of the vacuum, which we assumed at the end of the preceding section, will permit us to construct a family of eigenvectors of L_3 and $\boldsymbol{L}^2$ and we shall assume—without proof—that we can thus obtain all the eigenvectors of these two operators.

Let us in fact apply (3.72) to the vacuum; we obtain

$$L_3q_A^{(l, m)}(r, t)|0\rangle = mq_A^{(l, m)}(r, t)|0\rangle. \tag{3.73}$$

The eigenvectors of the component L_3 of the orbital angular momentum are therefore the vectors $q_A^{(l, m)}(r, t)|0\rangle$ and the eigenvalues are the numbers $m = 0, \pm 1, \pm 2, \ldots$, with $|m| \leqslant l$. By inserting the expansion (3.71) into (3.69) we should obtain by a similar calculation:

$$\boldsymbol{L}^2q_A^{(l, m)}(r, t)|0\rangle = l(l+1)q_A^{(l, m)}(r, t)|0\rangle, \tag{3.74}$$

an equation giving the eigenvectors and eigenvalues of the total orbital angular momentum.

We are interested in investigating the case of a particle with spin 0 in the Heisenberg picture; the eigenvalues of L_3 and $\boldsymbol{L}^2$ are good quantum numbers for, according to (3.47*b*), the operator $\boldsymbol{L}$ commutes with H. The state vector of the field, if it was initially chosen as an eigenvector common to L_3 and $\boldsymbol{L}^2$, remains the eigenvector of these observables and is characterized by the same eigenvalues as at the initial moment. It will, moreover, be readily seen that the eigenvectors of (3.74) are really independent of t. Indeed, we have seen [cf. (2.110)] that

$$Q(x) = e^{itH}Q(\boldsymbol{x}, 0)e^{-itH},$$

from which it follows that

$$q_A^{(l,m)}(r, t)|0\rangle = e^{itH} q_A^{(l,m)}(r, 0)|0\rangle \tag{3.75}$$

since $H|0\rangle = 0$. Since on the other hand $\int [L_j, H] = 0$, formulae (3.73) and (3.74) are obeyed by the vectors $q_A^{(l,m)}(r, 0)|0\rangle$ which are consequently independent of t.

If the field represents a non-zero spin particle, $\boldsymbol{L}$ is a function of the time in the Heisenberg picture. We can always define the state vector at the initial moment t_0 as an eigenvector of L_3 and $\boldsymbol{L}^2$ but, whereas the state vector remains constant, the eigenvectors of L_3 and $\boldsymbol{L}^2$ change in time; the expectation values of L_3 and $\boldsymbol{L}^2$ are therefore no longer equal to the eigenvalues of these operators at the time zero: these eigenvalues are therefore no longer good quantum numbers.

7. Parity of a field

The parity of a field is connected with the invariance of the action under the Lorentz transformations $L_-^\uparrow$ (cf. Chap. I, § 1); it concerns the behaviour of this field under a space symmetry†

$$\boldsymbol{x} \rightarrow \boldsymbol{x}' = -\boldsymbol{x}, \quad t \rightarrow t' = t \tag{3.76}$$

which by virtue of (3.7) transforms field variables as follows:

$$Q_A^{(\sigma)}(x') = \sigma_{AB} Q_B(x) = \sigma_{AB} Q_B(-\boldsymbol{x}', t'); \tag{3.77}$$

σ_{AB}, a matrix in the space of the indices A, is therefore a particular case of the matrix $\mathcal{J}_{AB}$ given by (3.7).

Let us now consider field variables as operators on the space $\mathcal{R}$. It is obvious that we cannot in this case appeal to the Schwinger–Feynman principle to affirm the existence of the operator U (2.158), corresponding to the symmetry concerned: the set of space-symmetries does not in fact contain the operator I. But we can conclude that such an operator exists by extending this principle; this extension is based essentially on the property that both field equations and commutation rules are inferred from the action and therefore remain invariant when the action is invariant.‡ We shall in fact be verifying this in the case of free fields (cf. for instance Chap. V, § 19); but we assume in any case that there exists a unitary operator Π such that

$$Q_A^{(\sigma)}(x) = \Pi Q_A(x) \Pi^{-1}. \tag{3.78}$$

† The study of the time reversal presents somewhat particular characteristics. We shall discuss it below: Chap. IV, § 20; Chap. VII, § 6d; and Chap. IX, § 4c.

‡ In other words, we shall admit, as on p. 92, that field equations and commutation rules determine in a unique way the algebra of the field variables. The non-conservation of parity for the neutrino field will be considered in Chap. VI, § 15.

Making use of (3.77), formula (3.78) can be written in the form

$$Q_A^{(\sigma)}(x) = \Pi Q_A(x)\Pi^{-1} = \sigma_{AB}Q_B(-\boldsymbol{x}, t); \tag{3.79}$$

then introducing (3.71) into (3.79):

$$\sum_{l,m} Y_{lm}(\theta, \varphi)\Pi q_A^{(l,m)}(r, t)\Pi^{-1} = \sum_{B,l,m} \sigma_{AB}Y_{lm}(\pi-\theta, \pi+\varphi)q_B^{(l,m)}(r, t)$$
$$= \sum_{B,l,m} (-1)^l\sigma_{AB}Y_{lm}(\theta, \varphi)q_B^{(l,m)}(r, t).$$

Spherical harmonics being orthogonal, we finally obtain the formula:

$$\Pi q^{(l,m)}(r, t) = (-1)^l\sigma q^{(t,m)}(r, t)\Pi, \tag{3.80}$$

which is expressed in matrix form.

On the other hand, let ε_0 be the parity of the vacuum

$$\Pi|0\rangle = \varepsilon_0|0\rangle; \tag{3.81}$$

by making (3.80) act on the vacuum, we get

$$\Pi q^{(l,m)}(r, t)|0\rangle = \varepsilon_0(-1)^l\sigma q^{l,m}(r, t)|0\rangle. \tag{3.82}$$

Suppose furthermore that there exists a unitary matrix S diagonalizing σ:

$$S\sigma S^{-1} = \delta. \tag{3.83}$$

δ is then a particular diagonal matrix; eqn. (3.82) is then written

$$\Pi Sq^{(l,m)}(r, t)|0\rangle = (-1)^l\varepsilon_0\delta Sq^{l,m}(r, t)|0\rangle, \tag{3.84}$$

and by setting

$$\bar{\omega}_A^{(l,m)}(r, t) = \sum_B S_{AB}q_B^{(l,m)}(r, t)|0\rangle, \tag{3.85}$$

eqn. (3.84) shows that we thus obtain the eigenvectors of Π:

$$\Pi\bar{\omega}_A^{(l,m)}(r, t) = (-1)^l\varepsilon_0\delta_A\bar{\omega}_A^{(l,m)}(r, t). \tag{3.86}$$

The parity of the state $\bar{\omega}_A^{(l,m)}(r, t)$ is therefore $(-1)^l\varepsilon_0\delta_A$; since Π has been assumed to be unitary, we also obtain $|\varepsilon_0\delta_A| = 1$. We see finally, that $\varepsilon_0\delta_A$ is the parity of the state $\bar{\omega}_A^{(0,0)}(r, t)$; we characterize in this way "the intrinsic parity" of the field.

Let us now consider the canonical transform by the operator Π of eqns. (2.154):

$$i\partial_\mu \Pi Q_A(x)\Pi^{-1} = \Pi[P_\mu, Q_A(x)]\Pi^{-1}.$$

Setting $P_\mu^{(\sigma)} = \Pi P_\mu \Pi^{-1}$ and making use of (3.79), the above relation may also be written

$$i\sigma_{AA'}\partial_\mu Q_{A'}(-\boldsymbol{x}, x_0) = \sigma_{AA'}[P_\mu^{(\sigma)}, Q_{A'}(-\boldsymbol{x}, x_0)]. \tag{3.87}$$

In terms of the variables x', the left side has the following components:

$$-i\sigma_{AA'}\partial'_j Q_{A'}(x') \quad \text{and} \quad +i\sigma_{AA'}\partial'_4 Q_{A'}(x').$$

Let us express these components afresh with the help of the canonical equations (2.154) using the variables x'; comparing the two sides of (3.87), we obtain†

$$\sigma_{AA'}[P_j^{(\sigma)}+P_j, Q_{A'}(x')] = 0,$$
$$\sigma_{AA'}[P_4^{(\sigma)}-P_4, Q_{A'}(x')] = 0,$$

from which it follows that $P_j^{(\sigma)} = -P_j$, $P_4^{(\sigma)} = P_4$. The energy–momentum observable has therefore the properties required by the correspondence principle; moreover,

$$[\Pi, P_j]_+ = 0, \quad [\Pi, P_4] = 0;$$

Π commutes therefore with P_0: it is a constant of motion in the Heisenberg picture.

Finally, the argument used at the end of the preceding section shows that the eigenvectors (3.87) do not depend on the time: $\overline{\omega}^{(l,m)}(r, 0)$ is an eigenvector of Π. The parity can therefore belong together with certain components of the operators P_μ and $M_{\mu\nu}$, to the complete set of observables defining the state vector in this picture.

The scalar field has the intrinsic parity of the vacuum ε_0. (which is chosen equal to $+1$); the pseudoscalar field has the parity $-\varepsilon_0$. A space-symmetry multiplies the three-space components of a vector field by $-\varepsilon_0$, and the fourth by ε_0, whereas for a pseudovector field the space components are multiplied by ε_0, and the time component by $-\varepsilon_0$. These two fields are said to have respectively $+1$ and -1 as their intrinsic parities.

For fields with half-integer spins, the δ_A's can be imaginary. For instance, in the case of the Dirac field, we shall see (6.1.14) that $\sigma=\lambda\gamma_4$ with $\lambda^2=\pm 1$, and consequently λ is one of the four numbers ± 1, $\pm i$.

8. Current four-vector and total charge of a field

Consider a Lagrangian density, a function of the field variables $Q(x)$ and of their Hermitian conjugates. To any "*gauge transformation of the first kind*"

$$Q(x) \rightarrow Q'(x) = e^{i\lambda}Q(x), \tag{3.88a}$$

where λ is an infinitesimal *c*-number independent of x, there corresponds the transformation

$$\tilde{Q}(x) \rightarrow \tilde{Q}'(x) = e^{-i\lambda}\tilde{Q}(x) \tag{3.88b}$$

† We have $\det \sigma \neq 0$.

and the variations

$$\delta Q(x) = i\lambda Q(x), \quad \delta \tilde{Q}(x) = -i\lambda \tilde{Q}(x). \tag{3.89}$$

Suppose that we carry out this transformation on a set of components of Q characterized by the index C; the variations δQ_c and $\delta \tilde{Q}_c$ are of the (2.139) type and it follows from formula (2.140) that we can readily write:

$$\delta \mathfrak{A}[\sigma, \sigma_0, Q, \tilde{Q}, \delta Q, \delta \tilde{Q}] = i\lambda \int_{\sigma(x)} \left(\frac{\partial \mathcal{L}}{\partial Q_{C,\mu}} Q_C(\xi) - \tilde{Q}_C(\xi) \frac{\partial \mathcal{L}}{\partial \tilde{Q}_{C,\mu}} \right) d\sigma_\mu. \tag{3.90}$$

Setting

$$j_\mu(x) = i \left(\frac{\partial \mathcal{L}}{\partial Q_{C,\mu}} Q_C(x) - \tilde{Q}_C(x) \frac{\partial \mathcal{L}}{\partial \tilde{Q}_{C,\mu}} \right) \tag{3.91a}$$

and

$$\mathcal{N}[\sigma] = \int_{\sigma(x)} j_\mu(\xi) d\sigma_\mu(\xi), \tag{3.91b}$$

the variation (3.90) of the action takes the form

$$\delta \mathfrak{A}[\ldots] = \lambda \mathcal{N}[\sigma]. \tag{3.91c}$$

With regard to the explicit form of $\delta \mathfrak{A}[\ldots]$, we must repeat what we said on p. 109: expression (3.91) is merely symbolic when the position of any operator in the given Langrangian density must be taken into account.

When the action is invariant under a transformation of this type one may show by an argument similar to that on p. 104 that $\delta \mathfrak{A}[\sigma, \sigma_0, \ldots]$ is independent of σ, and consequently that

$$\frac{\delta \mathcal{N}[\sigma]}{\delta \sigma(x)} = j_{\mu,\mu}(x) = 0. \tag{3.92}$$

A direct application of formula (2.130) expressing the Schwinger–Feynman variation principle then leads to

$$Q_C(x) = [\mathcal{N}, Q_C(x)], \tag{3.93a}$$

$$\tilde{Q}_C(x) = -[\mathcal{N}, \tilde{Q}_C(x)]. \tag{3.93b}$$

We shall now consider the particularly important case in which the field $Q_c(x)$ represents particles of absolute charge e; we define the current vector

$$J_\mu(x) = e j_\mu(x) \tag{3.94}$$

and the total charge operator

$$e_{\text{tot}} = \int_{\sigma(x)} J_\mu(\xi) d\sigma_\mu(\xi). \tag{3.95}$$

Note, furthermore, that in a non-quantized theory in which the field variables commute, the current $J_\mu(x)$ is zero if the wave functions $Q(x)$ are real. This property makes it possible to represent a neutral field by real wave functions, therefore by Hermitian operators after quantization. Any Hermitian field variable represents therefore a neutral field; the converse is not however true: the neutrino, for instance, a chargeless particle, is represented by a spinor.

The total charge is independent of σ (as is required by the invariance of the action), and this property is expressed, from (3.91), as follows:

$$J_{\mu,\mu}(x) = 0, \tag{3.96}$$

a formula which reflects the charge conservation.

Formulae (3.93) are in turn written:

$$eQ_C(x) = [e_{\text{tot}}, Q_C(x)], \tag{3.97a}$$
$$-e\tilde{Q}_C(x) = [e_{\text{tot}}, \tilde{Q}_C(x)] \tag{3.97b}$$

or again

$$e\delta^{(\sigma)}(x-x')Q_C(x) = n_\mu(x')[J_\mu(x'), Q_C(x)] \tag{3.97c}$$

with an analogous relation for (3.97*b*). These relations prove that the operators e_{tot} and $\mathcal{N}$ are the generating operators of the commutative gauge group. The physical reasons which led us to choose (3.95) and (3.97) as a definition of the charge will be examined in the final section of this chapter; we shall be showing there that one can define, from field variables, charge creation and annihilation operators which increase and diminish the eigenvalues of the total charge e_{tot} (3.95) by e.

In Chaps. V and VI particularly, it will be seen that in some very important cases, the operators P_μ and $M_{\mu\nu}$ are bilinear with respect to the wave functions. Taking into account the Hermitian (or anti-Hermitian) character we conclude that they are invariant under the transformation generated by e_{tot}. We therefore have in all these cases:

$$[e_{\text{tot}}, P_\mu] = 0 \tag{3.98}$$
$$[e_{\text{tot}}, M_{\mu\nu}] = 0. \tag{3.99}$$

These formulae prove therefore that in the cases we have previously considered, the eigenvalues of e_{tot} are good quantum numbers in the Heisenberg picture and that we can simultaneously measure the momentum and total charge of a field or its total angular momentum and total charge.

Within the framework of the corpuscular interpretation of the charge, the operator $\mathcal{N} = \frac{e_{tot}}{e}$ is the observable corresponding to the difference between the number of negatively charged particles and the number of positively charged ones.

Finally, in Chap. VII, §§ 3, 6 and 7, we shall define gauge transformations of a more general kind than those we have just been examining.

9. The set of eigenvectors of the total momentum P_μ and the total charge e_{tot}

It is convenient, in approaching this problem, to make use of momentum space

$$Q(x) = (2\pi)^{-4} \int e^{ikx} Q(k)\, d^4k, \tag{3.100a}$$

$$\tilde{Q}(x) = (2\pi)^{-4} \int e^{-ikx} \tilde{Q}(k)\, d^4k, \tag{3.100b}$$

$$\Pi(x) = (2\pi)^{-4} \int e^{-ikx} \Pi(k)\, d^4k. \tag{3.100c}$$

The canonical equations (2.84) and (2.86) concerning $Q(x)$ and $\Pi(x)$ take, after a Fourier transformation, the following form:

$$[Q(k), P_\mu] = \quad k_\mu Q(k), \tag{3.101a}$$

$$[\Pi(k), P_\mu] = -k_\mu \Pi(k). \tag{3.101b}$$

Similarly, we can verify that P_μ is a displacement operator for $\tilde{Q}(x)$ (note in this connection that P_0 is Hermitian and not P_4):

$$[P_\mu, \tilde{Q}(x)] = i\,\frac{\partial \tilde{Q}}{\partial x_\mu}. \tag{3.102}$$

Finally, on introducing the Fourier transform (3.100*b*) of $(\tilde{Q}x)$, we obtain

$$-[\tilde{Q}(k), P_\mu] = k_\mu \tilde{Q}(k). \tag{3.103}$$

Since the operator P_μ is an observable, it has a complete orthogonal set of eigenvectors $|\alpha, p\rangle$. The second argument denotes the eigenvalue of P_μ, whereas the first one symbolizes the set of quantum numbers relating to the observables which, together with P_μ, form a complete set of commuting observables. We thus have

$$P_\mu|p, \alpha\rangle = p_\mu|p, \alpha\rangle \tag{3.104}$$

and

$$\langle q, \beta\,|\,\alpha, p\rangle = \delta(q-p)\,\delta(\alpha, \beta), \tag{3.105}$$

where $\delta(\alpha, \beta)$ is the Kronecker symbol or the Dirac function according as α and β symbolize quantum numbers belonging to a discrete or continuous spectrum.

We take the matrix element of the canonical equation of $Q(x)$:

$$\frac{\partial}{\partial x_\mu}\langle p, \alpha\,|\,Q(x)\,|\,\beta, q\rangle = -i\langle p, \alpha\,|[P_\mu, Q(x)]\,|\,\beta, q\rangle. \quad (3.106)$$

This expression can be written, by virtue of (3.104):

$$\frac{\partial}{\partial x_u}\langle p, \alpha\,|\,Q(x)\,|\,\beta, q\rangle = -i(p_\mu - q_\mu)\langle p, \alpha\,|\,Q(x)\,|\,\beta, q\rangle.$$

We thus have

$$\langle p, \alpha\,|\,Q(x)\,|\,\beta, q\rangle = \langle p, \alpha\,|\,Q(0)\,|\,\beta, q\rangle e^{-i(p-q)x} \quad (3.107)$$

or by a Fourier transformation:

$$\langle p, \alpha\,|\,Q(k)\,|\,\beta, q\rangle = (2\pi)^4\langle p, \alpha\,|\,Q(0)\,|\,\beta, q\rangle\,\delta(p-q+k). \quad (3.108)$$

Consider the vector $Q_A(k)\,|\,\alpha, p\rangle$; since the vectors $|\,\beta, q\rangle$ form a complete orthonormal set, it can be expressed in the form

$$\begin{aligned} Q_A(k)\,|\,\alpha, p\rangle &= \int |\,\beta, q\rangle\langle q, \beta\,|\,Q_A(k)\,|\,\alpha, p\rangle\, dq\, d\beta \\ &= (2\pi)^4 \int |\,\beta, p-k\rangle\langle p-k, \beta\,|\,Q_A(0)\,|\,\alpha, p\rangle\, d\beta \end{aligned} \quad (3.109)$$

and $\tilde{Q}_A(k)$ can be expressed in an analogous form.

If, on the other hand, we apply eqn. (3.103) to the vector $|\,\alpha, p\rangle$, we verify that

$$P_\mu\{\tilde{Q}_A(k)\,|\,\alpha, p\rangle\} = (p_\mu + k_\mu)\{Q_A(k)\,|\,\alpha, p\rangle\}, \quad (3.110a)$$

Consequently, $\tilde{Q}_A(k)\,|\,\alpha, p\rangle$ is the eigenvector of the observable P_μ corresponding to the eigenvalue $p_\mu + k_\mu$.

Similarly, by computing $P_\mu\{Q_A(k)\,Q_B(k')\,|\,\alpha, p\rangle\}$, we obtain, by virtue of formulae (3.101) and (3.110):

$$P_\mu\{\tilde{Q}_A(k)\tilde{Q}_B(k')\,|\,\alpha, p\rangle\} = (p_\mu + k_\mu + k'_\mu)\{Q_A(k)Q_B(k')\,|\,\alpha, p\rangle\}. \quad (3.110b)$$

We again have an eigenvector of P_μ corresponding to the eigenvalue $p_\mu + k_\mu + k'_\mu$.

We obtain analogous formulae for the vector of the space $\mathcal{R}$ obtained by making an arbitrary number of field variables act on $|\,\alpha, p\rangle$.

Assume now that the field under consideration describes particles of charge $\pm e$, and that the operators e_{tot} and P_μ commute; they can therefore be simultaneously diagonalized: e_{tot} is consequently one of the operators which can form with P_μ a complete set of observables, and its eigenvalues can help determine the parameter β of the vectors $|\beta, q\rangle$. Then let $|\beta, e', p\rangle$ be an eigenvector of the total charge

$$e_{\text{tot}}|\beta, e', p\rangle = e'|\beta, e', p\rangle. \tag{3.111}$$

Starting from eqns. (3.97) and adopting the method which enabled us to obtain (3.110), we get

$$e_{\text{tot}}Q_A(x)|\beta, e', p\rangle = (e'+e)Q_A(x)|\beta, e', p\rangle, \tag{3.112a}$$
$$e_{\text{tot}}\tilde{Q}_A(x)|\beta, e', p\rangle = (e'-e)\tilde{Q}_A(x)|\beta, e', p\rangle; \tag{3.112b}$$

$\tilde{Q}(x)$ diminishes the total charge, whereas $Q(x)$ increases it.

Equations (3.112) exhibit the fundamental property of the charge observable: its spectrum is a discrete spectrum, whereas that of P_μ is continuous. The total charge of a field is a multiple of the elementary charge e: this is an essentially quantum property upon which is based the definition (3.95) of the charge, a definition which in itself appears arbitrary and artificial.

Note finally that formulae (3.112) will enable us to define, in the last section of this chapter, the creation and annihilation operators.

10. Charge conjugation

Consider an action invariant under the gauge transformations we examined in § 8; we shall be led to show further on (Chap. V, § 18) that there exists a transformation called "charge conjugation", a canonical transformation (under which field equations and commutation rules remain invariant) defined as follows:

$$Q_C^{(c)}(x) = \mathscr{C}_{CC'}\tilde{Q}_{C'}(x). \tag{3.113}$$

We shall be considering the existence and properties of the matrix $\mathscr{C}$ for various fields; note for the time being that the equations replacing (3.97) are, as may be easily verified:

$$\left.\begin{aligned} +e\tilde{Q}_C^{(c)}(x) &= [e_{\text{tot}}, \tilde{Q}_C^{(c)}(x)], \\ -eQ_C^{(c)}(x) &= [e_{\text{tot}}, Q_C^{(c)}(x)]. \end{aligned}\right\} \tag{3.114}$$

It therefore follows that in place of (3.113) we have

$$\left.\begin{aligned} e_{\text{tot}}Q_C^{(c)}(x)|\beta, e', p\rangle &= (e'-e)Q_C^{(c)}(x)|\beta, e', p\rangle, \\ e_{\text{tot}}\tilde{Q}_C^{(c)}(x)|\beta, e', p\rangle &= (e'+e)Q_C^{(c)}(x)|\beta, e', p\rangle. \end{aligned}\right\} \tag{3.115}$$

The field $Q^{(c)}(x)$ behaves in an opposite manner to the field $Q(x)$; $\tilde{Q}^{(c)}(x)$ increases the total charge, whereas $Q^{(c)}(x)$ diminishes it. We may therefore say that the field $Q(x)$ describes particles of opposite charge to those described by the field $Q^{(c)}(x)$ and that a theory invariant under charge conjugation describes particles and antiparticles. A well-known example of pairs of particles is supplied by electrons: the positron $+e$ has an opposite charge to that of the negatron $-e$; the positron is said to be the antiparticle of the negatron.

In a more general manner, the notion of particles and antiparticles may be extended to all non-Hermitian fields with an action invariant under gauge transformations; we need merely consider $\mathcal{N}$ as defined by eqns. (3.91) and the relations (3.93*a*) and (3.93*b*).

Note also that the fact that the charge conjugation—when it exists—must be canonical, implies a unitary representation Γ of this operation in the space $\mathcal{R}$ such that

$$Q^{(c)}(x) = \Gamma Q(x)\Gamma^{-1}; \qquad (3.116a)$$

Γ is the "charge parity" operator. Since, furthermore, the antiparticle of an antiparticle is the particle, it is necessary that

$$\Gamma^2 = I. \qquad (3.116b)$$

Consequently, Γ has the eigenvalues ± 1 which correspond to the two possible charge parities.

Finally, a calculation of exactly the same type as that performed for the space symmetry (and which will be used again in Chap. V, § 18) proves that if $P_\mu^{(c)}$ and $e_{\text{tot}}^{(c)}$ are the transformed total momentum and total charge observables, we have

$$P_\mu^{(c)} = P_\mu, \quad e_{\text{tot}}^{(c)} = -e_{\text{tot}}.$$

The total charge therefore changes its sign after a charge conjugation; this fact justifies us in interpreting this operation as an exchange of particles and antiparticles.

11. Creation and annihilation operators

The corpuscular interpretation of quantized fields is based on the introduction of creation and annihilation operators which, as their names suggest, increase or diminish the eigenvalue of certain observables such as the total energy–momentum of the field, its charge, etc. These operators are inferred from the field variables $Q(x)$ by separating them into parts of positive and negative frequency according to a process devised by Schwinger.†

In a theory covariant under the Lorentz group, it is legitimate to require that such a separation be intrinsic, in other words that a Lorentz transformat in transform the creation operators into themselves and the annihilation operators into themselves.

† J. Schwinger, *Phys. Rev.*, **75**, 651 (1949).

This condition implies restrictions with regard to the field equations [cf. the footnote on p. 123] so that finally only the free fields satisfy it. Nevertheless, in view of the formal nature of the discussions we shall be entering on and their close connection with the notions examined in the preceding sections, we have reserved them for the conclusion of this chapter.

Let the following two operators correspond to the operator $Q(x)$:

$$Q^{\pm}(x) = \frac{1}{2i\pi}\int_{C_+} Q(x\mp\varepsilon\tau)\frac{d\tau}{\tau}, \tag{3.117a}$$

the path C_+ is the real axis of the plane τ from $-\infty$ to $+\infty$ completed by a semi-circumference of infinitesimal radius below the point O (Fig. 3.1); ε is a time-like vector: $\varepsilon_\mu\varepsilon_\mu<0$ and $\varepsilon_0 = -i\varepsilon_4>0$, so that $x-\varepsilon\tau$ is the vector with components $x_\mu-\varepsilon_\mu\tau$. These two assumptions concerning ε_μ are covariant.

We can moreover write $Q^-(x)$ as follows:

$$Q^{-}(x) = \frac{1}{2i\pi}\int_{C_-} Q(x-\varepsilon\tau)\frac{d\tau}{\tau}. \tag{3.117b}$$

Here C_- denotes the axis of the plane τ from $-\infty$ to $+\infty$ with a semi-circumference of infinitesimal radius above the point $\tau=0$. The sum of these two paths gives a closed path and we have

$$Q^+(x)+Q^-(x) = \frac{1}{2i\pi}\oint Q(x-\varepsilon\tau)\frac{d\tau}{\tau} = Q(x).$$

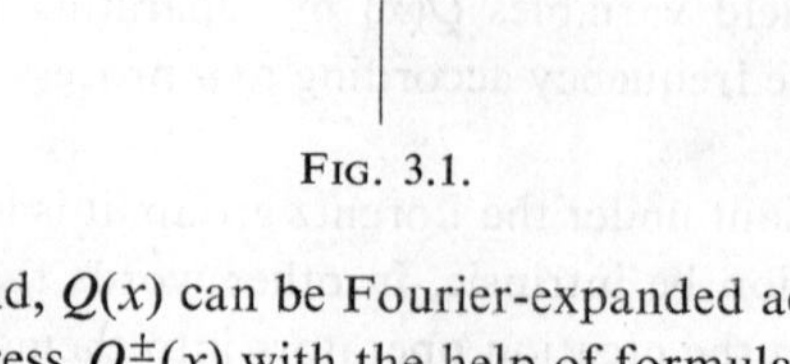

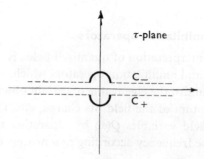

FIG. 3.1.

On the other hand, $Q(x)$ can be Fourier-expanded according to (3.100), so that we can express $Q^{\pm}(x)$ with the help of formula (3.114) as follows:

$$Q^{\pm}(x) = (2\pi)^{-4}\int e^{ikx}Q(k)\,d^4k\,\frac{1}{2i\pi}\int_{C+} e^{\mp ik\varepsilon\tau}\frac{d\tau}{\tau} \tag{3.118}$$

where

$$\frac{1}{2i\pi}\int_{C_+} e^{-ik\varepsilon\tau}\frac{d\tau}{\tau} = \begin{cases} 1 & \text{if} \quad k\varepsilon < 0, \\ 0 & \text{if} \quad k\varepsilon > 0, \end{cases}$$

so that finally we have

$$Q^{\pm}(x) = (2\pi)^{-4}\int_{k\varepsilon\left\{{<0 \atop >0}\right.} e^{ikx}Q(k)d^4k. \tag{3.119}$$

The two integration paths in the space k, characterized respectively by the sign of $-k\varepsilon$, are independent of ε_ν provided that ε and k are time-like vectors with a given sign for ε_0. It is in fact sufficient to observe that

$$-k\varepsilon = k_0\varepsilon_0\left(1-\frac{\boldsymbol{k}\,.\,\boldsymbol{\varepsilon}}{k_0\varepsilon_0}\right), \tag{3.120a}$$

with

$$\left|\frac{\boldsymbol{k}\,.\,\boldsymbol{\varepsilon}}{k_0\varepsilon_0}\right| < \left|\frac{\boldsymbol{k}}{k_0}\right|\cdot\left|\frac{\boldsymbol{\varepsilon}}{\varepsilon_0}\right| < 1, \tag{3.120b}$$

when ε and k are time-like.† The latter relation proves that the sign of $k_\mu\varepsilon_\mu$ is determined by that of $k_0\varepsilon_0$ and if furthermore we make the invariant assumption that ε_0 is positive, this sign will be that of k_0. $Q(x)$ is thus built up by a superposition of plane waves: the positive frequencies in Q^+ and the negative frequencies in Q^-.

We then set

$$Q^+(k) = Q(k), \quad Q^-(k) = Q(-k) \quad \text{for} \quad k_\mu\varepsilon_\mu < 0; \tag{3.121}$$

the two formulae (3.119) may be written

$$Q^{\pm}(x) = (2\pi)^{-4}\int_{k\varepsilon<0} e^{\pm ikx}Q^{\pm}(k)\,d^4k, \tag{3.122}$$

$$Q(x) = (2\pi)^{-4}\int_{k\varepsilon<0} (e^{ikx}Q^+(k)+e^{-ikx}Q^-(k))\,d^4k. \tag{3.123}$$

We also verify that for the particular choice $\varepsilon=0$, $\varepsilon_0>0$ the above integration is taken over all positive frequencies. On the other hand, let $\mathcal{C}(x-x')$ be the matrix in the space of the indices A whose element (AB) is‡

$$[Q_A(x), Q_B(x')]_{\pm} = \mathcal{C}_{AB}(x-x'). \tag{3.124}$$

† It is this assumption concerning k [leading to an invariant definition of $Q^{\pm}(k)$] which restricts the above definition to free fields: for such fields, indeed, the momentum vectors k_μ obey $k_\mu k_\mu+m^2 = 0$; they are time-like vectors [cf. (4.2)].

‡ This matrix may be identically zero, for instance in the case of the Dirac field, where the components of $Q(x)$ are $\psi_\alpha(x)$ and $\psi_\beta(x')$. $\mathcal{C}$ depends on $x-x'$ in all cases in which the action is invariant under translations.

The method of decomposition we considered above can also be applied to the right side of the commutation relations; we get

$$\mathcal{C}(x) = \mathcal{C}^{+}(x)+\mathcal{C}^{-}(x), \tag{3.125a}$$

$$\mathcal{C}^{\pm}(x) = (2\pi)^{-4}\int_{k\varepsilon<0} e^{\pm ikx}\mathcal{C}^{\pm}(k)\, d^4k. \tag{3.125b}$$

We shall now prove that the commutation (or anticommutation) relations of the $Q^{\pm}$'s are

$$[Q_A^{+}(x), Q_B^{+}(x')]_{\pm} = [Q_A^{-}(x), Q_B^{-}(x')]_{\pm} = 0, \tag{3.126a}$$

$$[Q_A^{+}(x), Q_B^{-}(x')]_{\pm} = \mathcal{C}_{AB}^{+}(x-x'), \tag{3.126b}$$

$$[Q_A^{-}(x), Q_B^{+}(x')]_{\pm} = \mathcal{C}_{AB}^{+}(x-x'). \tag{3.126c}$$

Let us prove, for instance, (3.126*a*); we have

$$[Q_A^{+}(x), Q_B^{+}(x')]$$

$$= \frac{1}{(2i\pi)^2}\int_{C_+} \mathcal{C}_{AB}(x-x'-\varepsilon(\tau-\tau'))\frac{d\tau}{\tau}\,\frac{d\tau'}{\tau'}$$

$$= \frac{1}{2\pi}\int_{-\infty}^{+\infty} d\alpha\,\mathcal{C}_{AB}(x-x'-\varepsilon\alpha)\frac{1}{2i\pi}\int_{C_+} e^{i\alpha\tau}\frac{d\tau}{\tau}\,\frac{1}{2i\pi}\int_{C_+} e^{-i\alpha\tau'}\frac{d\tau'}{\tau'} = 0,$$

where we have introduced the Fourier transforms with respect to the variable τ in $\mathcal{C}_{AB}(x-x'-\varepsilon\tau)$. By virtue of (3.118) the last two integrals cannot be simultaneously non-zero; thus we finally obtain the value of the commutator in question. The Fourier transforms of formulae (3.126) are also simple in form:

$$[Q_A^{+}(k), Q_B^{+}(k')]_{\pm} = [Q_A^{-}(k), Q_B^{-}(k')]_{\pm} = 0, \tag{3.127a}$$

$$[Q_A^{+}(k), Q_B^{-}(k')]_{\pm} = \mathcal{C}_{AB}^{+}(k)\,\delta(k-k'), \tag{3.127b}$$

$$[Q_A^{-}(k), Q_B^{+}(k')]_{\pm} = \mathcal{C}_{AB}^{-}(k)\,\delta(k-k'). \tag{3.127c}$$

One need, in fact, merely take the Fourier transforms of formulae (3.126), noting that the right side is a function of $x-x'$.

Analogous expressions may be written for

$$\tilde{Q}(x) = (2\pi)^{-4}\int e^{-ikx}\tilde{Q}(k)\, d^4k. \tag{3.128}$$

We have as before (but with different signs for $k_\mu\varepsilon_\mu$)

$$\tilde{Q}^{\pm}(x) = (2\pi)^{-4}\int_{k\varepsilon\left\{{>0 \atop <0}\right.} e^{-ikx}\tilde{Q}(k)\, d^4k. \tag{3.129}$$

Let us now define $\tilde{Q}^{-}(k)$ and $\tilde{Q}^{+}(k)$;

$$\tilde{Q}(k) = \tilde{Q}^{-}(k), \quad \tilde{Q}(-k) = \tilde{Q}^{+}(k) \quad \text{for} \quad k\varepsilon < 0; \tag{3.130}$$

going to x-space, one finally gets

$$\left.\begin{aligned}\tilde{Q}^+(x) &= (2\pi)^{-4}\int_{k\varepsilon<0} e^{ikx}\tilde{Q}^+(k)\,d^4k,\\ \tilde{Q}^-(x) &= (2\pi)^{-4}\int_{k\varepsilon<0} e^{-ikx}\tilde{Q}^-(k)\,d^4k.\end{aligned}\right\} \quad (3.131)$$

Comparing now (3.130) with (3.121), we easily obtain

$$\left.\begin{aligned}\{Q^+(k)\}^\sim &= \tilde{Q}^-(k),\\ \{Q^-(k)\}^\sim &= \tilde{Q}^+(k),\end{aligned}\right\} \quad (3.132a)$$

$$\left.\begin{aligned}\{Q^+(x)\}^\sim &= \tilde{Q}^-(x),\\ \{Q^-(x)\}^\sim &= \tilde{Q}^+(x).\end{aligned}\right\} \quad (3.132b)$$

Similarly, by denoting by $C(x-x')$ the commutator of the operator $Q(x)$ with its Hermitian conjugate $\tilde{Q}(x)$,

$$[Q_A(x), \tilde{Q}_B(x')] = C_{AB}(x-x'), \quad (3.133)$$

it is easy, by a Fourier transformation, to infer the following commutators:

$$[Q_A^-(k), \tilde{Q}_B^+(k')]_\pm = C_{AB}^+(k)\,\delta(k-k'), \quad (3.134a)$$

$$[\tilde{Q}_A^-(k), Q_B^+(k')]_\pm = C_{AB}^-(k)\,\delta(k-k'). \quad (3.134b)$$

Making use of the definition (3.117) of the operators $Q^\pm(x)$, one infers that P_μ is also a displacement operator for these operators; indeed

$$-i[P_\mu, Q_A^\pm(x)] = -\frac{1}{2\pi}\int_{C\pm}[P_\mu, Q(x-\varepsilon\tau)]\frac{d\tau}{\tau} \quad (3.135)$$

$$= \frac{\partial}{\partial x_\mu}\frac{1}{2i\pi}\int_{C\pm} Q(x-\varepsilon\tau)\frac{d\tau}{\tau} = \frac{\partial Q^\pm(x)}{\partial x_\mu}.$$

Consequently, after a Fourier transformation, one obtains

$$[Q^-(k), P_\mu] = -k_\mu Q^-(k), \quad (3.136a)$$

$$[Q^+(k), P_\mu] = \quad k_\mu Q^+(k), \quad (3.136b)$$

and similarly, by virtue of the definitions (3.130),

$$[\tilde{Q}^-(k), P_\mu] = -k_\mu \tilde{Q}^-(k), \quad (3.137a)$$

$$[\tilde{Q}^+(k), P_\mu] = \quad k_\mu \tilde{Q}^+(k). \quad (3.137b)$$

The matrix elements of $Q^\pm(k)$ are therefore given by formulae analogous to (3.108):

$$\langle p, \alpha|\,Q^\pm(k)\,|\beta, q\rangle = (2\pi)^4\langle p, \alpha|\,Q(0)\,|\beta, q\rangle\,\delta(p-q\pm k) \quad (3.138)$$

and [cf. (3.131)]:

$$\langle p, \alpha|\,\tilde{Q}^\pm(k)\,|\beta, q\rangle = (2\pi)^4\langle p, \alpha|\,\tilde{Q}(0)\,|\beta, q\rangle\,\delta(p-q\mp k). \quad (3.139)$$

We shall now consider the vector $Q_A^-(k)|\alpha, p\rangle$; it can be expressed in the following form, by virtue of the above relations:

$$Q_A^-(k)|\alpha, p\rangle = \int |\beta, q\rangle\langle q, \beta|Q_A^-(k)|\alpha, p\rangle\, dp\, d\beta \tag{3.140a}$$

$$= (2\pi)^4 \int |\beta, p+k\rangle\langle p+k, \beta|Q_A(0)|\alpha, p\rangle\, d\beta.$$

Similarly, by using (3.139),

$$\tilde{Q}_A^-(k)|\alpha, p\rangle = (2\pi)^4 \int |\beta, p+k\rangle\langle p+k, \beta|\tilde{Q}_A(0)|\alpha, p\rangle\, d\beta, \tag{3.140b}$$

and, in the same manner,

$$Q_A^+(k)|\alpha, p\rangle = (2\pi)^4 \int |\beta, p-k\rangle\langle p-k, \beta|Q_A(0)|\alpha, p\rangle\, d\beta, \tag{3.141a}$$

$$\tilde{Q}_A^+(k)|\alpha, p\rangle = (2\pi)^4 \int |\beta, p-k\rangle\langle p-k, \beta|\tilde{Q}_A(0)|\alpha, p\rangle\, d\beta. \tag{3.141b}$$

The following relations may also be easily inferred from eqns. (3.137) and (3.138):

$$P_\mu\{Q_A^\pm(k)|\alpha, p\rangle\} = (p_\mu \mp k_\mu)\{Q_A^\pm(k)|\alpha, p\rangle\}, \tag{3.142a}$$

$$P_\mu\{\tilde{Q}^\pm(k)|\alpha, p\rangle\} = (p_\mu \mp k_\mu)\{\tilde{Q}^\pm(k)|\alpha, p\rangle\}. \tag{3.142b}$$

Formulae (3.142) show that the operators $Q^-(k)$ and $\tilde{Q}^-(k)$ transform a state characterized by the value p of the total energy–momentum into another state in which this energy–momentum of the field is $p+k$. These operators, which are composed solely of negative frequencies (3.122) and (3.129) are "creation operators". For an analogous reason, formulae (3.142) prove that the operators $Q^+(k)$ and $\tilde{Q}^+(k)$ are "annihilation operators". We shall see moreover when we come to define the "number operator" in the case of free fields that the creation (or annihilation) operator transforms a state with N particles into a state with $N+1$ (or $N-1$) particles.

Consider once more the vacuum state as introduced in (3.60) and (3.61):

$$P_\mu|0\rangle = 0, \quad \langle 0|0\rangle = 1. \tag{3.143}$$

Since the vacuum state represents the state of minimal energy of the system, it can be said in accordance with the above interpretation of the annihilation operators:

$$Q_A^+(k)|0\rangle = 0, \quad \tilde{Q}_A^+(k)|0\rangle = 0. \tag{3.144}$$

We may also construct from the vacuum the eigenvectors of P_μ by letting a number of creation operators act on $|0\rangle$: this set will be written in explicit form in the case of free fields (cf. Chap. V, § 15).

Let us finally return to the total charge operator e_{tot}; according to (3.112) and adopting the method we have just used for P_μ, we can easily infer

$$e_{tot}\{Q^{\pm}_{\bar{A}}(k)\,|\,\beta, e', p\rangle\} = (e'+e)\{Q^{\pm}_{\bar{A}}(k)\,|\,\beta, e', p\rangle\}, \tag{3.145a}$$

$$e_{tot}\{\tilde{Q}^{\pm}_{\bar{A}}(k)\,|\,\beta, e', p\rangle\} = (e'-e)\{\tilde{Q}^{\pm}_{\bar{A}}(k)\,|\,\beta, e', p\rangle\}. \tag{3.145b}$$

Comparing eqns. (3.145) and (3.142), we see that the creation and annihilation operators can also be interpreted as follows:

	Type of operator	Momentum	Charge
$Q^+(k)$	Annihilation	k	$-e$
$\tilde{Q}^+(k)$	Annihilation	k	$+e$
$Q^-(k)$	Creation	k	$+e$
$\tilde{Q}^-(k)$	Creation	k	$-e$

It can be seen from this table that $Q^-(k)$ and $\tilde{Q}^-(k)$ respectively create particles of charge $+e$ and $-e$, whereas $\tilde{Q}^+(k)$ and $Q^+(k)$ respectively annihilate particles of charge $+e$ and $-e$; so that in accordance with formulae (3.113), $\tilde{Q}(k)$ diminishes the total charge (by creation of $-e$ or annihilation of $+e$) and $Q(k)$ increases the total charge (by creation of $+e$ or annihilation of $-e$).

Examples will be given in Chaps. V and VI of the observables we have built up in this chapter. The reader will see that, in the case of free fields in which the definition of the number of particles presents no difficulty, the observables P_μ and e_{tot} do in fact give the sum of the energy–momenta and the sum of the charges of the particles present at the time at which the field is considered.

CHAPTER IV

Linear wave equations and invariants of first order

1. Introduction

We noted in the last chapter the particularly important part played by the Lorentz group.

It enabled us to introduce the "ten generating operators": the four components P_μ of the energy–momentum vector operator and the six components $M_{\mu\nu}$ of the total angular momentum (skew-symmetrical tensor) operator. We also saw that these operators contribute to the definition of the complete set of commuting observables which determines a state completely.

A second group, the gauge group of the first kind, enabled us to define another operator which contributes to the definition of a state: the total charge e_{tot} of the field.

To return to the Lorentz group, we also noted the importance of another mathematical concept: that of "group representation". The investigation and classification of the irreducible unitary representations of the inhomogeneous Lorentz group were carried out by E. Wigner. Because of the non-compactness of this group, these representations are infinite-dimensional ones; each of them can be characterized by giving two numbers m and s.

The number m can be defined as follows: as we shall prove in Part 2 of this chapter, the operator $p_\mu p_\mu$ [formula (3.13a)] is a multiple of the operator unity I; we shall therefore set

$$p_\mu p_\mu = -m^2, \tag{4.1}$$

where m^2 is a constant. Consider free particles, i.e. non-interacting particles which are subject to no external field; the above formula is simply the well-known relation between their momentum, energy and mass.

The mass of the free particles appears therefore as a number characterizing a representation of the Lorentz group. In Part 2 of this chapter we shall prove that the second number s characteristic of the group is the spin (cf. Chap. III, § 4) of the field. We are thus led to the following definition of free particles: they correspond to unitary irreducible representations of the inhomogeneous Lorentz group, unitary in order that the norm of the vectors be preserved, and irreducible since no mixtures of particles should be involved (cf. Chap. III, § 4).

We shall also prove in the next chapter that in the case of free fields, a one-particle state (Heisenberg picture) can be defined by making a field operator act on the vacuum state.

The investigation of the evolution of one-particle states amounts therefore in the last analysis to the investigation of the evolution of field variables. It will therefore be necessary to investigate the type of functional equations they obey.

We shall assume that the free field variables obey differential equations of a finite order: they can therefore always be reduced to systems of equations of first order by defining as new field variables a number of their derivatives.

The canonical field variables (cf. p. 51) generally belong to this set; we have, indeed, seen that the canonical equations (2.59) are of first order in time variable.

On the other hand, taking into account the well-known correspondence between the energy–momentum vector and the differential operators ∂_μ, we shall include the energy–momentum relation (4.1) in our picture, by assuming that each of the components of the wave function $Q_A(x)$ must obey the equation of second order (4.2*a*) commonly known as the "Klein–Gordon equation". In particular, any free field described by a one-component variable—scalar particle—obeys a wave equation of second order

$$\{\Box - m^2\} Q_A(x) = 0, \tag{4.2a}$$

which will reduce to a differential system of first order and whose quantization will be discussed in the next chapter.† Note also that if the field under consideration corresponds to a mixture of particles with different masses $m_1, \ldots, m_N$, (4.2*a*) can be readily extended as follows:

$$\prod_1^N \{\Box - m_j^2\} Q_A(x) = 0. \tag{4.2b}$$

† See Chap. V, § 7: the field will in fact be quantized from a differential system (5.24) and (5.44), which will be of first order with respect to t.

We shall only occasionally consider equations of this kind which as a rule do not lead to a positive definite energy.

Supposing therefore that the quantization of a field described by a linear differential equation of first order with constant coefficients amounts to the quantization of eqn. (4.2*a*), we shall advance the following postulate:

REDUCTION POSTULATE. A. *The wave equations of the free particles form a system of partial differential equations of first order with constant coefficients:*

$$\mathcal{F}_{AB}(\partial)Q_B(x) = 0 \tag{4.3a}$$

or in matrix form:

$$\mathcal{F}(\partial)Q(x) = 0 \tag{4.3b}$$

$\mathcal{F}_{AB}(\partial)$ *is a differential polynomial.*

B. *These equations are covariant (i.e. keep the same form) under the Lorentz group.*

C. *There exists a differential operator* $\mathcal{D}(\partial)$ *such that*

$$\mathcal{F}(\partial)\mathcal{D}(\partial) = \Box - m^2. \tag{4.4}$$

These three postulates, however general they may be in character, impose considerable restrictions as to the possible forms of $\mathcal{F}(\partial)$, as will be seen below.

We shall divide this chapter into three parts:

(a) In Part 1 we shall run over the entire list of known particles: a list which has lengthened considerably during the last few years.

Readers will thus be able to note that the fields investigated in the following chapters do in fact correspond to actually observed particles.

(b) Part 2 will recapitulate the essential features of the Wigner–Bargman classification of elementary particles according to their masses and spins. We shall also consider the restrictions imposed by the preceding postulates on the forms of the differential equations of the wave functions.

(c) Finally, in Part 3 we shall give an explicit way of building up the field equations of particles with arbitrary masses and spins by de Broglie's method, which is known as "méthode de la fusion".

PART 1

THE EXPERIMENTAL FACTS

2. General considerations

An elementary particle is, first of all, characterized by its mass, i.e. its energy at rest. We know that for the photon γ and the neutrinos ν_e and ν_μ there are good reasons to believe that their masses are zero. The electron has a mass around 0·5 MeV, while certain resonances have masses around 3000 MeV. The mass spectrum of elementary particles is thus extremely wide: its lower bound is zero, and we do not know if there is any upper bound.

We are led furthermore to suppose that a particle may have a structure. In order to get some idea about its spatial extension, we may use Heisenberg's relation $\Delta x \cdot \Delta p \approx \hbar$. Taking for Δp its maximum value mc, one sees that the length scale in the description of a particle is of the order

$$\Delta x \approx \frac{\hbar}{mc},$$

which is the reduced Compton wavelength. Consider, for instance, a particle with 300 times the mass of the electron (let us say a π-meson); such a particle will have a spatial extension of the order of 1 Fermi (10^{-13} cm). The second Heisenberg relation $\Delta E \, \Delta t \approx \hbar$ may shed some light on the time scale in the world of particles; taking $\Delta E = mc^2$, then

$$\Delta t \approx \frac{\hbar}{mc^2}.$$

For the same particle as above, the time scale is of the order of 10^{-24} sec, a very short time indeed.

The *reaction time* is another useful concept in the theory of elementary particles. Suppose a particle crossing a nucleus whose spatial extension is of the order of $r_0 = 10^{-13}$ cm, the reaction time will be of the order

$$\frac{r_0}{c} \approx 10^{-23} \text{ sec}.$$

This concept may be linked with the *strength of the interaction*, expressed by a dimensionless number called the *coupling constant*, which is, roughly speaking, inversely proportional to the reaction time (up to a dimensional factor equal to 1).

For a decay process lasting 10^{-23} sec there will correspond a *strong interaction* characterized by a coupling constant of order 1. There are other processes occurring through photons, e.g. photo-disintegration, and generally *electromagnetic interactions* which are 1/100 times slower than the strong ones; their coupling constant will be of the order 1/100. This number may be compared with the electromagnetic coupling constant

$$\frac{e^2}{\hbar c} = \frac{1}{137}.$$

We also have decays with reaction times of the order 10^{-9} sec; their coupling constants should thus be of the order 10^{-14}, this kind of decays occurring through *weak interactions*. Let us finally quote the gravitational field whose quanta are the gravitons, particles with mass 0 and spin 2.

Among all the known particles, there are some which are stable when free, as is the case, for instance, for the neutrinos, the photon, the electron and the proton. The neutron decays within 10^3 sec through β-decay.

From a theoretical point of view, we may look for connections between the theory of elementary particles and the Lorentz group: it has been shown by Bargmann and Wigner (cf. Part 2 of this chapter) that a particle corresponds to a unitary irreducible representation of the Lorentz group and is defined by its mass m and its spin s. Its state is characterized by an eigenvector of a complete set of commuting observables: number of particles, momentum and total charge, etc.; certain "good quantum numbers" are therefore associated with its momentum, charge, and total angular momentum.

Coming back to the disintegration of a given particle through certain decay schemes, we may add that conservation laws (energy-momentum, angular momentum, charge, parity, etc.) severely restrict the various possibilities of decay.† For instance, the stability of the electron with respect to the emission of a photon, in other words the fact that the decay $e \to e' + \gamma$ is impossible, is a simple result of the conservation of energy–momentum. It is indeed easy to see that any transformation $A \to A' + B$, where A and A' have the same mass, is impossible: one need merely adopt a frame of reference in which A (or B) is at rest according as $m_A \neq 0$ (or $m_B \neq 0$).

Similarly, the stability of the "free" photon with respect to the emission of a pair of electrons, in other words the impossibility of the decay

$$\gamma \to e^+ + e^-$$

is a particular case of the following reaction:

$$A \to B + B',$$

† We thus have simple examples of selection rules; for more complex examples, cf. Chap. VII (§§ 2, 7, 8 and 9).

where it is assumed that B and B' have the same mass m_B and that $m_A = 0$.

The other conservation theorems play a similar role; for instance, the introduction of the neutrino into the β-decay:

$$n \to p + e^- + \bar{\nu}$$

is required by the fact that in $n \to p + e^-$ the negatron emitted could not have an energy spectrum but a well-defined energy fixed by that of the neutron and of the proton and that there would be no conservation of the angular momenta. It is also very easy to find decays which would be excluded by the charge conservation.

Finally, we have already seen that a particle and its antiparticle have the same properties except that their charges are opposite; for instance, the positron is the antiparticle of the negatron. Apart moreover from any question of charge, particles and antiparticles are always said to be charge conjugates.

3. Classification of elementary particles

Particles are normally classified according to their masses,† falling into several families. The *hadrons* are heavy particles capable of strong interactions: hadrons with half-integer spin (obeying Fermi–Dirac statistics) are called *baryons*, hadrons with integer spin (obeying Bose–Einstein statistics) are called *mesons*. Among the baryons the lightest one is the proton (around 938 MeV); among the mesons the lightest one is the Π° (around 135 MeV) with a life-time of 10^{-15} sec.

The family of *leptons* includes the neutrinos, the electron and the muon. Finally, there are the photon and the graviton, quanta of the electromagnetic and gravitational fields, respectively.

Let us finally note that all particles obeying Fermi–Dirac statistics are called *fermions* while all particles obeying Bose–Einstein statistics are called *bosons*. To the particles described in the table of p. 134 we have to add a very large number indeed of resonances which may be considered as particles with an extremely short life-time or as excited states of a given particle.

Let us quote some resonances: among the baryons, the family of resonances N^* consists of excited states $N^* = N + \Pi$ with an isospin equal to 1/2 or 3/2. Their mass spectrum starts at 2800 m_e, the Ω^- particle of 3350 m_e has a charge $-e$ and a decay scheme: $\Xi + \Pi$ or $\Lambda^\circ + K$.

† The figures in the table on p. 134 are given as a mere indication and may not agree exactly with the results of the latest experiments. In the same way, we do not claim to have given an exhaustive list of all the known types of decay.

	Charge		Mass (m)	Spin	Parity	Decay scheme	Lifetime (s)
Photon	γ	0	0	1		Stable	∞
Leptons	ν_s	0	$<5\times10^{-4}m_e$	1/2		Stable	∞
	ν_μ	0	$<4\ m_e$	1/2		Stable	∞
	e^-	$-e$	m_e	1/2		Stable	∞
	e^+	$+e$	m_e	1/2		$e^++e^-\to n\gamma;\ n=2, 3, \ldots$	$1{\cdot}5\times10^{-7}$ (*)
	$\mu^\pm$	$\pm e$	$206{\cdot}8\ m_e$	1/2		$\mu^\pm\to e^\pm+\nu+\nu$	$2{\cdot}22\times10^{-6}$
Mesons π or pions	Π^+	$+e$	$273{\cdot}3\ m_e$	0	$-$	$\Pi^+\to\mu^++\nu$	$2{\cdot}56\times10^{-8}$
	Π^-	$-e$	$273{\cdot}3\ m_e$	0	$-$	$\Pi^-\to\mu^-+\nu$	$2{\cdot}56\times10^{-8}$
	Π^0	0	$264{\cdot}3\ m_e$	0	$-$	$\Pi^0\to\gamma+\gamma$	$\sim10^{-15}$
						$\to e^++e^-+\gamma$	
Mesons K or kaons	K^0	0	$965\ m_e$	0		$K_1^0\to\Pi^++\Pi^-$	$\sim10^{-10}$
						$\to\Pi^0+\Pi^0$	
						$K_2^0\to\Pi^++\Pi^-+\Pi^0$	$\sim10^{-8}$
						$\to e^\pm+\Pi^\mp+\nu$	
						$\to\mu^\pm+\Pi^\mp+\nu$	
	$K^\pm$	$\pm e$	$967\ m_e$	0		$K^+\to\mu^++\nu$	$1{\cdot}3\times10^{-8}$
						$\to\Pi^++\Pi^0$	
						$\to\Pi^++\Pi^++\Pi^-$	
						$\to\Pi^++\Pi^0+\Pi^0$	
						$\to e^++\nu+\Pi^0$	
						$\to\mu^++\nu+\Pi^0$	
Nucleons	p	$+e$	$1836{\cdot}12\ m_e$	1/2		Stable	∞
	n	0	$1838{\cdot}65\ m_e$	1/2		$n\to p+e^-+\nu$	$1{\cdot}11\times10^3$ s
Hyperons	Λ_0	0	$2181{\cdot}5\ m_e$	1/2		$\Lambda^0\to p+\Pi^-$	3×10^{-10}
						$\to n+\Pi^0$	
	Σ^0	0	$2330\ m_e$	1/2		$\Sigma^0\to\Lambda^0+\gamma$	$\sim10^{-12}$
	Σ^+	$+e$	$2328\ m_e$	1/2		$\Sigma^+\to p+\Pi^0$	$\sim10^{-10}$
						$\to n+\Pi^+$	
	Σ^-	$-e$	$2341\ m_e$	1/2		$\Sigma^-\to n+\Pi^-$	$\sim10^{-10}$
	Ξ^0	0	$2630\ m_e$	1/2		$\Xi^0\to\Lambda^0+\Pi^0$	$\sim10^{-10}$
	Ξ^-	$-e$	$2586\ m_e$	1/2		$\Xi^-\to\Lambda^0+\Pi^-$	$\sim10^{-10}$

(*) In the ^3S state of the positronium.

$$m_e = 9{\cdot}1090\times10^{-28}\ \text{g}.$$

Note that the positon e^+ is the antiparticle of the negaton e^- and that μ^+ is the antiparticle of μ^-; the antiproton $\bar{p}$ and the antineutron $\bar{n}$ are respectively the antiparticles of the proton and the neutron. The mesons π^- and K^- are the antiparticles of the mesons π^+ and K^+. The meson K^0 and the hyperon Λ^0 are their own antiparticles (cf. Chap. III, § 10). Finally, the neutrino ν is also considered to

Among the meson resonances we may quote: the η of 1097 m_e, spin 0, negative parity, which decays into 3π or 2γ; the ϱ's of 1550 m_e, spin 1, negative parity, which decay essentially into 2π.

A detailed description of the experiments which made it possible to determine the characteristics of these particles would be beyond the scope of this book: we shall confine ourselves to a few remarks.

First of all, the proton and the neutron are to be considered as two states of the same particle (the nucleon) which differ in their isobaric spin or isospin (Chap. VII, § 7), but a difference in mass to the extent of 2·53 m_e in favour of the neutron may also be noted. It is reasonable to suppose that this difference in mass is due to the electromagnetic action of these particles upon themselves. Nevertheless, as will be seen below (Chap. X, § 7), all the numerical results expressing the action of a particle upon itself contain infinite quantities, the interpretation of which is fraught with difficulties.

We shall investigate further on (Chap. VII, § 8) some attempts to introduce new quantum numbers, particularly in connection with the charge of the particles. These quantum numbers enable us to define new selection rules excluding certain decays which have not been observed and which are not excluded by the selection rules inferred from the Lorentz and gauge groups.† We finally quote some recent and successful attempts to introduce abstract groups (SU(3), SU(6)...) in the classification of elementary particles.

PART 2

WAVE EQUATIONS AND LORENTZ GROUP

4. Elementary particles and representations of the Lorentz group. Polarization

As we stated in our introduction, to each irreducible representation of the proper inhomogeneous Lorentz group there corresponds a kind of elementary particle.

We shall now give a brief outline of the Bargman–Wigner method‡ of

† A full bibliography for the study of elementary particles would be extremely long: we shall simply refer to a few full-length works in which lists of original papers will be found: J. D. Jackson, *The Physics of Elementary Particles*, Princeton University Press, 1958; J. Hamilton, *The Theory of Elementary Particles*, Oxford, 1959; P. Roman, *Theory of Elementary Particles*, North Holland, 1960; D. B. Lichtenberg, *Meson and Baryon Spectroscopy*, Springer Verlag, 1965; S. Gasiorowicz, *Elementary Particle Physics*, Wiley.

‡ V. Bargman and E. P. Wigner, *Proc. Nat. Acad. Sc.*, **34**, 211 (1948).

obtaining these representations and thereby classifying elementary particles.

The method is a simple one: it consists in showing that each irreducible representation $D(L)$ [formula (3.44)] of the Lorentz group is characterized by two numbers m and s by proving that there exist two operators P and W, in close relations with m and s and which commute for each irreducible representation with the generating operators P_λ and $M_{\mu\nu}$ and are consequently multiples of the operator identity.†

In the following stages of the method, we note, first of all, that since the particles under consideration are free, $\mathcal{R}$ is a Hilbert space, as will be seen from the examples given in Chaps. V and VI. It is sufficient to consider the generators p_μ and $m_{\mu\nu}$ of formulae (3.13) as operators in a Hilbert space of functions. In other words, instead of discussing quantum field theory, we can confine ourselves to non-quantized field theory.

We shall therefore introduce the vector operator:

$$w_1 = w_{234}, \quad w_2 = w_{341}; \quad w_3 = w_{412}, \quad w_4 = w_{123}, \tag{4.5a}$$

with

$$w_{\lambda\mu\nu} = p_\lambda m_{\mu\nu} + p_\mu m_{\nu\lambda} + p_\nu m_{\lambda\mu} = m_{\mu\nu} p_\lambda + m_{\nu\lambda} p_\mu + m_{\lambda\mu} p_\nu; \tag{4.5b}$$

making use of the relations (3.47), we can therefore write:‡

$$w_\lambda = \frac{1}{2}\, \varepsilon_{\lambda\mu\nu\varrho} p_\mu m_{\nu\varrho}. \tag{4.5c}$$

We then verify by virtue of these same relations (3.47):

$$[m_{\mu\nu}, w_\lambda] = -i(\delta_{\mu\lambda} w_\nu - \delta_{\nu\lambda} w_\mu), \tag{4.6a}$$

$$[p_\mu, w_\lambda] = 0. \tag{4.6b}$$

It is easy to prove from these same relations and from (4.6) that the two operators

$$P = p_\mu p_\mu, \tag{4.7a}$$

$$W = -w_\mu w_\mu = \frac{1}{6}\, w_{\lambda\mu\nu} w_{\lambda\mu\nu} \tag{4.7b}$$

commute with all the operators p_λ and $m_{\mu\nu}$ and that they are therefore

† Remember that the operators $D(L)$ form an irreducible representation of the Lorentz group if, given the vector space $\mathcal{E}$ in which they operate, the space 0 and the space $\mathcal{E}$ itself are the only two subspaces of $\mathcal{E}$ which remain invariant; it is also known that any operator commuting with all the $D(L)$'s is a multiple of the identity.

‡ Remember that $\varepsilon_{\lambda\mu\nu\varrho}$, a completely skew-symmetrical tensor of fourth order, has the value ± 1 according to the parity of the permutation of the numbers 1, 2, 3, 4 represented by the indices λ, μ, ν, ϱ and is equal to 0 if two of these indices are equal.

multiples of the identity for any irreducible representation of the Lorentz group.

On the other hand, formulae (3.13) have led us to introduce the operators

$$l_{\mu\nu} = i\left\{k_\nu \frac{\partial}{\partial k_\mu} - k_\mu \frac{\partial}{\partial k_\nu}\right\}, \qquad s_{\mu\nu} = -iJ_{\mu\nu}.$$

We may also verify that $l_{\mu\nu}$ does not contribute to the expression (4.5*b*) of $w_{\lambda\mu\nu}$, so that we have

$$w_{\lambda\mu\nu} = p_\lambda s_{\mu\nu} + p_\mu s_{\nu\lambda} + p_\nu s_{\lambda\mu}, \tag{4.8a}$$

$$[s_{\mu\nu}, p_\lambda] = 0. \tag{4.8b}$$

It is equally easy to see that the components w_μ are finally written

$$w_1 = p_2 s_{34} + p_3 s_{42} + p_4 S_1, \quad w_2 = p_3 s_{14} + p_1 s_{43} + p_4 S_2,$$

$$w_3 = p_1 s_{24} + p_2 s_{41} + p_4 S_3, \quad w_4 = -\boldsymbol{p} \cdot \boldsymbol{S};$$

S_1, S_2, S_3 are the spin vector components (3.62). If we also set

$$S_1' = s_{14}, \quad S_2' = s_{24}, \quad S_3' = s_{34},$$

the preceding formulae may be written in condensed form:

$$\boldsymbol{w} = \boldsymbol{p} \wedge \boldsymbol{S}' + p_4 \boldsymbol{S}, \quad w_4 = -\boldsymbol{p} \cdot \boldsymbol{S}. \tag{4.9}$$

Finally, we may verify the important formula:

$$w_\mu p_\mu = 0. \tag{4.10}$$

Elementary particles may now be classified as follows:

1. *Particles of mass m, spin s*. Since the particles are free, $P = -m^2$. In a frame of reference in which the particle is at rest ($\boldsymbol{p} = 0$), with $p_0^2 = m^2$, we have

$$W = -p_4^2 \boldsymbol{S}^2 = m^2 \boldsymbol{S}^2. \tag{4.11}$$

Therefore the eigenvalues of the operator W/m^2 are the numbers $s(s+1)$ where $s=0, \frac{1}{2}, 1, \ldots$. The operator W will allow us to define the helicity of the particles.†

2. *Particles of mass* 0, *spin s*. We first look for representations which are the limits of the above representations when m^2 becomes zero.

We thus obtain with $P=0$ the equality inferred from (4.11): $W=0$, whence $w_\mu w_\mu = 0$, $p_\mu p_\mu = 0$, $p_\mu w_\mu = 0$, from which it follows that the vectors p_μ and w_μ are collinear: $w_\mu = \pm s p_\mu$‡ and by continuity (from the case $m \neq 0$)

† Polarization and helicity have been taken hitherto as synonyms; later on we shall use the word polarization to refer to a beam of particles.

‡ We infer indeed from the above three equalities that whatever the scalar λ may be, $(\boldsymbol{w}+\lambda \boldsymbol{p})^2 - (w_0 + \lambda p_0)^2 = 0$, which gives $\boldsymbol{w} = \dfrac{w_0}{p^0} \boldsymbol{p}$.

s represents the spin of the particle. The representation is thus characterized by the operators $P=0$, $w_\mu w_\nu = s^2 p_\mu p_\nu$.

If $W \neq 0$, we refer readers to the above-mentioned paper by Bargman and Wigner who investigate particles of mass 0, the spin of which can take a continuous series of values, and those of mass 0 and infinite spin: such particles have not been observed.

The pseudo-vector [cf. (4.5*c*)] whose components are the Hermitian operators w_j and w_0 represents an observable which is the "polarization" of the field investigated.

According to (4.11), the spin observable is then:

$$\frac{W}{m^2} = -\frac{w_\mu w_\mu}{m^2},$$

and the helicity of the field in the direction specified by the unitary vector u_μ is†

$$\frac{w_\mu u_\mu}{m}.$$

A particle of non-zero mass and of spin s is therefore described in a $2s+1$ dimensional spin–space, in other words the wave function $Q(x)$ of such a field has $2s+1$ linearly independent components.

The case of particles of mass zero calls for more delicate treatment; their spin—whatever it may be—is proved to be capable of only two alignments (parallel or antiparallel) with respect to an arbitrary direction corresponding to its two polarization states: such for instance is the case of the photon, a particle of spin 1.

We can consider a complete and oriented orthonormal basis of space–time such that the vectors $\boldsymbol{e}^{(1)}$ and $\boldsymbol{e}^{(2)}$ are orthogonal to the vector $\boldsymbol{p}$, ($e_4^{(1)} = e_4^{(2)} = 0$), the vector $\boldsymbol{e}^{(3)}$ is parallel to it $\left(e_j^{(3)} = \dfrac{p_0 p_j}{m|\boldsymbol{p}|},\ e_4^{(3)} = \dfrac{i|\boldsymbol{p}|}{m} \right)$ and $e_\mu^{(4)} = \dfrac{p^\mu}{m}$.

The vector w_μ then has only three non-zero components: those corresponding to the first three vectors; we may therefore speak of longitudinal and transversal helicity.

It can be proved, moreover, than the pseudovector with components $(i/m) w_\mu e_\mu^{(\alpha)}$, the square of whose length is W/m^2, does in fact represent the spin of the field, since it obeys the relations (3.63). The computation in-

† It will be noted that according to (4.9), $\boldsymbol{w}\dfrac{\boldsymbol{p}}{|\boldsymbol{p}| \cdot p_4}$ gives the observable $\boldsymbol{S}\dfrac{\boldsymbol{p}}{|\boldsymbol{p}|}$.

volved, which makes use of the commutation rules (4.6*a*) and the skew-symmetrical properties of $\varepsilon_{\lambda\mu\nu\varrho}$, presents no difficulty but is a trifle lengthy.

The preceding definition of the spin enables us therefore to extract it in a covariant manner from the expression of the operator $m_{\mu\nu}$.

It will be noted finally that the four operators p_j and p_0, the total spin operator W/m^2 and the helicity operator in a certain direction $(W_\mu u_\mu)/m$ form a complete set of six commuting Hermitian operators. They form a complete set of commuting observables capable of describing the field without any ambiguity with respect to the Lorentz group.†

5. Relativistic invariance of the wave equations

According to the reduction postulate (p. 130) $\mathcal{F}(\partial)$ is a differential polynomial of first order; it can therefore be written in the form

$$\mathcal{F}_{AB}(\partial) = (\varrho_\mu)_{AB}\partial_\mu + (\varrho)_{AB}m,$$

so that the wave function $Q(x)$ obeys equation

$$\{\varrho_\mu\partial_\mu + \varrho m\}Q(x) = 0. \tag{4.12a}$$

By exhibiting the indices, the former equation can also be written in the following form:

$$\{(\varrho_\mu)_{AA'}\partial_\mu + m\varrho_{AA'}\}Q_{A'}(x) = 0, \tag{4.12b}$$

the ϱ_μ's are matrices and the wave function $Q(x)$ is a column matrix in the space of the indices A.

We shall show in the next section (by virtue of the same reduction postulate) that ϱ is not a singular matrix, so that by multiplying eqn. (4.12*a*) by ϱ^{-1} it takes the form

$$\{\varrho^{-1}\varrho_\mu\partial_\mu + m\}Q(x) \equiv \{\beta_\mu\partial_\mu + m\}Q(x) = 0. \tag{4.12c}$$

We propose to examine first to what extent the relativistic invariance of the preceding equations restricts the algebra of the β matrices. We shall therefore suppose that under a homogeneous proper Lorentz transformation ($x' = Lx$) such that

$$Q(x) \to Q'(x') = \mathcal{J}(L)Q(x) \tag{4.13}$$

eqn. (4.12*c*) is invariant, in other words that this differential system is turned into another one with the same coefficients, consequently with the same β_μ's:

$$\left\{\beta_\mu \frac{\partial}{\partial x'_\mu} + m\right\} Q'(x') = 0, \tag{4.14}$$

† L. Michel, *Supplemento Nuovo Cimento*, **14**, series 10, 95 (1959).

and we shall prove that $Q'(x')$ obeys (4.14) provided that the β_μ's satisfy condition (4.15). Indeed, since $\dfrac{\partial}{\partial x'_\mu}$ is a vector with respect to these transformations, we have

$$\frac{\partial}{\partial x'_\mu} = L_{\mu\nu}\frac{\partial}{\partial x_\nu}$$

so that eqn. (4.14) is written

$$\left\{L_{\mu\nu}\beta_\mu\frac{\partial}{\partial x_\nu}+m\right\}\mathcal{J}Q(x) = 0$$

or multiplying the two sides on the left by $\mathcal{J}^{-1}$:

$$\left\{L_{\mu\nu}\mathcal{J}^{-1}\beta_\mu\mathcal{J}\frac{\partial}{\partial x_\nu}+m\right\}Q(x) = 0.$$

Identifying with (4.12*c*) we have

$$\mathcal{J}^{-1}\beta_\mu\mathcal{J} = L_{\mu\nu}\beta_\nu. \tag{4.15}$$

Suppose now that the Lorentz transformation is infinitesimal:

$$L_{\mu\nu} = \delta_{\mu\nu}+\varepsilon_{\mu\nu},$$

$$\mathcal{J}(L) = 1+\frac{1}{2}\varepsilon_{\mu\nu}J_{\mu\nu}.$$

From (1.70*b*), the relation (4.15) takes the form

$$\frac{1}{2}\varepsilon_{\lambda\varrho}[J_{\lambda\varrho}, \beta_\nu] = \varepsilon_{\mu\nu}\beta_\mu = \frac{1}{2}(\delta_{\nu\varrho}\varepsilon_{\lambda\varrho}\beta_\lambda-\delta_{\nu\lambda}\varepsilon_{\lambda\varrho}\beta_\varrho)$$

or again

$$[J_{\lambda\varrho}, \beta_\nu] = \beta_\lambda\delta_{\varrho\nu}-\beta_\varrho\delta_{\nu\lambda}. \tag{4.16}$$

Together with formula (3.22) which we rewrite

$$[J_{\mu\nu}, J_{\varrho\sigma}] = -\delta_{\mu\varrho}J_{\nu\sigma}+\delta_{\mu\sigma}J_{\nu\varrho}-\delta_{\nu\sigma}J_{\mu\varrho}+\delta_{\nu\varrho}J_{\mu\sigma}, \tag{4.17}$$

formula (4.16) represents the restrictions on the algebra of the β_μ's, restrictions which are due to the relativistic invariance of (4.12*c*).

The solution $J_{\mu\nu}$ of eqns. (4.16) and (4.17) is unique.

Proof: Let $J_{\mu\nu}$ and $J'_{\mu\nu}$ be two solutions of these equations; according to (4.17), their difference commutes with β_λ:

$$[J_{\mu\nu}-J'_{\mu\nu}, \beta_\lambda] = 0,$$

consequently, it commutes with any function of the β_μ's, and therefore with $J_{\varrho\sigma}$ and $J'_{\varrho\sigma}$. More specifically, we have

$$[J_{\mu\nu}, J'_{\varrho\sigma}-J_{\varrho\sigma}] = 0, \quad [J'_{\mu\nu}-J_{\mu\nu}, J'_{\varrho\sigma}] = 0, \tag{4.18}$$

whence

$$[J'_{\mu\nu}, J'_{\varrho\sigma}]-[J_{\mu\nu}, J_{\varrho\sigma}] = 0.$$

On the other hand, making use of (4.18), we can also write

$$[J'_{\mu\nu}, J'_{\varrho\sigma}]-[J_{\mu\nu}, J_{\varrho\sigma}] = 2(J'_{\nu\sigma}-J_{\nu\sigma}),$$

then comparing the last two equations, we do in fact obtain $J'_{\nu\sigma} = J_{\nu\sigma}$.

The calculation of the cross-sections necessitates, as we shall see further on, the calculation of the spurs of products of β_μ. The foregoing considerations of a general nature enable us to point out an interesting fact: take the spur (in the space of the indices A) of formula (4.15); we have

$$\mathrm{Sp}\,\{\mathfrak{J}^{-1}\beta_\mu\mathfrak{J}\} = \mathrm{Sp}\,\{\beta_\mu\} = L_{\mu\nu}\,\mathrm{Sp}\,\{\beta_\nu\}$$

in accordance with the property $\mathrm{Sp}(AB) = \mathrm{Sp}(BA)$. Now, det $\{I-L\}$ is non-zero,† therefore the only solution of the system $\mathrm{Sp}\ \beta_\mu = L_{\mu\nu}\,\mathrm{Sp}\,\beta_\nu$ is

$$\mathrm{Sp}\beta_\mu = 0.$$

The extension of this type of argument to a product of an odd number of β_μ makes it possible to write‡

$$\mathrm{Sp}\,\{\beta_{\mu_1}\ldots\beta_{\mu_{2n+1}}\} = 0.$$

We shall conclude with a few remarks on the possibility of determining the algebra of the β_μ's.

The simplest form of $J_{\mu\nu}$ that can be tested is

$$J_{\mu\nu} = g^2[\beta_\mu, \beta_\nu], \tag{4.19}$$

and its introduction into eqns. (4.16) and (4.17) gives relations which the algebra of the β_μ's must obey. We shall see when we come to investigate particles of spin $\frac{1}{2}$ and 1 that this form of $J_{\mu\nu}$ is in fact the solution whose uniqueness was established above. For particles of spin more than 1, it can be shown that this solution leads to particles with a mass spectrum.

In a general manner, the β_μ's are matrices which have to obey a number of algebraic relations: they can thus have—as in the well-known case of the Dirac γ_μ matrices—several representations. For obvious physical reasons, the various representations are connected together. Indeed, an observable is generally of the form $\int_{t=\text{const.}} \tilde{Q}(x)\,O(\beta_\mu)Q(x)\,d^3\boldsymbol{x}$ where $O(\beta_\mu)$ is a

† By taking L to be infinitesimal, $I-L$ is a 4×4 skew-symmetrical matrix; its determinant is therefore a positive definite number.

‡ For an investigation of the algebra of these matrices, consult Harisch Chandra, *Phys. Rev.*, **71**, 793 (1947); W. A. Hepner, *Phys. Rev.*, **84**, 744 (1951), or the following books: E. M. Corson, *Introduction to Tensors, Spinors and Relativistic Wave Equations*, Blackie; O. Costa de Beauregard, *Théorie synthétique de la relativité restreinte et des quanta*, Gauthier-Villars, 1957; P. Roman, *op. cit.*; H. Umezawa, *Quantum Field Theory*, North Holland, 1956.

function of the β_μ's which must not change when one representation β_μ is replaced by another, namely β'_μ.

The above holds true when there exists a unitary matrix T (as is in fact the case for the Dirac γ_μ's) such that we have

$$\beta'_\mu = T\beta_\mu T^{-1}.$$

In that case, the observables as given in the form under consideration are in fact independent of the representation of the β_μ's (again provided that a scalar product in L^2 has been defined such that the expression of the observable is $\langle Q, O(\beta_\mu)\, Q\rangle$).

We shall have occasion to expand the foregoing remarks in the case of the Dirac field. We are now going to examine the precise consequences of the reduction postulate of p. 130.

6. Reduction postulate and field equation†

The reduction postulate informs us that there exists a differential operator $\mathcal{D}(\partial)$ such that if $\mathcal{F}(\partial)Q(x)$ are the wave equations, we have

$$\mathcal{F}(\partial)\mathcal{D}(\partial) = \Box - m^2. \tag{4.20a}$$

Suppose that $\mathcal{F}(\partial)$ is defined as follows:

$$\mathcal{F}(\partial) = -\{\beta_\mu \partial_\mu + m\} \tag{4.20b}$$

with a minus sign in its definition.

Consider first in a more general manner the field equation in the form (4.12*a*); we shall prove that the existence of $\mathcal{D}(\partial)$ implies that the matrix ϱ is a non-singular matrix. Write $\mathcal{D}(\partial)$ in the form of a differential polynomial (or an infinite series)

$$\mathcal{D}(\partial) = \alpha_0 + \alpha_{\sigma_1}\partial_{\sigma_1} + \alpha_{\sigma_1\sigma_2}\partial_{\sigma_1}\partial_{\sigma_2} + \ldots \equiv \sum_{n=0} \alpha_{\sigma_1 \ldots \sigma_n}\partial_{\sigma_1}\ldots\partial_{\sigma_n}, \tag{4.21}$$

where the coefficients $\alpha_{\sigma_1 \ldots \sigma_n}$ are matrices in the space of the indices A.

Identifying in (4.20*a*) the terms which have the same number of differential operators, we obtain as the first term

$$\varrho m \alpha_0 = m^2,$$

the product det ϱ. Det α_0 is non-zero, therefore both det ϱ and det α_0 must be non-zero.

Considering therefore the field equation in the form (4.12*c*) and taking $\mathcal{D}(\partial)$ as expressed in (4.21), we obtain by identifying the terms of the two

† H. Umezawa and A. Visconti, *Nucl. Phys.*, **1**, 20 (1956).

sides of (4.20a):†

$$\left.\begin{aligned}\alpha_0 &= +m, \quad \alpha_\sigma = -\beta_\sigma,\\ \alpha_{\sigma_1\sigma_2} &= -\frac{1}{m}(\delta_{\sigma_1\sigma_2}-\beta_{\sigma_1}\beta_{\sigma_2}),\\ \alpha_{\sigma_1\sigma_2\sigma_3} &= +\frac{1}{m^2}\beta_{\sigma_3}(\delta_{\sigma_2\sigma_1}-\beta_{\sigma_2}\beta_{\sigma_1}),\\ &\dots\dots\dots\dots\dots\dots,\\ \alpha_{\sigma_1\dots\sigma_n} &= \left(-\frac{1}{m}\right)^{n-1}\beta_{\sigma_n}\dots\beta_{\sigma_3}(\delta_{\sigma_2\sigma_1}-\beta_{\sigma_2}\beta_{\sigma_1}),\end{aligned}\right\} \tag{4.22}$$

and consequently the following expression of $\mathcal{D}(\partial)$:

$$\begin{aligned}\mathcal{D}(\partial) &= +m-\beta_\sigma\partial_\sigma-\frac{1}{m}\{\Box-\beta_{\sigma_1}\beta_{\sigma_2}\partial_{\sigma_1}\partial_{\sigma_2}\}+\dots\\ &\quad+\left(-\frac{1}{m}\right)^{n-1}\{\Box-(\beta_\sigma\partial_\sigma)^2\}\{\beta_\sigma\partial_\sigma\}^{n-2}+\dots.\end{aligned} \tag{4.23}$$

If therefore we bring in no other considerations, the expansion of $\mathcal{D}(\partial)$ forms an infinite series, but this operator must obey an important restriction due to the relativistic aspect of the theory.

We shall see in the next chapter that

$$[Q_A(x), \bar{Q}_B(x')] = i\mathcal{D}_{AB}(\partial)\,\Delta(x-x') \quad \text{[formula (5.103)]},$$

where $\Delta(x)$ is a scalar distribution with respect to Lorentz transformations. Let us carry out a rotation of the Euclidean space and let $\mathcal{D}_S$ be its representation [$Q'(x')=\mathcal{D}_S Q(x)$, where $\mathcal{D}_S$ is a $(2S+1)\times(2S+1)$ matrix]. The representation generated by the commutator splits according to the Clebsch–Gordan theorem into irreducible representations: $D_{2S}, D_{2S-1}, \dots$. The right side of the above commutator is expressed as products of $\Box\Delta(x-x')$ which are invariant and of terms of the form $(\beta_\sigma\partial_\sigma)^n\Delta(x-x')$. Now, under a Lorentz transformation (a particular case of which is the rotation under consideration), the term $\beta_\sigma\partial_\sigma\Delta(x)$ is transformed into

$$\beta_\sigma\partial'_\sigma\,\Delta(x) = L_{\sigma\varrho}\beta_\sigma\partial_\varrho\,\Delta(x),$$

since under this transformation $\mathcal{F}(\partial)\to\mathcal{F}(\partial')$ from which it follows that $\mathcal{D}(\partial)\to\mathcal{D}(\partial')$ and one may prove moreover $\Delta'(x')=\Delta(x)$. This term

† The same result may be obtained by writing

$$\begin{aligned}\mathcal{D}(\partial) &= -\{\Box-m^2\}\{\beta_\mu\partial_\mu+m\}^{-1}\\ &= -\frac{1}{m}\{\Box-m^2\}\left\{1+\frac{1}{m}\beta_\mu\partial_\mu\right\}^{-1}\end{aligned}$$

and by formally expanding the third factor of the product with respect to $\beta_\mu\partial_\mu$.

$\beta_\sigma \partial_\sigma \Delta(x)$ induces a representation of degree 1 of the rotation under consideration.

Consequently, in accordance with the same Clebsch–Gordan theorem, the term of nth order of $\mathcal{D}(\partial)$ gives a representation of maximal order n of this group. The terms of order $l > 2S$ must therefore have the form

$$\alpha_{\sigma_1 \ldots \sigma_l} = a_{\sigma_1 \ldots \sigma_{2S}} \{\Box\}^{\frac{l}{2}-S} \partial_{\sigma_1 \ldots \sigma_{2S}}.$$

Take the first term which follows $\alpha_{\sigma_1 \ldots \sigma_{2S}}$ and which corresponds to $l = 2S+1$: this term is obviously zero $\left(\dfrac{l}{2}-2S \text{ is not an integer}\right)$ and by virtue of the recurrence formula (4.22) giving $\alpha_{\sigma_1 \ldots \sigma_l}$ from the single term $\alpha_{\sigma_1 \ldots \sigma_{l-1}}$, the terms of order $2S+2, 2S+3, \ldots$, are also zero.

We thus have the theorem:

If the wave equation represents particles of spin S, the differential operator $\mathcal{D}(\partial)$ is a polynomial derivative of 2Sth order.

Reverting now to the expression (4.21) of $\mathcal{D}(\partial)$: since the operator $\partial_{\sigma_1} \ldots \partial_{\sigma_n}$ is symmetric with respect to the indices $\sigma_1, \ldots, \sigma_n$, only the symmetric part of $\alpha_{\sigma_1 \ldots \sigma_n}$ contributes to (4.21) as follows:

$$\mathcal{D}(\partial) = \sum \alpha^{(\mathrm{sym})}_{\sigma_1 \ldots \sigma_n} \partial_{\sigma_1} \ldots \partial_{\sigma_n}$$

but $\alpha^{(\mathrm{sym})}_{\sigma_1 \ldots \sigma_n}$ is proportional to

$$\mathsf{S}\alpha_{\sigma_1 \ldots \sigma_n},$$

where S denotes a summation over all the $\alpha_{\sigma_1 \ldots \sigma_n}$ obtained by permutations of $\sigma_1, \ldots, \sigma_n$.

Hence, we can readily see from the above that we have

$$\mathsf{S}\alpha_{\sigma_1 \ldots \sigma_{2S+1}} = 0,$$

or

$$\mathsf{S}\beta_{\sigma_{2S+1}} \ldots \beta_{\sigma_3}(\delta_{\sigma_2\sigma_1} - \beta_{\sigma_2}\beta_{\sigma_1}) = 0. \tag{4.24}$$

Consider a half-integer spin field; formula (4.24) gives the commutation relations of the Dirac matrices:

$$\mathsf{S}(\delta_{\sigma_2\sigma_1} - \beta_{\sigma_2}\beta_{\sigma_1}) = 2\,\delta_{\sigma_2\sigma_1} - [\beta_{\sigma_1}, \beta_{\sigma_2}]_+ = 0 \tag{4.25}$$

and formula (4.23) gives the following expression of $\mathcal{D}(\partial)$:

$$\mathcal{D}(\partial) = -(\beta_\mu \partial_\mu - m). \tag{4.26}$$

Consider now a field of spin 1; (4.24) is written

$$\mathsf{S}\beta_\mu(\beta_\lambda\beta_\nu - \delta_{\lambda\nu}) = 0$$

or again

$$\mathsf{S}(\beta_\mu\beta_\lambda\beta_\nu + \beta_\nu\beta_\lambda\beta_\mu - \beta_\mu\,\delta_{\lambda\nu} - \beta_\nu\,\delta_{\mu\lambda}) = 0. \tag{4.27}$$

The expression of $\mathcal{D}(\partial)$ is then

$$\mathcal{D}(\partial) = -\left\{\frac{1}{m}\{\Box - m^2\} + \beta_\mu\partial_\mu - \frac{1}{2m}[\beta_\nu, \beta_\nu]_+\partial_\mu\partial_\nu\right\}. \quad (4.28a)$$

In particular, the β_μ's will obey (4.27) if they obey the Petiau–Duffin–Kemmer relation:†

$$\beta_\mu\beta_\lambda\beta_\nu + \beta_\nu\beta_\lambda\beta_\mu - \beta_\mu\,\delta_{\lambda\mu} - \beta_\nu\,\delta_{\mu\lambda} = 0 \quad (4.28b)$$

which characterizes the algebra of the matrices representing particles of spin 1. This restriction tends to suggest that the relation (4.24) does not necessary imply that the algebra of the β_μ's is finite: we can prove that for this to be so it is necessary to add further conditions to (4.24).‡

Taking all the coefficients σ to be equal in (4.24) we obtain an equation which may characterize the β_μ's:

$$(\beta_\mu)^{2S-1}(\beta_\mu^2 - I) = 0, \quad (4.29)$$

an equation which shows in particular that the β_μ's cannot be Hermitian when $S>1$. Indeed, one of the β_μ's, say β_λ, could be diagonalized and in accordance with (4.29) its eigenvalues would be 0 and +1. But in that case the equation which the matrix β_λ obeys would be

$$\beta_\lambda(\beta_\lambda^2 - I) = 0 \quad (4.30)$$

and not (4.29).

We have also recognized that eqn. (4.12c) can represent states with different masses. Each component of the wave function $Q(x)$ then obeys eqn. (4.2b) which we can write

$$\{\Box - m_1^2\}\{\Box - m_2^2\}\ldots\{\Box - m_N^2\}Q_A(x) = 0. \quad (4.31)$$

Define $\mathcal{D}(\partial)$ as before by

$$\{\beta_\mu\partial_\mu + m\}\mathcal{D}(\partial) = \prod_{j=1}^{N}\{\Box - m_j^2\}. \quad (4.32)$$

Since the argument which fixed the degree of the polynomial $\mathcal{D}(\partial)$ is still valid, the above formula shows that we have

$$2N \leqslant 2S+1. \quad (4.33)$$

We therefore see that for particles of half-integer spin: $2N \leqslant 2$, whereas for particles of spin $1: 2N \leqslant 3$; $N = 1$ is the value of N common to these two fields. Only a single mass theory accounts therefore for these two cases.

† G. Petiau, *Thèse*, Masson, 1936; R. J. Duffin, *Phys. Rev.*, **54,** 1114 (1938); N. Kemmer, *Proc. Roy. Soc.*, A **173,** 91 (1939). For particles of spin 3/2 see W. Rarita and J. Schwinger, *Phys. Rev.*, **60,** 61 (1941); S. N. Gupta, *Phys. Rev.*, **95,** 1334 (1954).

‡ Harish Chandra, *Phys. Rev.*, **71,** 793 (1947).

For particles of spin more than 1, $\mathcal{D}(\partial)$ will be determined by the same identification method as before. It can be noted first of all that

$$\prod_{j=1}^{N}\{\Box-m_j^2\}=\sum_{0}^{N}(-1)^{N-n}a_{N-n}\Box^n, \tag{4.34a}$$

where

$$a_N=\prod_{1}^{N}m_j^2,\quad a_{N-1}=a_N\sum_{1}^{N}\frac{1}{m_j^2},\quad a_{N-2}=a_N\sum{}'\frac{1}{m_j^2m_k^2},\quad \ldots, \tag{4.34b}$$

as can be proved by a simple calculation.

By identifying the two sides of (4.34*a*) we can write the general formula as follows:

$$\left.\begin{aligned}\alpha_{\sigma_1\ldots\sigma_{2p}}&=\frac{(-1)^N}{m^{2p+1}}(a_N\beta_{\sigma_{2p}}\ldots\beta_{\sigma_1}-m^2a_{N-1}\beta_{\sigma_{2p}}\ldots\beta_{\sigma_3}\delta_{\sigma_2\sigma_1}\\&\quad+\ldots+(-1)^pm^{2p}a_{N-p}\,\delta_{\sigma_{2p}\sigma_{2p-1}}\ldots\delta_{\sigma_2\sigma_1}),\\\alpha_{\sigma_1\ldots\sigma_{2p+1}}&=\frac{(-1)^{N+1}}{m^{2p+2}}\beta_{\sigma_{2p+1}}\alpha_{\sigma_1\ldots\sigma_{2p}}.\end{aligned}\right\} \tag{4.35}$$

Taking into account that $\mathcal{D}(\partial)$ is of degree $2S$ and making use of the symmetry condition concerning the $\alpha_{\sigma_1\ldots\sigma_m}$, we can obtain relations characterizing the β_μ's [as in (4.24) in the case of a single mass] and these results confirm similar results obtained by Bhabha.†

7. Hermitian character of the action and wave equation

Throughout this chapter we have generally adopted the framework of a non-quantized theory and have treated field variables as wave functions.

The detailed investigation of the "operators" $Q(x)$ will be reserved for the next chapter, but we shall outline in this section a few properties of the β_μ's obtained by assuming the action to be Hermitian (noting all the time that the same conclusions would be arrived at if one were to start from a real *c*-number action).

The wave equation $\mathcal{F}(\partial)Q(x)=0$ can be inferred from the following action:

$$\begin{aligned}\mathfrak{A}&=\int\tilde{Q}_A(x)\eta_{AA'}\mathcal{F}_{A'B}(\partial)Q_B(x)\,d^4x\\&=\int\tilde{Q}_A(x)\eta_{AA'}\mathcal{F}_{A'B}(-\overleftarrow{\partial})Q_B(x)\,dx,\end{aligned} \tag{4.36}$$

† H. J. Bhabha, *Rev. Mod. Phys.*, **17**, 300 (1945) and **21**, 451 (1949).

where the integral is taken over the entire space–time and the second formula has been obtained by a partial integration of the first.†

On the other hand, η is a non-singular matrix chosen such that the equations inferred from (4.36) by independent variations of $\tilde{Q}(x)$ and $Q(x)$ are compatible. The Hermitian conjugate (or the adjoint) of the equation

$$\eta\mathcal{F}(\partial)Q(x) = 0 \tag{4.37}$$

obtained by varying $\tilde{Q}(x)$, will have to be identical with the equation obtained by varying $Q(x)$:

$$\tilde{Q}(x)\eta\mathcal{F}(-\overleftarrow{\partial}) = 0. \tag{4.38}$$

The adjoint of (4.38) is

$$\eta\mathcal{F}(\partial) = \{\eta\mathcal{F}(-\overleftarrow{\partial})\}^{\sim} = \tilde{\mathcal{F}}(-\partial)\tilde{\eta} \tag{4.39}$$

and it is easy to see that this last equation expresses the Hermitian character of $\mathfrak{A}$.‡

Let us now return to the case $\mathcal{F}(\partial) = -\{\beta_\mu\partial_\mu+m\}$; the condition (4.39) takes the form

$$\eta\{\beta_\mu\partial_\mu+m\} = -\{\beta_\mu\partial_\mu+m\}^{\sim}\tilde{\eta} = \{-\tilde{\beta}_j\partial_j+\tilde{\beta}_4\partial_4+m\}\tilde{\eta},$$

from which we obtain the relations

$$\left.\begin{aligned} \tilde{\eta} = \eta, \quad \eta\beta_j &= -\tilde{\beta}_j\eta = -(\eta\beta_j)^{\sim}, \\ \eta\beta_4 &= \tilde{\beta}_4\eta = (\eta\beta_4)^{\sim}, \end{aligned}\right\} \tag{4.40a}$$

which can be written in the form

$$\tilde{\beta}_j = -\eta\beta_j\eta^{-1}, \quad \tilde{\beta}_4 = \eta\beta_4\eta^{-1}. \tag{4.40b}$$

† Let $F\left(\frac{d}{dx}\right) = \sum a_n \frac{d^n}{dx^n}$ be a differential polynomial with a single variable; we have indeed

$$\int f(x)F\left(\frac{d}{dx}\right)g(x)\,dx = \sum a_n \int f(x)\frac{d^n}{dx^n}g(x)\,dx$$
$$= \sum a_n \int (-1)^n \frac{d^n f}{dx^n} g(x)\,dx = \int f(x)F\left(-\frac{\overleftarrow{d}}{dx}\right)g(x)\,dx.$$

‡ The Hermitian conjugation in (4.39) for $\mathcal{F}(\partial)$ is an adjunction for the matrix with elements $\mathcal{F}_{AB}$, the symbol ∂ being treated as a c-number, in other words

$$\{\tilde{\mathcal{F}}(-\partial)\}_{AB} = \{\mathcal{F}_{BA}(-\partial)\}^*.$$

Note also that it is not the Lagrangian density $\mathcal{L} = -\bar{Q}(x)\{\beta_\mu\partial_\mu+m)\}Q(x)$ which is Hermitian, but only the action. This hermiticity is due to the fact that we integrate over the entire space–time and that the field variables vanish at infinity (hence, the possibility of integrating by parts without introducing the integrated terms). The above does not hold when the domain of integration extends between two surfaces $\sigma_{\rm I}$ and $\sigma_{\rm II}$; the Hermitian character of $\mathfrak{A}[\sigma_{\rm II}, \sigma_{\rm I}, Q]$ implies that of $\mathcal{L}$. One would then have to choose in place of the density under consideration that of (5.138), which is Hermitian.

It is equally easy to verify that

$$\bar{Q}_A(x) \equiv \tilde{Q}_{A'}(x)\eta_{A'A} \tag{4.41}$$

obeys the equation

$$\bar{Q}(x)\{-\beta_\mu \overleftarrow{\partial}_\mu + m\} = 0. \tag{4.42}$$

Note on the other hand that the Fourier transform $Q(k)$ of $Q(x)$ $[Q(x) = (2\pi)^{-4} \int e^{ikx} Q(k)\, d^4k]$ obeys the equation

$$(i\beta_\mu k_\mu + m)Q(k) = 0, \tag{4.43}$$

whereas if we set

$$\bar{Q}(x) = (2\pi)^{-4} \int e^{-ikx} \bar{Q}(k)\, d^4k, \tag{4.44a}$$

we find that $\bar{Q}(k)$ also obeys the equation

$$\bar{Q}(k)(i\beta_\mu k_\mu + m) = 0 \tag{4.44b}$$

and furthermore that by virtue of the relations (4.40), the matrix

$$\eta(i\beta_\mu k_\mu + m)$$

is Hermitian.

We also have†

$$\frac{\partial}{\partial t} \int \bar{Q}(x)\beta_4 Q(x)\, d^3\boldsymbol{x} = \frac{\partial}{\partial t} i \int_{\sigma(x)} \bar{Q}(x)\beta_\mu Q(x)\, d\sigma_\mu = 0, \tag{4.45}$$

taking into account the Hermitian character of $\eta(\beta_\mu \partial_\mu + m)$.

It is particularly important to observe how $\bar{Q}(x)$ varies under a proper Lorentz transformation.

Since the field equation of $\bar{Q}(x)$ must remain invariant under such a transformation, we have‡

$$\tilde{Q}'(x')\eta'(\beta_\mu \overleftarrow{\partial}'_\mu - m) = 0.$$

But

$$\tilde{Q}'(x') = \tilde{Q}(x)\tilde{\mathfrak{I}}(L),$$

† Note that $\bar{Q}(x)\beta_\mu Q(x)$ is proportional to the current vector $J_\mu(x)$ when the field is charged.

‡ It must be clearly understood that the dash of η' does not mean that η' is the transform of η under the Lorentz transformation: $\eta' = \mathfrak{I}\eta\mathfrak{I}^{-1}$. The matrix η' must be considered simply as defining $\bar{Q}'(x')$ according to (4.47a) which is itself defined by the relations (4.40) and (4.46). Readers may take as an example the Dirac equation with the standard representation of the γ_μ's, where $\eta = \eta' = \gamma_4$.

so that the above equation is written

$$\tilde{Q}(x)\tilde{\mathcal{J}}\eta'\mathcal{J}\{\mathcal{J}^{-1}\beta_\mu\mathcal{J}L_{\mu\nu}\overleftarrow{\partial_\nu}-\mathcal{J}^{-1}m\mathcal{J}\} = 0$$

or

$$\tilde{Q}(x)\tilde{\mathcal{J}}\eta'\mathcal{J}\{\beta_\mu\overleftarrow{\partial_\mu}-m\} = 0.$$

The latter equation shows that η' must obey not only conditions of the form (4.40) but also

$$\tilde{\mathcal{J}}\eta'\mathcal{J} = \eta. \tag{4.46}$$

We finally obtain the law of transformation of $\bar{Q}(x)$

$$\bar{Q}'(x') \equiv \tilde{Q}'(x')\eta' = \tilde{Q}(x)\eta\mathcal{J}^{-1} = \bar{Q}(x)\mathcal{J}^{-1}. \tag{4.47a}$$

We shall verify below (5.104) that the commutation relations are also invariant: the proper Lorentz transformations are therefore canonical transformations.

It should be noted that the right side of (4.46) is only defined up to a phase factor. This factor only becomes important if one considers the full Lorentz Group (in particular the time reversals). Indeed, if we wished to represent the time reversals for free fields by a linear transformation (instead of an antilinear one as we shall do in Chap. V, §§ 11 and 20), we should be obliged, in order to make the commutation relations invariant and since $\Delta(x)$ is odd for the variable t, to choose:

$$\bar{Q}'(x') = -\bar{Q}(x)\mathcal{J}^{-1} \tag{4.47b}$$

corresponding to $\tilde{\mathcal{J}}\eta\mathcal{J} = -\eta$.

Let us now consider more closely the observables of the field; since under a proper Lorentz transformation, the density $\mathcal{L}$ becomes

$$\mathcal{L}' = -\bar{Q}'(x')\{\beta_\mu\partial'_\mu+m\}Q'(x');$$

$\mathcal{L}$ is thus a scalar and we again verify that the equations obeyed by Q and $\bar{Q}$ are invariant.

Now, the observables of the transformed field can therefore be considered as being given by the same expressions as those of the initial field in which Q is to be replaced by Q' and $\bar{Q}$ by $\bar{Q}'$: in other words they are field co-variants. We shall endeavour to determine their variance.

Let us prove for instance that $\bar{Q}(x)Q(x)$ is a scalar; we have indeed under a Lorentz transformation:

$$\bar{Q}(x)Q(x) = \bar{Q}'(x')\mathcal{J}\mathcal{J}^{-1}Q'(x') = \bar{Q}'(x')Q'(x'). \tag{4.48a}$$

Similarly, we can prove that $\bar{Q}(x)\beta_\mu Q(x)$ is a vector; indeed

$$\bar{Q}(x)\beta_\mu Q(x) = \bar{Q}'(x')\mathcal{J}^{-1}\beta_\mu\mathcal{J}Q'(x') = L_{\mu\lambda}\bar{Q}'(x')\beta_\lambda Q'(x'). \tag{4.48b}$$

This method will be of particular service to us in forming the invariants of the Dirac field.†

We shall investigate in conclusion a few properties of the field of spin $S \leqslant 1$: we have seen that it is possible to find Hermitian representations of the β_μ's [text following formula (4.29)].

Formulae (4.40) are then written simply

$$[\eta, \beta_j]_+ = 0, \quad [\eta, \beta_4]_- = 0. \tag{4.49}$$

If we set $k_4 = ik_0$, the wave equation (4.43) takes the form

$$k_0\beta_4 Q(k) = (i\beta_j k_j + m)Q(k).$$

Take the adjoint and use the Hermitian character of the β_μ's:

$$k_0\tilde{Q}(k)\beta_4 = \tilde{Q}(k)(-i\beta_j k_j + m).$$

By multiplying the first of these two equations on the left by $\tilde{Q}(k)$, the second on the right by $Q(k)$ and adding, we obtain

$$\tilde{Q}(k)\beta_4 Q(k) = \frac{m}{k_0}\tilde{Q}(k)Q(k), \tag{4.50}$$

a formula which will enable us to carry out an interesting normalization of Dirac wave functions.

It will be noted in conclusion that we have confined ourselves in this part to purely formal considerations in that we have not proved the existence of a field $Q(x)$ which obeys all the postulates stated and that we have scarcely attempted to investigate their compatibility.

A relatively simple example of a case in which the existence and non-contradiction of the postulates can be verified, is the Dirac field, which we shall be studying in some detail in Chap. VI.

For the moment, we propose to consider in the next part the description of a field of given spin and single mass by de Broglie's "méthode de la fusion".

† It will obviously also help us form the transformed expectation values, as will be seen in detail in the next chapter; if we consider for instance:

$$j_\mu(x) = \bar{Q}(x)\beta_\mu Q(x),$$

we obtain according to formula (3.1*b*)

$$\begin{aligned}\langle j_\mu(x')\rangle' &= \langle \Psi | D^{-1}(L)\bar{Q}(x')\beta_\mu Q(x')D(L) | \Psi \rangle \\ &= \langle \Psi | \bar{Q}'(x')\beta_\mu Q'(x') | \Psi \rangle = \langle \Psi | \bar{Q}(x)\mathcal{J}^{-1}\beta_\mu \mathcal{J} Q(x) | \Psi \rangle = L_{\mu\nu}\langle j_\nu(x)\rangle.\end{aligned}$$

The picture adopted is that of Heisenberg; the field under investigation $Q(x)$ is therefore free since it obeys (4.12c).

PART 3

DE BROGLIE FUSION METHOD†

The basic idea underlying the fusion method is to assume that any particle of spin more than $\frac{1}{2}$ is formed by the "fusion" of particles of spin $\frac{1}{2}$; the fusion of the particles represents an operation whereby the wave function of a particle with a spin more than $\frac{1}{2}$ is obtained by applying certain mathematical processes to wave functions describing particles of spin $\frac{1}{2}$.

This theory, when the wave functions are subjected to no symmetry condition, leads to the definition of particles with a given maximal spin: for instance, the equation relative to a spin 1 describes in this method equally well a particle of spin 0 as one of spin 1: the former corresponds to the singlet state of the system (antiparallel spins) of spin 0 and the latter to the triplet state (parallel spins) of spin 1. In addition, the theory describes a field of spin $S/2$ no longer by a matrix equation of the type (4.12*c*) but by a system of S equations of this type.

In place of the markedly intuitive approach adopted by de Broglie in the exposition of his theory, we shall employ, for lack of space, a more abstract method but at the same time a more rapidly effective one, using concepts from group theory.

Note also that this chapter ought logically to follow the one devoted to the Dirac equation, but the results used are sufficiently elementary for us to assume them to be known by the reader, who can, moreover, refer to Chap. VI, Part 1.

8. Formation of wave equations for a particle of spin S/2

Consider S spinors $\psi(x) = u(p)e^{ipx}$ describing plane waves with the same momenta. The components of the function $\psi^{S/2}(x)$ will by definition be

$$\psi^{S/2}_{\sigma_1 \ldots \sigma_s}(x) \equiv \psi_{\sigma_1}(x)\psi_{\sigma_2}(x)\ldots\psi_{\sigma_s}(x) = u_{\sigma_1}(p)e^{ipx}\ldots u_{\sigma_s}(p)e^{ipx} = u_{\sigma_1}(p)\ldots u_{\sigma_s}(p)e^{iSpx}, \tag{4.51}$$

each of the indices σ runs from 1 to 4: this function therefore has 4^S components and each of these components is symmetric with respect to

† L. de Broglie, *Théorie générale des particules à spin*, Gauthier-Villars, 1943. The following papers may also be consulted: P. A. M. Dirac, *Proc. Roy. Soc.*, A **155**, 447 (1936); M. Fierz, *Helv. Phys. Acta*, **12**, 3 (1939); M. Fierz and W. Pauli, *Proc. Roy. Soc.*, A **173**, 211 (1939).

$\sigma_1, \ldots, \sigma_S$. We now clearly have

$$\sum_{\lambda} \{\gamma_\mu \partial_\mu + M\}_{\sigma_p \lambda} \psi^{S/2}_{\sigma_1 \ldots \sigma_{p-1}, \lambda, \sigma_{p+1} \ldots \sigma_s}(x) = 0, \tag{4.52a}$$

where $M = Sm$ and where the index λ can take the place of any one of the σ's.

The preceding equation can also be expressed in the form

$$\sum_{\lambda_1 \ldots \lambda_S} \{\delta_{\sigma_1 \lambda_1} \ldots \delta_{\sigma_{p-1} \lambda_{p-1}} (\gamma_\mu)_{\sigma_p \lambda_p} \delta_{\sigma_{p+1} \lambda_{p+1}} \ldots \delta_{\sigma_s \lambda_s} \partial_\mu + M \delta_{\sigma_1 \lambda_1} \ldots \delta_{\sigma_s \lambda_s}\} \psi^{S/2}_{\lambda_1 \ldots \lambda_s}(x) = 0; \tag{4.52b}$$

the first term of the bracket represents the tensor product of S factors:

$$\Gamma^{p,\, S/2}_\mu \equiv I \otimes I \otimes \ldots \otimes \gamma_\mu \otimes \ldots \otimes I, \tag{4.53}$$

where γ_μ occupies the pth rank. Equation (4.52) can then be reduced to a condensed form:

$$\{\Gamma^{p,\, S/2}_\mu \partial_\mu + M\} \psi^{S/2}(x) = 0. \tag{4.54}$$

We clearly have S equations of this form: $p = 1, 2, \ldots, S$, corresponding to the various positions of γ_μ in (4.53).

It can easily be verified that the $\Gamma^{pS/2}_\mu$'s obey the following relations:

$$[\Gamma^{p,\, S/2}_\mu, \Gamma^{q,\, S/2}_\nu]_+ = 2\, \delta_{pq}\, \delta_{\mu\nu}, \tag{4.55a}$$

$$[\Gamma^{p,\, S/2}_\mu \partial_\mu + M, \Gamma^{q,\, S/2}_\nu \partial_\nu + M] = 0, \tag{4.55b}$$

$$\{\Gamma^{p,\, S/2}_\mu \partial_\mu + M\}\{\Gamma^{p,\, S/2}_\nu \partial_\nu - M\} = \Box - M^2. \tag{4.55c}$$

When the wave functions are not made symmetric with respect to the indices σ, it can be shown that eqns. (4.54) represent a particle of maximal spin $S/2$. Indeed, a proper Lorentz transformation transforms $\psi^{S/2}(x)$ as follows:

$$\psi^{S/2}(x) \to \psi^{(S/2)'}(x') = \mathcal{J}^{S/2} \psi^{S/2}(x),$$

but since by the very definition of $\psi^{S/2}(x)$, the matrix $\mathcal{J}^{1/2}$ corresponds to a particle of spin $\frac{1}{2}$, we have

$$\mathcal{J}^{S/2} = \mathcal{J}^{1/2} \otimes \mathcal{J}^{1/2} \otimes \ldots, \tag{4.56}$$

a product with S factors: then the Clebsch–Gordan theorem indicates that the representation of $\psi^{S/2}$ splits into a sum of irreducible representations and that the highest order one meets with is $S/2$.

We shall see nevertheless (end of § 10) that the symmetry condition imposed on the wave functions makes it possible to describe a field of spin $S/2$ by eqns. (4.54).

It can be proved just as easily that the system of equations is invariant under Lorentz transformations: this proof starts out from the relativistic invariance of the Dirac equations from which we set out.

The set of the $S \cdot 4^S$ equations can be expressed in various forms. For instance, by setting

$$\beta_\mu = \frac{1}{S} \sum_{p=1}^{S} \Gamma_\mu^{p, S/2}, \tag{4.57}$$

we obtain the following systems of S equations: adding first eqns. (4.54), one has

$$\{\beta_\mu \partial_\mu + M\}\psi^{S/2}(x) = 0. \tag{4.58a}$$

Now by subtracting these same equations by pairs, we get the following $S-1$ equations:

$$(\Gamma_\mu^{1, S/2} - \Gamma_\mu^{p, S/2})\, \partial_\mu \psi^{S/2}(x) = 0 \quad (p=2, \ldots, S). \tag{4.58b}$$

Equations (4.58*a*) are known as the evolution equations and eqns. (4.58*b*) as the auxiliary equations.

It can be shown in fact that if the auxiliary equations are obeyed at a particular time t_0, they continue to be obeyed at any subsequent time t and that it is possible, by investigating plane waves, to find a system of solutions such that all these equations (4.58) are compatible at any time t_0. For all these questions, the reader is referred to the work of de Broglie (*op. cit.*, p. 138 *et seq.*). Note simply in conclusion that while the wave equation implies the auxiliary equation for particles of spin 1, this is no longer the case for particles of spin more than 1; this condition has to be postulated.

9. Tensor components of the wave functions of integer spin fields

The fusion method also enables us to prove readily that an integer spin field can be represented by a set of variables each of which has a well-defined tensor character.

Consider therefore the wave function of a particle of integer spin N: $\psi^N(x)$ is labelled by $2N$ indices: $\psi_{\sigma_1 \ldots \sigma_N \sigma_1' \ldots \sigma_N'}$. On the other hand, to the particle of spin $N/2$ are attached the matrices $\Gamma_\mu^{p, N/2}$ given by formula (4.53); a total of $4N$ matrices each of whose elements has $2N$ indices. Then consider the set of matrices

$$\left.\begin{array}{l} I, \\ \Gamma_1^{1, N/2} \ldots \Gamma_4^{N, N/2}, \\ \Gamma_1^{1, N/2}\Gamma_2^{1, N/2}, \quad \Gamma_1^{1, N/2}\Gamma_3^{1, N/2}; \quad \ldots, \quad \Gamma_3^{N, N/2}\Gamma_4^{N, N/2}, \\ \ldots\ldots\ldots\ldots\ldots\ldots\ldots\ldots\ldots\ldots\ldots\ldots, \\ \Gamma_1^{1, N/2}\Gamma_2^{1, N/2}\Gamma_3^{1, N/2} \ldots \Gamma_4^{N, N/2}. \end{array}\right\} \tag{4.59}$$

We have thus formed

$$1 + C_{4N}^1 + \ldots + C_{4N}^{4N} = 16^N \tag{4.60}$$

new matrices and it can be easily verified that their squares are equal to 1 (matrix unity) and their spurs to 0 (except, of course, matrix 1).†

Denote by $\Gamma_A^{N/2}$ any one of these quantities (A represents a collection of indices); we can expand $\psi^N(x)$ in terms of the Γ_A's as follows:

$$\psi^N(x) = \sum_1^{16^N} \varphi_A(x)\Gamma_A^{N/2}, \tag{4.61a}$$

with

$$(\Gamma_A^{N/2})^2 = I, \quad Tr\Gamma_A^{N/2} = 0,$$

from which it follows that

$$\varphi_A(x) = \frac{1}{4^N} Tr\{\psi^N(x)\Gamma_A^{N/2}\}; \tag{4.61b}$$

the coefficients φ_A have precisely the required tensor character.

Consider as an application of these formulae the case of a particle of maximal spin 1; the above basis is the one formed by the sixteen matrices we come across in Dirac's theory (6.1.3) and eqn. (4.61) becomes

$$\psi^1(x) = \varphi_0 + \sum \varphi_\mu \gamma_u + \sum \varphi_{\mu\nu}\gamma_\mu\gamma_\nu + \dots$$

It has been shown (cf. L. de Broglie, *op cit.*, p. 107 *et seq.*) that we thus obtain a scalar φ_0, a four-vector φ_μ, the six components of a skew-symmetrical tensor of order 2; $\varphi_{\mu\nu}$, the four components of a completely skew-symmetrical tensor of order 3 (therefore a pseudovector); $\varphi_{\lambda\mu\nu}$ and the single component of a completely skew-symmetrical tensor of order 4 (pseudoscalar): $\varphi_{\lambda\mu\nu\varrho}$.

The $\varphi_{\mu\nu}$'s and φ_μ's obey equations we shall be meeting with again when we study the vector field:

$$M\varphi_{\mu\nu} = \varphi_{\nu,\,\mu} - \varphi_{\mu,\,\nu}, \quad \varphi_{\mu\nu,\,\mu} = M\varphi_\nu,$$

$$\varphi_{\mu,\,\mu} = 0, \quad \varphi_{\mu\nu,\,\varrho} + \varphi_{\nu\varrho,\,\mu} + \varphi_{\varrho\mu,\,\nu} = 0.$$

They are the Proca–de Broglie equations.

The relations between $\varphi_{\mu\nu\varrho\sigma}$ and $\varphi_{\mu\nu\sigma}$ are those of the pseudo-scalar field (Chap. V):

$$M\varphi_{\nu\varrho\sigma} = \varphi_{\mu\nu\varrho\sigma,\,\mu}, \quad \varphi_{\mu\nu\varrho,\,\mu} = 0$$

$$\varphi_{\mu\nu\varrho,\,\sigma} + \varphi_{\nu\varrho\sigma,\,\mu} + \varphi_{\varrho\sigma\mu,\,\nu} + \varphi_{\sigma\mu\nu,\,\varrho} = M\varphi_{\mu\nu\varrho\sigma}.$$

For a detailed study of these equations, we refer readers to the above-mentioned work by de Broglie. Another example dealt with in the same work is that of a particle of maximal spin 2: it is well known that the particles describing the gravitation field (the gravitons) obey the equations

† The same calculations will be made in connection with the Dirac equation; for details refer to Chap. VI, Part 1.

of the particle of spin 2, which represent a linear approximation of the equations of the general theory of relativity.†

Finally, the fusion theory enables us to discuss in an elegant manner certain aspects of the classical electromagnetic theory by defining the photon as a particle of vanishing mass. For an investigation of these matters, readers are referred to *La Mécanique ondulatoire du photon et la théorie quantique des champs* by L. de Broglie (Chap. VI).

10. The spin

Following the same method as in Chap. II, the definition of the spin will be based on the operator $\mathcal{J}^{S/2}$ given by (4.56).

We shall see (6.1.9*a*) for infinitesimal Lorentz transformations that the operator $\mathcal{J}^{1/2}$ is expressed as follows:

$$\mathcal{J}^{1/2} = I + \frac{1}{8}\varepsilon_{\mu\nu}[\gamma_\mu, \gamma_\nu] = I + \sum_{\mu\neq\nu}\frac{1}{4}\varepsilon_{\mu\nu}\gamma_\mu\gamma_\nu,$$

consequently,

$$\mathcal{J}^{S/2} = \left(I + \frac{1}{4}\varepsilon_{\mu\nu}\gamma_\mu\gamma_\nu\right)\otimes\left(I + \frac{1}{4}\varepsilon_{\lambda\varrho}\gamma_\lambda\gamma_\varrho\right)\otimes\ldots,$$

with a total of S factors. Placing ourselves in the S-dimensional space, we can then easily see that

$$\mathcal{J}^{S/2} = I + \frac{1}{4}\varepsilon_{\mu\nu}\sum_{p=1}^{S}\Gamma_\mu^{p,\,S/2}\Gamma_\nu^{p,\,S/2}. \tag{4.62}$$

We thus obtain the operators $J_{\mu\nu}$ and, by using (3.62), the three components of the spin operator.

For a detailed study and classification of the spin states, readers should consult de Broglie, *op. cit.*, pp. 161–74.

We shall show, in conclusion, that when the components $\psi^{(S/2)}_{\sigma_1\ldots\sigma_S}(x)$ of the wave functions are symmetrized with respect to the indices σ_n, the wave function $\psi^{S/2}(x)$ has $S+1$ independent components, so that the field described is indeed one of spin $S/2$.

Assume γ_4 to be diagonal, with 1, 1, -1, -1 as diagonal elements and write the wave equation in the momentum space:

$$(i\Gamma_\mu^{j,\,S/2}p_\mu + M)\psi^{S/2}(p) = 0 \quad (j = 1, \ldots, S). \tag{4.63}$$

Adopting a frame in which the particle is at rest with $p_0 = m$, this equation then shows that the only components of $\psi^{S/2}$ which are non-zero are those which correspond to the first two rows of the γ_4's, which is the jth factor in the expression (4.53) of $\Gamma_\mu^{j,\,S/2}$. There are 2^S components of this

† M. A. Tonnelat, *Thèse*, Masson, 1942.

form; the remaining components, $4^S - 2^S$ in number, vanish. Furthermore, the components which do not vanish are not all independent. In fact, the symmetry of $\psi^{S/2}$ with respect to its indices implies the equality of the components of $\psi^{S/2}$ for which the same number m of the S indices σ corresponds to the first row of the γ_μ's in the expression of the Γ's, the $S-m$ other indices corresponding to the second row. Since m runs from 0 to S, there are $S+1$ independent components. The same considerations apply in the case $p_0 = -m$, where the last two rows of γ_4 play the part which the first two rows played in the preceding case.†

11. Probability density and sign of the energy

Let us define the matrix

$$\Gamma^{S/2} = \frac{1}{S}\sum_p \Gamma_4^{S,\, S/2}\Gamma_4^{S-1,\, S/2} \ldots \Gamma_4^{p-1,\, S/2}\Gamma_4^{p+1,\, S/2} \ldots \Gamma_4^{1,\, S/2}; \quad (4.64)$$

it can be shown that

$$\frac{\partial \varrho}{\partial t} = 0, \quad (4.65a)$$

where

$$\varrho = \int \tilde{\psi}^{S/2}(x)\Gamma^{S/2}\psi^{S/2}(x)\, d^3\boldsymbol{x}. \quad (4.65b)$$

One sees, therefore, that $\Gamma^{S/2}$ is the analogue of the matrix η introduced in (4.37). Furthermore, in the x representation of wave mechanics $\tilde{\psi}^{S/2}\Gamma^{S/2}\psi^{S/2}$ can play the part of a probability density. It should be noted, however, that, as in the case of the Klein–Gordon equation, this quantity is not always positive definite.‡

To see this it is convenient to consider the space p: a solution $\psi^{S/2}(p)$ of (4.63) also obeys the Klein–Gordon equation (4.2a), which may be written

$$(p_\mu p_\mu + M^2)\psi^{S/2}(p) = 0. \quad (4.66)$$

We can then set

$$\psi^{S/2}(p) = \delta(p^2 + M^2)A^{S/2}(p), \quad (4.67)$$

and the function

$$\psi^{S/2}(x) = (2\pi)^{-4}\int e^{i\boldsymbol{p}.\boldsymbol{x} - ip_0 t}\, \delta(p^2+M^2)A^{S/2}(p)\, d^4p \quad (4.68)$$

is a solution of (4.54) when $A^{S/2}(p)$ is a solution of (4.63).

† V. Bargman and E. P. Wigner, *op. cit.*, p. 218.

‡ With the definition (4.57) of the β_μ's, the density of the corresponding probability current is:

$$J_k(x) = \tilde{\psi}^{S/2}(x)\Gamma^{S/2}\beta_k\psi^{S/2}(x).$$

Now by virtue of formula (1.104f),

$$\delta(p^2+M^2) = \frac{1}{2\omega(\boldsymbol{p})}\left(\delta(p_0-\omega)+\delta(p_0+\omega)\right),$$

$$\omega(\boldsymbol{p}) = +\sqrt{\boldsymbol{p}^2+M^2},$$

such that

$$\delta(p^2+M^2)A^{S/2}(p) = \frac{1}{2\omega}\left(A^{S/2}(\boldsymbol{p},\omega)\,\delta(p_0-\omega)+A^{S/2}(\boldsymbol{p},-\omega)\,\delta(p_0+\omega)\right).$$

The Fourier transform (4.68) of $\psi^{S/2}(x)$ can be expressed in the form

$$\psi^{S/2}(x) = \frac{(2\pi)^{-4}}{2}\int e^{i\boldsymbol{p}.\boldsymbol{x}-ip_0t}A^{S/2}(p)\frac{d^3\boldsymbol{p}}{|p_0|}, \tag{4.69}$$

where the integral is taken over the two sheets of the hyperboloid $p_\mu p_\mu+M^2 = 0$ and the density ϱ defined by (4.65b) may be written

$$\varrho = \int \tilde{\psi}^{S/2}(\boldsymbol{x},0)\Gamma^{S/2}\psi^{S/2}(\boldsymbol{x},0)\,d^3\boldsymbol{x}$$

$$= \frac{(2\pi)^{-5}}{4}\int \tilde{A}^{S/2}(p)\Gamma^{S/2}A^{S/2}(p)\frac{d^3\boldsymbol{p}}{|p_0|^2}. \tag{4.70}$$

On the other hand, we can easily prove the following formula:

$$p_0^n\Gamma_4^{n,S/2}\ldots\Gamma_4^{1,S/2}A^{S/2}(p) = (M^n+B_{(n)})A^{S/2}(p) \quad (n\leqslant S), \tag{4.71}$$

where $B_{(n)}$ is an anti-Hermitian matrix; to prove this, we have only to proceed recurrently, verifying first of all that it is true for $n=1$.

Suppose (4.71) to be true and multiply it by $p_0\Gamma_4^{n+1,S/2}$; we get

$$p_0^{n+1}\Gamma_4^{n+1,S/2}\ldots\Gamma_4^{1,S/2}A^{S/2}(p) = (M^np_0\Gamma_4^{n+1,S/2}+p_0\Gamma_4^{n+1,S/2}B_{(n)})A^{S/2}(p)$$

$$= M^{n+1}+(iM^n\Gamma_j^{n+1,S/2}p_j+p_0\Gamma_4^{n+1,S/2}B_{(n)})A^{S/2}(p)$$

and we have indeed an expression of the form (4.71) since the expression between brackets is in fact anti-Hermitian.

In particular, let us therefore multiply (4.71) on the left by $\tilde{A}^{S/2}(p)$ and make $n=S$; we get

$$p_0^S\tilde{A}^{S/2}\Gamma^{S/2}A^{S/2} = M^S|A^{S/2}|^2+\tilde{A}^{S/2}B_{(S)}A^{S/2}; \tag{4.72}$$

the last term is purely imaginary and must be zero since all the other terms are real.

On the other hand, a direct extension of this theorem shows that the expression (4.70) of ϱ can be written in the form

$$\varrho = \frac{(2\pi)^{-5}}{4}\int\left(\frac{M}{p_0}\right)^{S-1}|A^{S/2}(p)|^2\frac{d^3\mathbf{p}}{|p_0|^2}$$

$$= \frac{(2\pi)^{-5}}{4}\frac{1}{M^2}\int\left(\frac{M}{p_0}\right)^{S+1}|A^{S/2}(p)|^2\, d^3\mathbf{p}. \qquad (4.73)$$

Consequently, if we consider a negative-energy solution of the wave equations of a particle of spin $S/2$, ϱ is only positive definite if the spin $S/2$ is half-integer.

Let us now find the expectation value of the energy (in wave mechanics); we have

$$\langle\psi(\mathbf{p}, p_0)|p_0|\psi(\mathbf{p}, p_0)\rangle = (2\pi)^{-3}\frac{I}{M}\int\left(\frac{M}{p_0}\right)^S|A^{S/2}(p)|^2\, d^3\mathbf{p}. \quad (4.74)$$

It is apparent that the energy is only positive definite if the spin $S/2$ is integer.

It is worth pointing out in this connection that the coefficient M/p_0 is equal to the Lorentz contraction factor $\sqrt{1-\beta^2}$: this fact might therefore suggest that a particle of spin $S/2$ is in fact a particle composed of elementary particles of spin $\frac{1}{2}$ which suffers in its internal structure the Lorentz contraction.

The investigation of the tensor $T_{\mu\nu}$ would lead to conclusions identical to those we have just presented.

In classifying the representations of the inhomogeneous Lorentz group, Wigner and Bargmann defined the norm of a wave function $\psi(x)$ as follows:

$$\langle\psi, \psi\rangle = \int\left|\frac{M}{p_0}\right|^S|A^{S/2}(p)|^2\frac{d^3\mathbf{p}}{|p_0|}, \qquad (4.75)$$

where the integral continues to be taken over the two sheets of the hyperboloid $p_\mu p_\mu + M^2 = 0$. This expression is invariant under the full Lorentz group and is positive definite.

Making use of formula (4.72), we can easily verify that it can be expressed in the form†

$$\langle\psi, \psi\rangle = \int|\bar{A}^{S/2}(p)\Gamma_4^{S, S/2}\ldots\Gamma_4^{1, S/2}A^{S/2}(p)|\frac{d^3\mathbf{p}}{|p_0|}. \qquad (4.76)$$

† For theories making use of the abstract theory of spinors, we refer readers to K. J. Le Couteur, *Proc. Roy. Soc.*, A **202**, 395 (1950); R. Potier, *Journal de Phys.*, **16,** 688 (1955). For a general survey, consult H. Umezawa, *Quantum Field Theory*, North Holland, Chaps. IV and V.

CHAPTER V

Quantization of linear field equations with constant coefficients

1. Introduction

The previous chapter was devoted to the study of certain systems of partial differential equations which will be used in the next chapter for the description of the fields of particles actually observed. The properties inferred have generally been obtained without reference to the operator character of the field variables, in other words the fields investigated were not quantized.

Their quantization provides us with further information which we propose to investigate in this chapter. We shall see that the spin 0 field is going to play a fundamental role, for we shall show that the quantization of a field of arbitrary spin is in the last analysis reduced (by virtue of the reduction postulate, Chap. IV, § 1) to the quantization of that field. The first part of this chapter will be therefore concerned with the spin 0 field and its quantization, while Parts 2 and 3 will be devoted to the study of fields of arbitrary spin.

In the course of the investigation we shall be carrying out, it is important to keep constantly in mind the double role which the wave equations we shall be studying can play:

(a) in the case of interacting fields, there exists a picture, that of interaction (Chap. II, § 12), in which the wave equations are linear equations with constant coefficients and in which the interaction only affects the state vector;

(b) in the case of free particles (in other words those which are not interacting and are subject to no external action), the wave equations describe the behaviour of those particles completely, provided one chooses the Heisenberg picture (constant state vector in the space $\mathcal{R}$).

We could therefore have entitled this chapter "Quantization of free fields", but have preferred a more abstract title, since we shall not as a rule find it necessary to specify which picture is used.

PART 1

SCALAR AND PSEUDO-SCALAR FIELDS

2. Wave equation and non-relativistic approximation

The wave equation of particles of spin 0 is, as we have already seen:

$$\{\square - m^2\}\Phi(x) = 0, \tag{5.1}$$

with

$$\square \equiv \partial_\mu \partial_\mu = \nabla^2 - \frac{\partial^2}{\partial t^2}. \tag{5.2}$$

The form of its non-relativistic approximation (cf. footnote on p. 176, for instance) is well known:

$$i\frac{\partial \Phi(n, r)}{\partial t} = -\frac{1}{2m}\nabla^2 \Phi(n, r). \tag{5.3}$$

3. Elementary solutions

Consider the function with one or more determinations $\Delta^{(\mathscr{C})}(x)$ defined by the following integral:

$$\Delta^{(\mathscr{C})}(x) = \frac{1}{(2\pi)^4}\int_{\mathscr{C}} \frac{e^{ikx}}{k^2+m^2}\, d^4k = -\frac{1}{(2\pi)^4}\int d^3\boldsymbol{k}\, e^{i\boldsymbol{k}.\boldsymbol{x}} \int_{\mathscr{C}} \frac{e^{-ik_0x_0}}{k_0^2-\omega^2}\, dk_0, \tag{5.4}$$

where

$$\omega^2 = \boldsymbol{k}^2 + m^2, \tag{5.5}$$

k_0 is a variable of the complex plane, the first integral is taken over the entire Euclidean space and the second one is taken along a path $\mathscr{C}$ of the complex plane k_0.†

† It will be noted in the first place that the mathematical entity we have called a function is actually a distribution, and secondly that the above method is an extension of the so-called Laplace method for differential equations.

Let $\mathscr{L}\left\{\frac{d}{dx}\right\}$ be a linear differential operator with constant coefficients; make it act on the function

$$u(x) = \frac{1}{2\pi}\int_{\mathscr{C}} e^{ipx} U(p)\, dp,$$

where $\mathscr{C}$ is a path to be determined. We have

$$\mathscr{L}\left\{\frac{d}{dx}\right\} u(x) = \frac{1}{2\pi}\int_{\mathscr{C}} e^{ipx} \mathscr{L}(ip) U(p)\, dp.$$

(Continued on p. 161)

Make the differential operator $\{\Box - m^2\}$ act on (5.4):

$$(\Box - m^2)\Delta^{(\mathscr{C})}(x) = -\frac{1}{(2\pi)^4}\int_{\mathscr{C}} e^{ikx}\, d^4k. \tag{5.6}$$

The value of the right side of the preceding relation depends essentially on the path $\mathscr{C}$. We shall show that according to the choice of the path $\mathscr{C}$ it will be possible to obtain either solutions of (5.1) or its elementary solutions (Green functions).

The paths we shall be considering are those formed by the four semi-straight lines A_-, A_+, R_-, R_+ (Fig. 5.1).

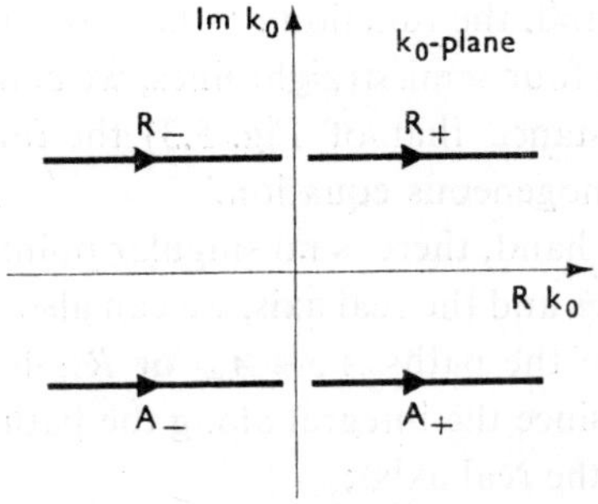

FIG. 5.1.

They will help us to construct all the paths we shall need, for instance the paths

$$A_- + R_+, \quad A_- - R_-$$

corresponding to the paths respectively open and closed in Figs. 5.2 and 5.3.

Choose

$$U(p) = \frac{1}{\mathscr{L}(ip)};$$

we then have

$$\mathscr{L}\left\{\frac{d}{dx}\right\} u(x) = \frac{1}{2\pi}\int_{\mathscr{C}} e^{ipx}\, dp.$$

If $\mathscr{C}$ is any closed path without points at infinity, $u(x)$ is a solution of the homogeneous differential equation

$$\mathscr{L}\left\{\frac{d}{dx}\right\} u(x) = 0.$$

If $\mathscr{C}$ is the straight line $-\infty, +\infty$, $u(x)$ is an elementary solution of the differential operator $\mathscr{L}\left\{\frac{d}{dx}\right\}$ (cf. Appendix I).

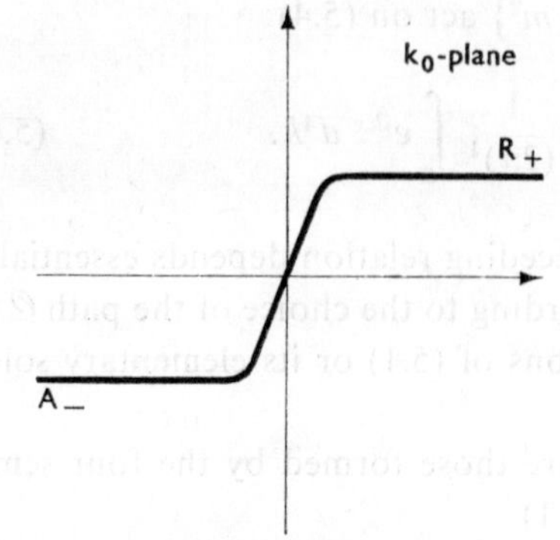

FIG. 5.2.

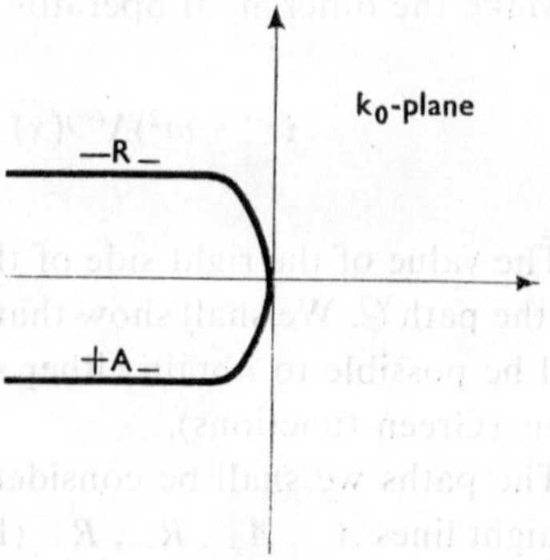

FIG. 5.3.

Since on the one hand, the function e^{ikx} has no singular point inside the path formed by these four semi-straight lines, we can say that: "when $\mathscr{C}$ is a closed path (for instance that of Fig. 5.3) the function thus defined is a solution of the homogeneous equation."

Since, on the other hand, there is no singular point of e^{ikx} between these four semi-straight lines and the real axis, we can also say that:

"When $\mathscr{C}$ is one of the paths $A_- + A_+$ or $R_- + R_+$, the right side of (5.6) equals: $-\delta(x)$ (since the integral along the path thus formed is equal to the integral along the real axis);

"When $\mathscr{C}$ is one of the semi-straight lines A_- or R_-, the right side of this same equation is $-\delta(\boldsymbol{x})\, \delta_+(x_0)$;

"If finally $\mathscr{C}$ is one of the semi-straight lines A_+ or R_+, the right side is $-\delta(\boldsymbol{x})\, \delta_-(x_0)$."

The integration relative to k_0 can generally be carried out by using the residues theorem. We must then complete the path $\mathscr{C}$ by one of the semi-circumferences Γ centred at the origin, with infinite radius and chosen such that the integrals

$$\int_\Gamma \frac{e^{ikx}}{k^2+m^2}\, d^4k, \quad \int_\Gamma e^{ikx}\, d^4k$$

be zero. It is obvious that this condition is satisfied by taking the lower semi-circumference if $x_0 > 0$ or the upper semi-circumference if $x_0 < 0$.

The singular points which can be enclosed within these paths are the simple poles

$$\pm\omega(k) = \pm\sqrt{\boldsymbol{k}^2+m^2}.$$

The various functions which can be obtained in this way are summed up in the following table. The symbols in the first column denote the functions considered: in the second column, the corresponding path is described with the aid of the semi-straight lines $A_{\pm}$, $R_{\pm}$. Finally the last column gives various expressions of the functions $\Delta^{(\mathscr{C})}(x)$.

TABLE OF THE FUNCTIONS $\Delta^{(\mathscr{C})}(x) = (2\pi)^{-4}\int d^3\boldsymbol{k}\int_{\mathscr{C}} dk_0 \frac{e^{ikx}}{\boldsymbol{k}^2-k_0^2-m^2}$:

x).	A_-.	R_-.	A_+.	R_+.	
	+	−			$= -\frac{i}{(2\pi)^3}\int_{+k\varepsilon>0} e^{+ikx}\varepsilon(k)\delta(k^2+m^2)\,d^4k$ (†) $= +\frac{i}{(2\pi)^3}\int e^{ikx}\theta_-(k_0)\delta(k^2+m^2)\,d^4k$ $= +\frac{i}{(2\pi)^3}\int e^{-i(\boldsymbol{k}.\boldsymbol{x}-\omega x_0)}\frac{d^3\boldsymbol{k}}{2\omega}$ $\Delta^-(-x) = -\Delta^+(x);\quad \Delta^{-*}(x) = \Delta^+(x);\quad \{\square - m^2\}\Delta^-(x) = 0.$

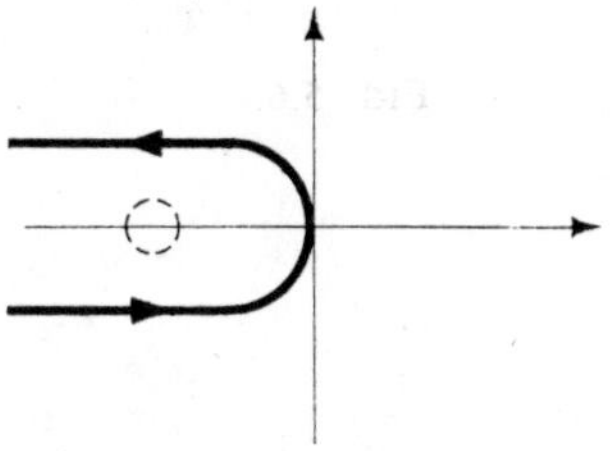

FIG. 5.4.‡

x).	A_-.	R_-.	A_+.	R_+.	
			+	−	$= -\frac{i}{(2\pi)^3}\int_{k\varepsilon<0} e^{ikx}\varepsilon(k)\delta(k^2+m^2)\,d^4k$ (†) $= -\frac{i}{(2\pi)^3}\int e^{ikx}\theta_+(k)\delta(k^2+m^2)\,d^4k$ $= -\frac{i}{(2\pi)^3}\int e^{i(\boldsymbol{k}.\boldsymbol{x}-\omega x_0)}\frac{d^3\boldsymbol{k}}{2\omega}$ $\Delta^+(-x) = -\Delta^-(x);\quad \Delta^{+*}(x) = \Delta^-(x);\quad \{\square - m^2\}\Delta^+(x) = 0$

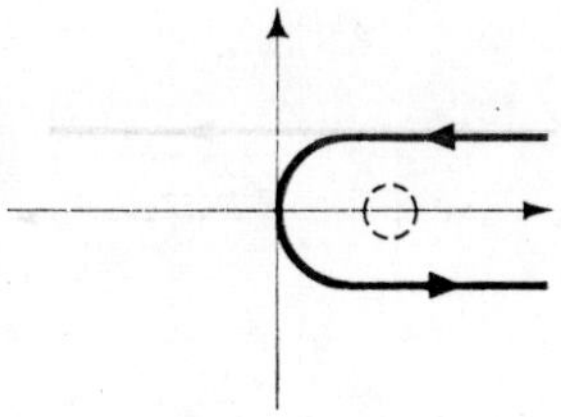

FIG. 5.5.

† ε is a time-like vector such that $\varepsilon_0>0$. These functions are obtained by applying (3.106).

‡ The paths in Figs. 5.4 to 5.11 are drawn in the complex k_0 plane.

$\Delta^{(C)}(x)$.	A_-.	R_-.	A_+.	R_+.	
Δ_{adv}	$+$		$+$		$+\theta_-(x_0)\Delta(x) = \begin{cases} \Delta(x) & \text{if } x_0 < 0 \ (\dagger), \\ 0 & \text{if } x_0 > 0; \end{cases}$ $\{\square - m^2\}\Delta_{adv}(x) =$ $= \overline{\Delta}(x) + \frac{1}{2}\Delta(x); \quad \Delta_{adv}(-x) = \Delta_{ret}(x)$ $\Delta^*_{adv}(x) = \Delta_{adv}(x)$

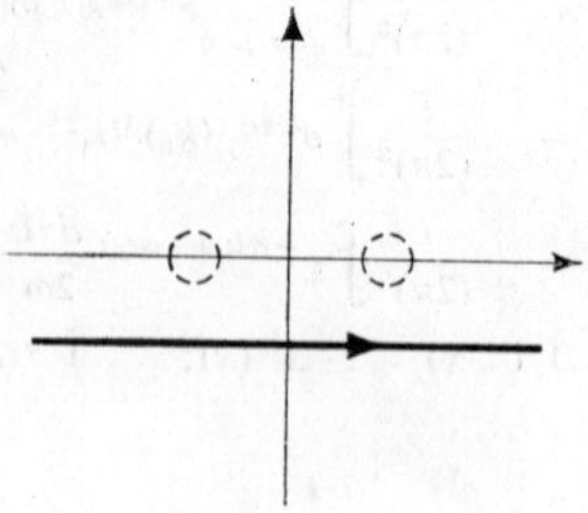

FIG. 5.6.

$\Delta^{(C)}(x)$.	A_-.	R_-.	A_+.	R_+.	
Δ_{ret}		$+$		$+$	$-\theta_+(x)\Delta(x) = \begin{cases} 0 & \text{if } x_0 < 0 \ (\dagger), \\ -\Delta(x) & \text{if } x_0 > 0; \end{cases}$ $\{\square - m^2\}\Delta_{ret}(x) =$ $= \overline{\Delta}(x) - \frac{1}{2}\Delta(x); \quad \Delta_{ret}(-x) = \Delta_{adv}(x)$ $\Delta^*_{ret}(x) = \Delta_{ret}(x)$

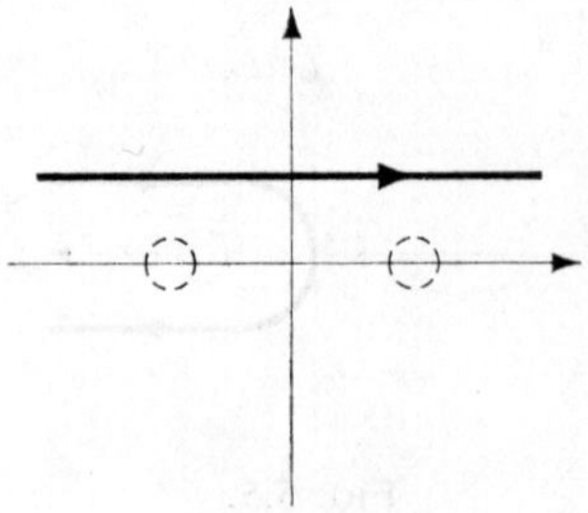

FIG. 5.7.

† Cf. also formulae (7.122).

(x).	A_-.	R_-.	A_+.	R_+.
	+	−	+	−

$$= \begin{Bmatrix} +\Delta_{\text{adv}}(x) & \text{if} & x_0 < 0 \\ -\Delta_{\text{ret}}(x) & \text{if} & x_0 > 0 \end{Bmatrix} = -\frac{i}{(2\pi)^3}\int e^{ikx}\varepsilon(k)\delta(k^2+m^2)\,d^4k$$

$$= -\frac{1}{(2\pi)^3}\int e^{i\boldsymbol{k}.\boldsymbol{x}}\frac{\sin\omega x_0}{\omega}\,d^3\boldsymbol{k}$$

$$= \Delta_{\text{adv}}(x) - \Delta_{\text{ret}}(x) = \Delta^+(x) + \Delta^-(x); \qquad \{\Box - m^2\}\Delta = 0$$

$$\Delta(x) = -\Delta(-x); \qquad \Delta(\boldsymbol{x}, x_0) = -\Delta(\boldsymbol{x}, -x_0)$$

$$\Delta(\boldsymbol{x}, x_0) = \Delta(-\boldsymbol{x}, x_0); \qquad \Delta^*(x) = \Delta(x)$$

$$\Delta(x)\Big]_{x_0=0} = 0; \qquad \frac{\partial\Delta(x)}{\partial x_j}\Big]_{x_0=0} = 0; \qquad \frac{\partial\Delta(x)}{\partial x^4}\Big]_{x_0=0} = +i\delta(\boldsymbol{x})$$

Where L is a proper Lorentz transformation: $\Delta(x) = \Delta(Lx)$; it follows that $\Delta(x) = 0$ for $x_\mu x_\mu > 0$ since any space-like vector x can be transformed into a vector x' such that $x_0' = 0$. We thus see that $\Delta(x)$ vanishes outside the light cone.

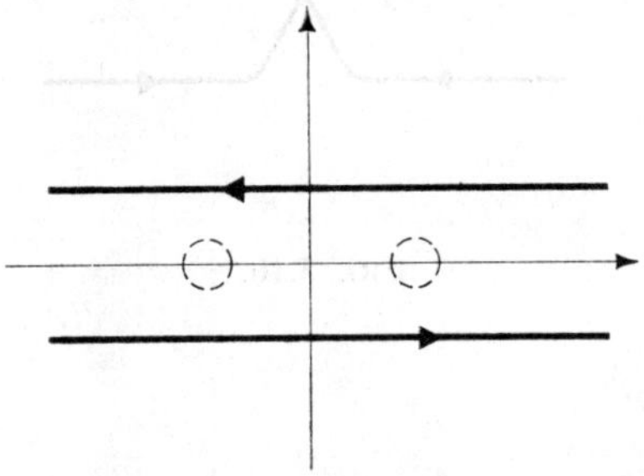

FIG. 5.8.

)(x).	A_-.	R_-.	A_+.	R_+.
	+	+	+	+

$$= -\frac{1}{2}\varepsilon(x_0)\Delta(x) = \frac{1}{2}(\Delta_{\text{ret}}(x) + \Delta_{\text{adv}}(x))$$

$$= \frac{P}{(2\pi)_4}\int \frac{e^{ikx}}{k^2+m^2}\,d^4k; \qquad \{\Box - m^2\}\overline{\Delta} = -\delta(x)$$

$$\overline{\Delta}(-x) = \overline{\Delta}(x)$$

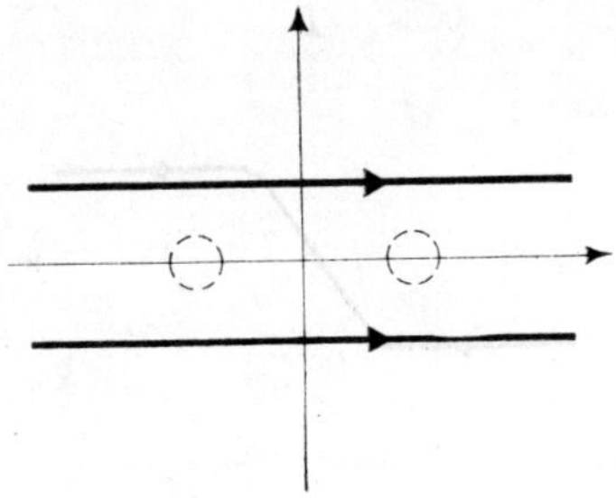

FIG. 5.9.

$\Delta^{(\mathscr{C})}(x)$.	A_-.	R_-.	A_+.	R_+.	
$i\Delta^{(1)}$	+	−	−	+	$\Delta^{(1)}(x) = \frac{1}{(2\pi)^3}\int e^{ikx}\delta(k^2+m^2)\,d^4k$ $= \frac{1}{(2\pi)^3}\int e^{i\mathbf{k}.\mathbf{x}}\frac{\cos\omega x_0}{\omega}\,d^3\mathbf{k}$ $= i(\Delta^+(x)-\Delta^-(x));\quad \{\Box-m^2\}\Delta^{(1)}(x) = 0$ $\Delta^{(1)}(x) = \Delta^{(1)}(-x);\quad \left.\frac{\partial\Delta^{(1)}(x)}{\partial x_4}\right]_{x_4=0} = 0$ Does not vanish outside the light cone.

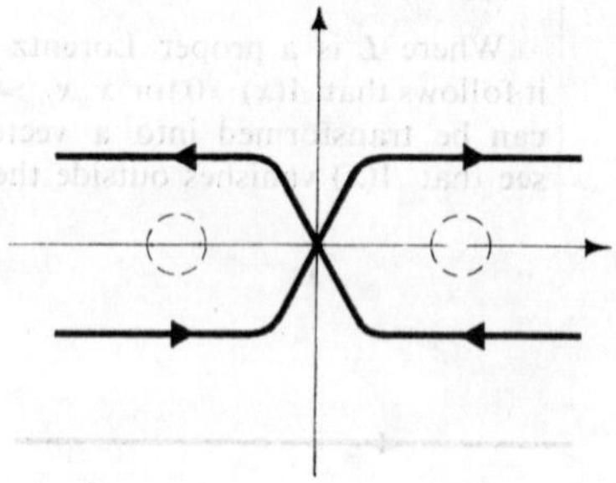

FIG. 5.10.

$\Delta^{(\mathscr{C})}(x)$.	A_-.	R_-.	A_+.	R_+.	
$\frac{i}{2}\Delta_C$	+			+	$\Delta_C(x) = +\frac{2}{(2\pi)^3}\int e^{ikx}\delta_-(k^2+m^2)\,d^4k$ $= \lim_{\varepsilon=0} -\frac{2i}{(2\pi)^4}\int d^4k\,\frac{e^{ikx}}{k^2+m^2-i\varepsilon}$ $= \left.\begin{cases} -2i\Delta^-(x) & \text{if } x_0<0 \\ 2i\Delta^+(x) & \text{if } x_0>0 \end{cases}\right\} = \Delta^{(1)}(x)-2i\bar{\Delta}(x)$ $\Delta_C(-x) = \Delta_C(x);\quad \{\Box-m^2\}\Delta_C(x) = 2i\delta(x)$

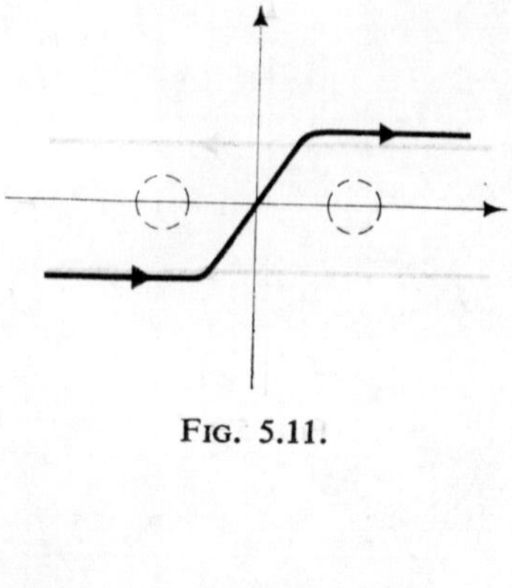

FIG. 5.11.

Notes. 1. The functions Δ_{adv}, Δ_{ret}, $\bar{\Delta}$ are expressed and can be defined with the aid of the function Δ and the step functions $\theta_{\pm}$, ε.

2. The functions Δ, $\Delta^{(1)}$, Δ_C can be expressed with the aid of Δ^+ and Δ^-.

3. The differential equations obeyed by Δ_{adv}, Δ_{ret}, $\bar{\Delta}$ can be established by using (1.108). The differential equation concerning Δ_C is inferred from the relation $\Delta_C = \Delta^{(1)} - 2i\bar{\Delta}$. Finally, Δ and $\Delta^{(1)}$ are solutions of the homogeneous equation since their integration paths in the complex plane are closed.

When $m = 0$, the functions $\Delta^{(\mathscr{C})}(x)$ are denoted by $D^{(\mathscr{C})}(x)$; a table of these functions is given below:

TABLE OF THE FUNCTIONS $D^{(\mathscr{C})}(x)$:

$$\text{Case of a mass} \quad m = 0 : D^{(\mathscr{C})}(x) = (2\pi)^{-4} \int d^3\boldsymbol{k} \int_{\mathscr{C}} dk_0 \frac{e^{ikx}}{\boldsymbol{k}^2 - k_0^2}.$$

With $r = |\boldsymbol{x}|$, $K = |\boldsymbol{k}| = \omega$, we have

$$D(x) = -\frac{1}{(2\pi)^3} \int e^{i\boldsymbol{k}.\boldsymbol{x}} \frac{\sin K x_0}{K} d^3\boldsymbol{k} = \frac{2\pi}{(2\pi)^3} \int_0^{\infty} dK \frac{\sin K x_0}{ir} \int_0^{\pi} e^{iKr\cos\theta} d(iKr\cos\theta)$$

$$= \frac{\pi}{(2\pi)^3 r} \int_{-\infty}^{+\infty} (e^{iK(x_0+r)} - e^{iK(x_0-r)}) dK$$

$$= \frac{1}{4\pi r}(\delta(x_0+r) - \delta(x_0-r)) = \begin{cases} +\dfrac{1}{4\pi r}\delta(x_0+r) & \text{if} \quad x_0 < 0 \\ -\dfrac{1}{4\pi r}\delta(x_0-r) & \text{if} \quad x_0 > 0 \end{cases}$$

$$= -\frac{1}{2\pi}\varepsilon(x)\delta(x_\mu x_\mu)$$

$$D_{ret}(x) = \frac{1}{4\pi r}\delta(x_0 - r)$$

$$D_{adv}(x) = \frac{1}{4\pi r}\delta(x_0 + r)$$

$$\bar{D}(x) = \frac{1}{8\pi r}(\delta(x_0+r) + \delta(x_0-r)) = \frac{1}{4\pi}\delta(x^2)$$

$$D^{(1)}(x) = \frac{P}{2\pi^2(r^2-t^2)} = \frac{P}{2\pi^2 x_\mu x_\mu}$$

$$D_C(x) = \frac{2}{(2\pi)^3}\int e^{ikx}\delta_-(k_\mu k_\mu)\, d^4k = \lim_{\varepsilon=0} -\frac{2i}{(2\pi)^4}\int \frac{e^{ikx}}{k_\mu k_\mu - i\varepsilon} d^4k$$

$$= D^{(1)}(x) - 2i\,\bar{D}(x) = -\frac{i}{\pi}\left[\frac{P}{2i\pi}\frac{1}{-x^2} + \frac{1}{2}\delta(-x^2)\right]$$

$$= -\frac{i}{\pi}\delta_-(-x^2) = -\frac{i}{\pi}\frac{1}{2r}[\delta_-(x_0-r) + \delta_-(-x_0-r)].$$

Let us show, for instance, how one may prove some of the results relative to $\Delta(x)$. First of all, one notes that the second expression of this function is obtained by a simple calculation of residues; then the first one [the one including a $\delta(\ldots)$] can be inferred as follows:

$$\Delta(x) = -\frac{1}{(2\pi)^3}\int e^{i\boldsymbol{k}.\boldsymbol{x}}\,\frac{\sin \omega x_0}{\omega}\,d^3\boldsymbol{k} \tag{5.7}$$

$$= \frac{i}{(2\pi)^3}\int d^3\boldsymbol{k}\int_{-\infty}^{+\infty} dk_0\,\frac{e^{ikx}}{2\omega}\,(\delta(\omega+k_0)-\delta(\omega-k_0))$$

$$= -\frac{i}{(2\pi)^3}\int e^{ikx}\varepsilon(k_0)\,\delta(\omega^2-k_0^2)\,d^4k.$$

All the factors in the above expression are Lorentz invariants, except the factor $\varepsilon(k_0)$ which seems to disturb the global invariance of $\Delta(x)$. In fact when $\varepsilon(k_0)$ is not invariant, in other words when k_μ is space-like, then $\delta(k^2+m^2)$ is always zero, since its argument does not vanish for any space-like vector k_μ.

Consequently, $\Delta(x)$ is an invariant and its value can therefore only depend on $x_\mu x_\mu$.

It can be verified (from its three-dimensional expression) that

$$\Delta(x)\Big]_{x_0=0} = 0, \tag{5.8a}$$

$$\frac{\partial\Delta(x)}{\partial x_4}\Big]_{x_0=0} = i\,\delta(\boldsymbol{x}) \tag{5.8b}$$

and also that

$$\frac{\partial\Delta^{(1)}(x)}{\partial x_4}\Big]_{x_0=0} = 0. \tag{5.9}$$

These formulae may be expressed in a covariant form by introducing a generalized distribution, namely $\delta_\mu^{(\sigma)}(x)$,

$$\int_\sigma \delta_\mu^{(\sigma)}(x-x')f(x')\,d\sigma_\mu = f(x) \tag{5.10}$$

which, when σ is a hyperplane, has the following components:

$$\delta_\mu^{(\sigma)}(x) = \begin{Bmatrix} 0 \\ i\,\delta(\boldsymbol{x}) \end{Bmatrix} \quad \begin{matrix} \text{if} & \mu = j, \\ \text{if} & \mu = 4. \end{matrix} \tag{5.11}$$

Formula (5.8*b*) is then written

$$\frac{\partial\Delta(x)}{\partial x_\mu} = \delta_\mu^{(\sigma)}(x). \tag{5.12}$$

Note also that the distribution $\delta^{(\sigma)}(x)$ introduced in (2.145) may be written in terms of $\delta_\mu^{(\sigma)}(x)$:

$$\delta^{(\sigma)}(x) = n_\mu(x)\,\delta_\mu^{(\sigma)}(x). \tag{5.13}$$

The singular functions $\Delta(x)$ and $\Delta^{(1)}(x)$ allow us to prove the Cauchy problem relative to (5.1); let us write

$$\Phi(x) = \frac{1}{(2\pi)^3}\int e^{i\boldsymbol{k}.\boldsymbol{x}}\frac{\sin\omega t}{\omega}f(\boldsymbol{k})\,d^3\boldsymbol{k}+\frac{1}{(2\pi)^3}\int e^{i\boldsymbol{k}.\boldsymbol{x}}\frac{\cos\omega t}{\omega}g(\boldsymbol{k})\,d^3\boldsymbol{k}$$

$$= -\int \Delta(\boldsymbol{x}-\boldsymbol{x}', t)f(\boldsymbol{x}')\,d^3\boldsymbol{x}'+\int \Delta^{(1)}(\boldsymbol{x}-\boldsymbol{x}', t)g(\boldsymbol{x}')\,d^3\boldsymbol{x}'. \qquad (5.14)$$

For $t = 0$, we have

$$\left.\begin{aligned}\Phi(x, 0) &= \frac{1}{(2\pi)^3}\int e^{i\boldsymbol{k}.\boldsymbol{x}}\frac{g(\boldsymbol{k})}{\omega(\boldsymbol{k})}\,d^3\boldsymbol{k},\\ \left[\frac{\partial\Phi(x)}{\partial t}\right]_{t=0} &= \frac{1}{(2\pi)^3}\int e^{i\boldsymbol{k}.\boldsymbol{x}}f(\boldsymbol{k})\,d^3\boldsymbol{k},\end{aligned}\right\} \qquad (5.15)$$

and by a Fourier inversion of the last two formulae, $\dfrac{g(\boldsymbol{k})}{\omega(\boldsymbol{k})}$ and $f(\boldsymbol{k})$ can be determined in terms of the initial data.

The same problem can moreover be solved with the exclusive use of the function $\Delta(x-x')$ and its time derivative.

It can indeed be easily verified that we have

$$\begin{aligned}\Phi(\boldsymbol{x}, t) &= -\int \frac{\partial\Delta(\boldsymbol{x}-\boldsymbol{x}', t-t')}{\partial t}\Phi(\boldsymbol{x}', t')\,d^3\boldsymbol{x}'\\ &\quad -\int \Delta(\boldsymbol{x}-\boldsymbol{x}', t-t')\frac{\partial\Phi(\boldsymbol{x}', t')}{\partial t'}\,d^3\boldsymbol{x}'\\ &\equiv \int \Delta(\boldsymbol{x}-\boldsymbol{x}')\,\frac{\overleftrightarrow{\partial}}{\partial t'}\,\Phi(x')\,d^3\boldsymbol{x}' \end{aligned}\qquad (5.16)$$

provided that $\Phi(\boldsymbol{x}', t')$ converges strongly enough to zero to counter-balance the strong singularity of $\left[\dfrac{\partial^2\Delta(\boldsymbol{x}-\boldsymbol{x}'\ t-t')}{\partial t^2}\right]_{t=t'}$. The right side gives the definition of the derivation operator $\dfrac{\overleftrightarrow{\partial}}{\partial t'}$.

The expression (5.16) can be readily written in covariant form by making use of (5.12):

$$\Phi(x) = \int_\sigma\left(\frac{\partial\Delta(x-x')}{\partial x_\mu}\Phi(x')+\Delta(x-x')\frac{\partial\Phi(x')}{\partial x'_\mu}\right)d\sigma'_\mu. \qquad (5.17)$$

Consider now the expression of $\Delta_C(x-x')$ in its three-dimensional form.

$$\Delta_C(x-x') = \left\{\begin{aligned}&\int\frac{1}{(2\pi)^3\omega}e^{-i[\omega(x'_0-x_0)-\boldsymbol{k}(\boldsymbol{x}-\boldsymbol{x}')]}\,d^3\boldsymbol{k} &&\text{if } x_0-x'_0 < 0,\\ &\int\frac{1}{(2\pi)^3\omega}e^{-i[\omega(x_0-x'_0)-\boldsymbol{k}(\boldsymbol{x}-\boldsymbol{x}')]}\,d^3\boldsymbol{k} &&\text{if } x_0-x'_0 > 0.\end{aligned}\right. \qquad (5.18)$$

As an elementary solution of a hyperbolic equation, it represents a propagation. We shall show that it characterizes the propagation of a signal such that it is only perceived after being sent out.

Let there be, in fact, two space points A and B separated by the distance $|\overline{AB}|=l$. The point B at the time t receives a signal emitted from A at the time t' when $t>t'$; we thus have a wave progressing from A to B, described by a function $f(t-t'-l/V)$, where V is the speed of propagation of the signal; $\Delta_C(x-x')$ therefore does in fact describe the above property when $x_0-x_0'>0$. Conversely, the point B at the time t emits a signal perceived by A at t' when $t<t'$; we thus have a wave progressing from B to A, described by a function $f(t'-t-l/V)$, which corresponds to the form of $\Delta_C(x-x')$ when $x_0-x_0'<0$. Finally since $\Delta_C(x-x') = \Delta_C(x'-x)$ we obtain a complete space–time symmetry. This is the characteristic of $\Delta_C(x)$ which distinguishes it from the other elementary solutions $\Delta_{\text{ret}}(x)$, $\Delta_{\text{adv}}(x)$, . . ., and which explains why the process of emission of particles of spin 0 followed by their absorption is in fact described by this function.†

Schwinger has devised a method of evaluating the functions $\Delta(x)$ in terms of Bessel functions. We shall give merely a few results without proofs.

Set $\lambda = -x_\mu x_\mu$, $r^2 = \boldsymbol{x}^2$:

$$\Delta(x) = \frac{1}{4\pi r}\frac{\partial}{\partial r}\begin{cases} J_0(m\sqrt{\lambda}) & \text{if} \quad t<-r \\ 0 & \text{if} \quad -r<t<+r \\ -J_0(m\sqrt{\lambda}) & \text{if} \quad t>r \end{cases}$$

$$= -\frac{1}{2\pi r}\,\varepsilon(x)\left(\delta(\lambda)+\frac{m^2}{4}\,\theta_+(-\lambda)+\ldots\right), \tag{5.19}$$

$$\Delta^{(1)}(x) = \frac{1}{4\pi r}\frac{\partial}{\partial r}\begin{cases} +N_0(m\sqrt{\lambda}) & \text{if} \quad |t|>r \\ -iH_0^{(1)}(im\sqrt{\lambda}) & \text{if} \quad |t|<r \end{cases}$$

$$= -\frac{1}{(2\pi)^2}\left(\frac{P}{\lambda}-\frac{m^2}{2}\log\left[\frac{\gamma}{2}\sqrt{m^2|\lambda|}\right]+\frac{m^2}{4}+\ldots\right), \tag{5.20}$$

where γ is the Euler constant.‡

† M. Fierz, *Helv. Phys. Acta*, **23**, 731 (1950); E. C. G. Stueckelberg and T. A. Green, *Helv. Phys. Acta*, **24**, 153 (1951).

‡ The singular functions were introduced by P. Jordan and W. Pauli, *Zeits. Phys.*, **47**, 151 (1928). They were investigated by J. Schwinger, *Phys. Rev.*, **75**, 651 (1949). Their definition as distributions has been the subject of numerous papers which we are unable to quote here, for lack of space.

4. Quantization and observables of the Hermitian scalar (or pseudo-scalar) field

We have already taken this field as an example (pp. 86 *et seq.)* and we have seen that its field equations could be derived from the Lagrangian density†

$$\mathcal{L} = -\frac{1}{2}(\partial_\mu\Phi\partial_\mu\Phi+m^2\Phi^2). \tag{5.21}$$

From this density we obtain the expression of the conjugate variable:

$$\Pi(x) = \frac{\partial\mathcal{L}}{\partial\dot{\Phi}} = \frac{\partial\Phi(x)}{\partial t}, \tag{5.22}$$

of the Hamiltonian density:

$$\Lambda(x) = \frac{1}{2}[(\Pi(x))^2+(\nabla\Phi)^2+m^2\Phi^2], \tag{5.23}$$

and of the Hamilton equations:

$$\frac{\partial\Phi(x)}{\partial t} = \Pi(x), \qquad \frac{\partial\Pi(x)}{\partial t} = \{\nabla^2-m^2\}\Phi(x). \tag{5.24}$$

The calculation of the commutator $[\Phi(x), \Phi(x')]$ can be carried out according to the following simple argument: let $C(x, x')$ be this commutator.

As a function of x and x', $C(x, x')$ has the following properties:

(a) $\{\Box-m^2\}C(x, x') = 0.$

(b) If $t = t'$, $C(x, x')$ is identically zero, since this field commutes with itself on an equitemporal surface.

(c) Differentiate the commutator with respect to t;

$$\left[\frac{\partial\Phi(x)}{\partial t}, \Phi(x')\right] = [\Pi(x), \Phi(x')] = \frac{\partial C(x, x')}{\partial t},$$

then taking $t=t'$, we have

$$\left.\frac{\partial C(x, x')}{\partial t}\right]_{t=t'} = -i\delta(\boldsymbol{x}-\boldsymbol{x}').$$

These three conditions determine $C(x, x')$ in a unique way (Cauchy's problem). Comparing $C(x, x')$ with $\Delta(x-x')$ of the preceding section, we get

$$C(x, x') = i\Delta(x-x'). \tag{5.25}$$

† The quantization of this field was first studied by W. Pauli and V. W. Weisskopf, *Helv. Phys. Acta*, 7, 709 (1934).

It will be noted moreover that the dependence of C with respect to $x-x'$ was to be expected since the theory is invariant under four-dimensional translations.

The same commutator can be calculated by using the method described in Chap. II, § 9. On the one hand, the commutation relations are given by formula (2.97) where the differential operator $W(\nabla)$ is defined by formula (2.93):

$$W(\nabla) \equiv \frac{\partial W^{(\Pi)}(t, \nabla)}{\partial t}\Big]_{t=0}.$$

On the other hand, we infer from formula (2.87) the expression of

$$\frac{\partial \Phi(x)}{\partial t} = \frac{\partial W^{(Q)}(t-t_0, \nabla)}{\partial t} Q(\boldsymbol{x}, t_0) + \frac{\partial W^{(\Pi)}(t-t_0, \nabla)}{\partial t} \Pi(\boldsymbol{x}, t_0). \quad (5.26)$$

Comparing the above expression with the first of the Hamilton equations (5.24), we obtain for $t = t_0$:

$$W(\nabla) = \frac{\partial W^{(\Pi)}(t-t_0, \nabla)}{\partial t}\Big]_{t=t_0} = 1, \quad (5.27)$$

and consequently the commutation relation

$$[\Phi(x), \Phi(x')] = i\Delta(x-x'). \quad (5.28)$$

Since the field is neutral, we can define the field variable $\Phi(x)$ as a Hermitian operator; consequently the operators $\Phi^+(k)$ and $\Phi^-(k)$ have a simple definition in the momentum space [formula (3.121)]:

$$\Phi^+(k) = \Phi(k), \qquad \Phi^-(k) = \Phi(-k), \; k_0 > 0 \quad (5.29)$$

and by virtue of (3.126*b*) we obtain as the commutator of Φ^+ and Φ^-:

$$[\Phi^+(x), \Phi^-(x')] = +i\Delta^+(x-x'). \quad (5.30)$$

The latter commutator makes it possible to compute vacuum-expectation values of products of field variables by a method whose importance will become clear further on; as an example, let us compute

$$\langle 0|\Phi(x)\Phi(x')|0\rangle = \langle 0|(\Phi^+(x)+\Phi^-(x))(\Phi^+(x')+\Phi^-(x'))|0\rangle. \quad (5.31)$$

Now by virtue of the properties (3.144) of the vacuum, the vacuum expectation values:

$$\langle 0|\Phi^+(x)\Phi^+(x')|0\rangle, \quad \langle 0|\Phi^-(x)\Phi^-(x')|0\rangle, \quad \langle 0|\Phi^-(x)\Phi^+(x')|0\rangle$$

are zero; there thus remains

$$\begin{aligned}\langle 0|\Phi(x)\Phi(x')|0\rangle &= \langle 0|\Phi^+(x)\Phi^-(x')|0\rangle \quad (5.32)\\ &= \langle 0|[\Phi^+(x), \Phi^-(x')]|0\rangle = i\Delta^+(x-x').\end{aligned}$$

Let us now turn to the computation of the energy–momentum tensor; we readily obtain its expression:

$$T_{\mu\nu} = \Phi_{,\mu}\Phi_{,\nu} - \frac{1}{2}\,\delta_{\mu\nu}(\Phi_{,\sigma}\Phi_{,\sigma} + m^2\Phi^2) \tag{5.33}$$

and the symmetrization of its first term gives the symmetrical energy–momentum tensor $\Theta_{\mu\nu}$

$$\Theta_{\mu\nu} = \frac{1}{2}\,[\Phi_{,\mu}\Phi_{,\nu}]_+ - \frac{1}{2}\,\delta_{\mu\nu}(\Phi_{,\sigma}\Phi_{,\sigma} + m^2\Phi^2). \tag{5.34}$$

The vacuum-expectation value of this tensor is not zero; let us in fact define this expectation value as follows:

$$\langle 0|T_{\mu\nu}|0\rangle = \lim_{x=x'}\langle 0|\Phi_{,\mu}(x)\Phi_{,\nu}(x') - \frac{1}{2}\,\delta_{\mu\nu}(\Phi_{,\sigma}(x)\Phi_{,\sigma}(x') + m^2\Phi(x)\Phi(x'))\,|0\rangle, \tag{5.35}$$

an expression which can also be written

$$\langle 0|T_{\mu\nu}(x)|0\rangle = \lim_{x=x'}\left\{\frac{\partial^2}{\partial x_\mu \partial x'_\nu} - \frac{1}{2}\,\delta_{\mu\nu}\left(\frac{\partial^2}{\partial x_\sigma \partial x'_\sigma} + m^2\right)\right\}\langle 0|\Phi(x)\Phi(x')\,|0\rangle,$$

or again

$$\langle 0|T_{\mu\nu}(x)|0\rangle = \lim_{x=x'} i\left\{\frac{\partial^2}{\partial x_\mu \partial x'_\nu} - \frac{1}{2}\,\delta_{\mu\nu}(-\Box_x + m^2)\right\}\Delta^+(x-x')$$

$$= \lim_{x=x'} i\,\frac{\partial^2 \Delta^+(x-x')}{\partial x_\mu \partial x'_\nu}. \tag{5.36}$$

Introducing the Fourier transform of Δ^+ and making $x = x'$, we get

$$\langle 0|T_{\mu\nu}(x)|0\rangle = -\frac{1}{(2\pi)^3}\int k_\mu k_\nu \theta_+(k)\,\delta(k^2+m^2)\,d^4k.$$

On the other hand, in order to respect the variance of $T_{\mu\nu}$, the preceding integral must be of the form

$$-\frac{1}{(2\pi)^3}\,\delta_{\mu\nu}\int k^2\theta_+(k)\,\delta(k^2+m^2)\,d^4k,$$

but we have

$$k^2\,\delta(k^2+m^2) = -m^2\,\delta(k^2+m^2).$$

So that finally

$$\langle 0|T_{\mu\nu}|0\rangle = \frac{m^2}{(2\pi)^3}\,\delta_{\mu\nu}\int \theta_+(k)\delta(k^2+m^2)\,d^4k = \frac{m^2}{2(2\pi)^3}\,\delta_{\mu\nu}\int \frac{d^3\boldsymbol{k}}{\sqrt{\boldsymbol{k}^2+m^2}}$$

$$= \frac{m^2}{(2\pi)^2}\,\delta_{\mu\nu}\int_0^\infty \frac{K^2\,dK}{\sqrt{K^2+m^2}}, \qquad K^2 = \boldsymbol{k}^2, \tag{5.37}$$

which is a quadratically divergent expression.

It can nevertheless be verified that the operator P_μ inferred from this tensor (and which itself has an infinite vacuum-expectation value) is indeed the displacement operator of this field; let us prove this for P_0. It suffices to calculate:

$$[\Phi(\boldsymbol{x}, t), \Lambda(\boldsymbol{x}', t)] = \frac{1}{2}[\Phi(\boldsymbol{x}, t), (\Pi(\boldsymbol{x}', t))^2] = i\,\delta(\boldsymbol{x}-\boldsymbol{x}')\Pi(\boldsymbol{x}', t),$$

$$[\Pi(\boldsymbol{x}, t), \Lambda(\boldsymbol{x}', t)] = \frac{1}{2}[\Pi(\boldsymbol{x}, t), (\nabla'\Phi(\boldsymbol{x}', t))^2 + m^2\Phi^2(\boldsymbol{x}', t)]$$
$$= i(\nabla\Phi(x)\nabla\delta(\boldsymbol{x}-\boldsymbol{x}') + m^2\Phi(x)\,\delta(\boldsymbol{x}-\boldsymbol{x}')).$$

By integrating over x' at a given time, we obtain for the first of the above relations:

$$[\Phi(x), P_0] = i\Pi(x) = i\frac{\partial\Phi(x)}{\partial t} \tag{5.38a}$$

and for the second of the same relations

$$[\Pi(x), P_0] = i\{\nabla^2 + m^2\}\Phi(x) = i\frac{\partial\Pi(x)}{\partial t}. \tag{5.38b}$$

The unphysical value of the vacuum expectation value of P_μ is thus due to the kinematics of the theory and not to its dynamics.

This infinity is a first example of the type of difficulty encountered in quantum field theory; the above difficulty is moreover only an apparent one. Indeed, the energy of a field (and consequently the number of particles of that field present in a given state) is only defined apart from an arbitrary additive constant. The above considerations show that this constant is infinite; it can be eliminated by a displacement (itself infinite, it is true!) of the zero of the energy.

At the end of the next section we shall be presenting another method, based on the definition of the ordered products, which permits a satisfactory solution of this difficulty.

5. Non-Hermitian (charged) scalar and pseudo-scalar fields

We have seen that in the formalism of non-Hermitian fields the field variable $\Phi(x)$ and its adjoint $\tilde{\Phi}(x)$ have to be treated as two independent operators (Chap. II, § 10) obeying the equations

$$(\Box - m^2)\Phi(x) = 0, \tag{5.39}$$
$$(\Box - m^2)\tilde{\Phi}(x) = 0, \tag{5.40}$$

which can be inferred from the Lagrangian density

$$\mathcal{L} = -(\tilde{\Phi}_{,\mu}\Phi_{,\mu} + m^2\tilde{\Phi}\Phi). \tag{5.41}$$

The conjugate momenta are then

$$\Pi(x) = \frac{\partial\tilde{\Phi}(x)}{\partial t}, \qquad \tilde{\Pi}(x) = \frac{\partial\Phi(x)}{\partial t} \tag{5.42}$$

and we can easily write down the Hamilton equations

$$\left.\begin{aligned} \frac{\partial\Phi(x)}{\partial t} &= \tilde{\Pi}(x), & \frac{\partial\tilde{\Phi}(x)}{\partial t} &= \Pi(x), \\ \frac{\partial\Pi(x)}{\partial t} &= (\nabla^2 - m^2)\tilde{\Phi}(x), & \frac{\partial\tilde{\Pi}(x)}{\partial t} &= (\nabla^2 - m^2)\Phi(x). \end{aligned}\right\} \tag{5.43}$$

By carrying out the quantization in a similar fashion to that of neutral particles, we obtain the commutation relations

$$[\Phi(x), \Phi(x')] = [\tilde{\Phi}(x), \tilde{\Phi}(x)'] = 0, \tag{5.44}$$

$$[\Phi(x), \tilde{\Phi}(x')] = i\,\Delta(x-x'). \tag{5.45}$$

On the other hand, let us separate the field variables $\Phi(x)$, $\tilde{\Phi}(x)$, into positive and negative frequency parts by the method of Chap. III, § 11, or in accordance with formulae (5.58) and (5.61). Referring to (3.142), we know that

$\Phi^+(x)$, $\tilde{\Phi}^+(x)$ annihilate particles respectively of charge $-e$ and $+e$,
$\Phi^-(x)$, $\tilde{\Phi}^-(x)$ create particles respectively of charge $+e$ and $-e$.

The commutators between the operators are all zero except the following ones [cf. (3.134)]:

$$[\Phi^+(x), \tilde{\Phi}^-(x')] = i\Delta^+(x-x'), \tag{5.46a}$$

$$[\Phi^-(x), \tilde{\Phi}^+(x')] = i\Delta^-(x-x'). \tag{5.46b}$$

The energy–momentum tensor operator can be readily expressed

$$\begin{aligned} T_{\mu\nu}(x) &= \tilde{\Phi}_{,\mu}(x)\Phi_{,\nu}(x) \\ &\quad + \tilde{\Phi}_{,\nu}(x)\Phi_{,\mu}(x) - \delta_{\mu\nu}(\tilde{\Phi}_{,\sigma}(x)\Phi_{,\sigma}(x) + m^2\tilde{\Phi}(x)\Phi(x)) \end{aligned} \tag{5.47}$$

and for the current $J_\mu(x)$ we have

$$J_\mu(x) = -ie(\tilde{\Phi}_{,\mu}(x)\Phi(x) - \tilde{\Phi}(x)\Phi_{,\mu}(x)) \tag{5.48}$$

or alternatively, making use of (5.45),

$$J_\mu(x) = -ie(\tilde{\Phi}_{,\mu}(x)\Phi(x) - \Phi_{,\mu}(x)\tilde{\Phi}(x) - \delta_{\mu 4} \lim_{\boldsymbol{x}=\boldsymbol{x}'} \delta(\boldsymbol{x}-\boldsymbol{x}')).$$

Neglecting as we said at the end of the above section the infinite term $-ie\,\delta_{\mu 4}\,\delta(0)$, we get another expression of the current:

$$J_\mu(x) = -ie(\tilde{\Phi}_{,\mu}(x)\Phi(x) - \Phi_{,\mu}(x)\tilde{\Phi}(x)). \tag{5.49}$$

The expression (5.49) is generally preferred to the expression (5.48) when the operator $\Phi(x)$ is split into Hermitian and anti-Hermitian parts; let us indeed set

$$\Phi(x) = \frac{1}{\sqrt{2}}(\Phi_1(x)+i\Phi_2(x)), \quad \tilde{\Phi}(x) = \frac{1}{\sqrt{2}}(\Phi_1(x)-i\Phi_2(x)) \tag{5.50a}$$

where the Hermitian operators Φ_1 and Φ_2 are defined as follows:

$$\Phi_1(x) = \frac{\Phi(x)+\tilde{\Phi}(x)}{\sqrt{2}}, \quad \Phi_2(x) = \frac{\Phi(x)-\tilde{\Phi}(x)}{i\sqrt{2}} \tag{5.50b}$$

and express the current J_μ in terms of Φ_1 and Φ_2:

$$J_\mu(x) = -e\left(\frac{\partial\Phi_2(x)}{\partial x_\mu}\Phi_1(x)-\frac{\partial\Phi_1(x)}{\partial x_\mu}\Phi_2(x)\right). \tag{5.51}$$

The formula thus obtained is of great importance when we consider the interaction of a nucleon field with a charged or neutral meson field.

From the current $J_\mu(x)$ we infer the number operator:

$$\begin{aligned}\mathcal{N} &= \frac{1}{e}\int J_\mu(x)\, d\sigma_\mu = i\int\left(\frac{\partial\tilde{\Phi}(x)}{\partial t}\Phi(x)-\tilde{\Phi}(x)\frac{\partial\Phi(x)}{\partial t}\right)d^3\boldsymbol{x}\\ &= i\int(\Pi(x)\Phi(x)-\tilde{\Phi}(x)\tilde{\Pi}(x))\, d^3\boldsymbol{x},\end{aligned} \tag{5.52}$$

which obeys the characteristic relations (3.93).

We have seen that the physical interpretation of the results is particularly simple in the momentum space. The Fourier transform $\Phi(k)$ of the wave function obeys the equation

$$(k^2+m^2)\Phi(k) = 0; \tag{5.53a}$$

$\Phi(k)$ is therefore of the form

$$\Phi(k) = A(k)\,\delta(k^2+m^2), \tag{5.53b}$$

where $A(k)$ is arbitrary (chosen, for instance, such that one gets in this way a solution to a Cauchy problem).

Let us now return to the expression of $\Phi(x)$ in terms of $A(k)$.†

$$\Phi(x) = (2\pi)^{-4}\int\delta(k^2+m^2)A(k)e^{ikx}\, d^4k, \tag{5.54}$$

† Let us find the non-relativistic form of $\Phi(k)$ given by (5.53*b*). The argument of $\delta(\ldots)$ only cancels if $k_0 = (m^2+\boldsymbol{k}^2)^{\frac{1}{2}} \approx m+\frac{1}{2m}\boldsymbol{k}^2$; consequently the expression of $\Phi^{(n,r)}(k)$ is

$$\Phi^{(n,r)}(k) = a(k)\,\delta\left(k_0-m-\frac{1}{2m}\boldsymbol{k}^2\right) \equiv a(k)\,\delta\left(E-\frac{1}{2m}\boldsymbol{k}^2\right)$$

and is a solution of eqn. (5.3).

making use of the computations which led to formulae (4.67) *et seq.*, we see that $\Phi(x)$, after integration with respect to k_0, can be written

$$\Phi(x) = (2\pi)^{-4} \int \frac{d^3\boldsymbol{k}}{2\omega}\, e^{i\boldsymbol{k}.\boldsymbol{x}} \big(A(\boldsymbol{k}, \omega)e^{-i\omega t} + A(\boldsymbol{k}, -\omega)e^{i\omega t}\big). \tag{5.55}$$

Set

$$A^+(\boldsymbol{k}) \equiv A(\boldsymbol{k}, \omega), \quad A^-(\boldsymbol{k}) = A(-\boldsymbol{k}, -\omega), \tag{5.56}$$

then change $\boldsymbol{k}$ into $-\boldsymbol{k}$ in the integrand involving $A(\boldsymbol{k}, -\omega)$; one finally gets

$$\Phi(x) = (2\pi)^{-4} \int \frac{d^3\boldsymbol{k}}{2\omega}\, e^{i(\boldsymbol{k}.\boldsymbol{x}-\omega t)} A^+(\boldsymbol{k}) + (2\pi)^{-4} \int \frac{d^3\boldsymbol{k}}{2\omega}\, e^{-i(\boldsymbol{k}.\boldsymbol{x}-\omega t)} A^-(\boldsymbol{k}). \tag{5.57}$$

We have thus separated $\Phi(x)$ into parts of positive $\Phi^+(x)$ and negative $\Phi^-(x)$ frequency with†

$$\Phi^{\pm}(x) = (2\pi)^{-4} \int_{k_0=\omega} \frac{d^3\boldsymbol{k}}{2\omega}\, e^{\pm ikx} A^{\pm}(\boldsymbol{k}). \tag{5.58}$$

The same analysis can be applied to $\tilde{\Phi}(x)$; we first have

$$\tilde{\Phi}(x) = (2\pi)^{-4} \int \delta(k^2+m^2)\tilde{A}(k)e^{-ikx}\, d^4k, \tag{5.59}$$

then setting

$$\tilde{A}^+(\boldsymbol{k}) = \tilde{A}(-\boldsymbol{k}, -\omega), \quad \tilde{A}^-(\boldsymbol{k}) = \tilde{A}(\boldsymbol{k}, \omega), \tag{5.60}$$

we obtain

$$\tilde{\Phi}(x) = \tilde{\Phi}^+(x) + \tilde{\Phi}^-(x) \tag{5.61a}$$

$$\tilde{\Phi}^{\pm}(x) = (2\pi)^{-4} \int_{k_0=\omega} \frac{d^3\boldsymbol{k}}{2\omega}\, e^{\pm ikx} \tilde{A}^{\pm}(\boldsymbol{k}), \tag{5.61b}$$

and the relations

$$\tilde{A}^+(\boldsymbol{k}) = [A^-(\boldsymbol{k})]^{\tilde{}}, \quad \tilde{A}^-(\boldsymbol{k}) = [A^+(\boldsymbol{k})]^{\tilde{}}. \tag{5.62}$$

We can easily verify that the results thus obtained are the same as those given by the Schwinger method as defined in Chap. II, § 13. We now express the commutators (5.46*a* and *b*) with the help of the Fourier transforms

† By introducing the factor $\delta(k_0-\omega)$ into (5.58) (as into the three-dimensional expression of $\Delta^{\pm}(x)$ on p. 163) the three-dimensional integrals can be transformed into four-dimensional ones. By taking into account formula (1.105), we see that

$$\Phi^{\pm}(x) = \frac{i}{2\pi} \int \Delta^{\pm}(x-\xi) \mathrm{A}^{\pm}(\xi)\, d^4\xi$$

where $A^{\pm}(\xi)$ is the Fourier transform of $A^{\pm}(\pm\boldsymbol{k})$. See also Chap. IX, § 16, Note (2).

(5.58) and (5.61*b*) of the operators $\Phi(x)$ and $\tilde{\Phi}(x)$; the left side of the first of these commutators takes the form

$$[\Phi^+(x), \tilde{\Phi}^-(x')] = (2\pi)^{-3} \int \frac{d^3\boldsymbol{k}}{2\omega}\, d^3\boldsymbol{k}' \tag{5.63a}$$

$$\times e^{i(kx-k'x')}\, (2\pi)^{-5}(2\omega')^{-1}[A^+(\boldsymbol{k}), \tilde{A}^-(\boldsymbol{k}')],$$

while its right side can be written as

$$i\Delta^+(x-x') = (2\pi)^{-3} \int_{k_0=\omega} \frac{d^3\boldsymbol{k}}{2\omega}\, e^{ik(x-x')}. \tag{5.63b}$$

Identifying thus the expression (5.63*a* and *b*), we obtain

$$[A^+(\boldsymbol{k}), \tilde{A}^-(\boldsymbol{k}')] = (2\pi)^5 2\omega\delta(\boldsymbol{k}-\boldsymbol{k}'). \tag{5.64}$$

By introducing the following "reduced operators":

$$a(\boldsymbol{k}) = (2\pi)^{-\frac{5}{2}}\, (2\omega)^{-\frac{1}{2}}\, A(\boldsymbol{k}), \tag{5.65}$$

the relation (5.64) then takes the simpler form

$$[a^+(\boldsymbol{k}), \tilde{a}^-(\boldsymbol{k}')] = \delta(\boldsymbol{k}-\boldsymbol{k}'). \tag{5.66a}$$

The commutation relation between the operators $a(\boldsymbol{k})$ and $\tilde{a}(\boldsymbol{k})$ can be easily obtained from the definitions (5.56) and (5.62) of the operators $A(\boldsymbol{k})$ and $\tilde{A}(\boldsymbol{k})$. We find

$$[\tilde{a}^+(\boldsymbol{k}), a^-(\boldsymbol{k}')] = \delta(\boldsymbol{k}-\boldsymbol{k}'), \tag{5.66b}$$

and

$$[a(\boldsymbol{k}), \tilde{a}(\boldsymbol{k}')] = \delta(\boldsymbol{k}-\boldsymbol{k}'), \tag{5.67}$$

The reader may find in § 6 the representation of the field operators by infinite matrices [formula (5.129) *et seq.*].

Finally, in terms of the reduced operators, the field variables are written as follows:

$$\Phi^{\pm}(x) = (2\pi)^{-\frac{3}{2}} \int_{k_0=\omega} \frac{d^3\boldsymbol{k}}{\sqrt{2\omega}}\, e^{\pm ikx} a^{\pm}(\boldsymbol{k}), \tag{5.68a}$$

$$\tilde{\Phi}^{\pm}(x) = (2\pi)^{-\frac{3}{2}} \int_{k_0=\omega} \frac{d^3\boldsymbol{k}}{\sqrt{2\omega}}\, e^{\pm ikx} \tilde{a}^{\pm}(\boldsymbol{k}). \tag{5.68b}$$

Formulae similar to (3.136) and (3.137) can be easily established for the operators $a^+(\boldsymbol{k})$, $a^-(\boldsymbol{k})$, ... It is clearly sufficient to express the field variables in (3.136) and (3.137) in terms of these operators; we obtain the following formulae

$$[a^-(\boldsymbol{k}), P_\mu] = -k_\mu a^-(\boldsymbol{k}), \quad (5.69a)$$

$$[\tilde{a}^-(\boldsymbol{k}), P_\mu] = -k_\mu \tilde{a}^-(\boldsymbol{k}), \quad (5.69b)$$

$$[a^+(\boldsymbol{k}), P_\mu] = k_\mu a^+(\boldsymbol{k}), \quad (5.69c)$$

$$[\tilde{a}^+(\boldsymbol{k}), P_\mu] = k_\mu \tilde{a}^+(\boldsymbol{k}) \quad (5.69d)$$

and similar formulae for the total charge operator.

It can thus be stated that the operators $a^+(\boldsymbol{k})$ and $\tilde{a}^+(\boldsymbol{k})$ respectively annihilate particles of charge $-e$ and $+e$, whereas the operators $a^-(\boldsymbol{k})$ and $\tilde{a}^+(\boldsymbol{k})$ respectively create particles of charge $+e$ and $-e$, all these particles having moreover $\boldsymbol{k}$ as a common momentum: the names "annihilation and creation operators" are generally reserved for the reduced operators; $a^-(\boldsymbol{k})$ and $\tilde{a}^+(\boldsymbol{k})$ are also said to create and annihilate particles (or antiparticles), whereas $\tilde{a}^-(\boldsymbol{k})$ and $a^+(\boldsymbol{k})$ create and annihilate antiparticles (or particles).†

We now wish to express the physical quantities of the field in terms of these operators. Taking first the operator $\mathcal{N}$, it will be easily seen or directly verified from formulae (3.91) that this operator can be written:

$$\begin{aligned}\mathcal{N} &= -\int d^3\boldsymbol{k}\left[\frac{\tilde{A}^-(\boldsymbol{k})}{\sqrt{(2\pi)^5 2\omega}}\,\frac{A^+(\boldsymbol{k})}{\sqrt{(2\pi)^5 2\omega}} - \frac{\tilde{A}^+(\boldsymbol{k})}{\sqrt{(2\pi)^5 2\omega}}\,\frac{A^-(\boldsymbol{k})}{\sqrt{(2\pi)^5 2\omega}}\right]\\ &= -\int d^3\boldsymbol{k}(\tilde{a}^-(\boldsymbol{k})a^+(\boldsymbol{k}) - \tilde{a}^+(\boldsymbol{k})a^-(\boldsymbol{k}))\\ &= -\int d^3\boldsymbol{k}(\mathcal{N}_-(\boldsymbol{k}) - \mathcal{N}_+(\boldsymbol{k})) = \mathcal{N}_+ - \mathcal{N}_-. \end{aligned} \quad (5.70)$$

The number operator $\mathcal{N}_-$ ("number of negative particles") is a positive definite operator and its expectation value in the vacuum state is zero:

$$\langle 0|\mathcal{N}_-|0\rangle = 0 \quad (5.71)$$

since $a^+(\boldsymbol{k})|0\rangle = 0$. The number operator $\mathcal{N}_+$ ("number of positive particles") is also positive definite, but its vacuum-expectation value is infinite. Now, by virtue of (5.66), we have

$$\tilde{a}^+(\boldsymbol{k})a^-(\boldsymbol{k}') = a^-(\boldsymbol{k}')\tilde{a}^+(\boldsymbol{k}) - \delta(\boldsymbol{k}-\boldsymbol{k}'). \quad (5.72)$$

Consequently,

$$\mathcal{N}_+ = \int a^-(\boldsymbol{k})\tilde{a}^+(\boldsymbol{k})\, d^3\boldsymbol{k} - \delta(0)\int d^3\boldsymbol{k}. \quad (5.73)$$

† To decide on the suitable term to be applied to each type of particle, it is vital to consider the selection rules in the various *decays* which concern it. For instance, in the case of the electron field, the negatron $(-e)$ is the particle, and the positron $(+e)$ the antiparticle. In the case of the proton field, the proton $(+e)$ is the particle and the antiproton $(-e)$, as its name suggests, is the antiparticle.

Apart from the infinite constant $\delta(0)\int d^3\boldsymbol{k}$, we can therefore define a number operator $\mathcal{N}_+ = \int a^-(\boldsymbol{k})\tilde{a}+(\boldsymbol{k})\,d^3\boldsymbol{k}$ such that

$$\langle 0|\mathcal{N}_+|0\rangle = 0. \tag{5.74}$$

To evaluate the energy–momentum operator, we must carry out some relatively easy computations; we find†

$$P_j = \int d^3\boldsymbol{k}k_j(\mathcal{N}_+(\boldsymbol{k})+\mathcal{N}_-(\boldsymbol{k})), \tag{5.75a}$$

$$P_4 = i\int d^3\boldsymbol{k}\omega(\boldsymbol{k})\,(\mathcal{N}_+(\boldsymbol{k})+\mathcal{N}_-(\boldsymbol{k})). \tag{5.75b}$$

The corpuscular interpretation can be readily obtained, and we see that apart from an infinite term introduced by redefining $\mathcal{N}_+$, the operator

$$E = -iP_4$$

is indeed positive definite.‡

It should be noted in passing that this problem of infinite expectation values in the vacuum state can be solved in a formally satisfactory manner by the use of "ordered products", which will be studied in detail in Chap. IX, § 7.

† The following relations are of use in these computations:

$$(kk')\,\delta(\boldsymbol{k}-\boldsymbol{k}') = -m^2\,\delta(\boldsymbol{k}-\boldsymbol{k}')$$
$$(kk')\,\delta(\boldsymbol{k}+\boldsymbol{k}') = (m^2-2\omega^2)\,\delta(\boldsymbol{k}+\boldsymbol{k}'), \quad \text{with} \quad k_0 = \omega.$$

‡ In the case of the Hermitian scalar field, all the above formulae can be simplified; we have

$$\Phi^\pm(x) = (2\pi)^{-\frac{3}{2}}\int_{k_0=\omega}\frac{d^3\boldsymbol{k}}{\sqrt{2\omega}}\,e^{\pm ikx}a^\pm(\boldsymbol{k}), \tag{5.68'}$$

$$a^+(\boldsymbol{k}) = a^-(\boldsymbol{k}))^\sim, \quad a^-(\boldsymbol{k}) = (a^+(\boldsymbol{k}))^\sim, \tag{5.62'}$$

$$[a^+(\boldsymbol{k}), a^-(\boldsymbol{k}')] = \delta(\boldsymbol{k}-\boldsymbol{k}'), \tag{5.66'}$$

$$\mathcal{N} = \int d^3\boldsymbol{k}a^-(\boldsymbol{k})a^+(\boldsymbol{k}), \tag{5.70'}$$

$$P_\mu = \int d^3\boldsymbol{k}k_\mu a^-(\boldsymbol{k})a^+(\boldsymbol{k}). \tag{5.75'}$$

Verifying the characteristic commutation relations of the observable P_μ with the field variables $\Phi^\pm x$, $\Phi(x)$, it will be noted that we have

$$[\mathcal{N}, \Phi^\pm(x)] = \mp\Phi^\pm(x)$$

and

$$[P_0, \mathcal{N}] = 0.$$

By definition, the ordered product of two field variables $\Phi(x)$ and $\Phi(x')$ is in the case of bosons:

$$\begin{aligned} :\Phi(x)\Phi(x'): &\equiv \Phi(x)\Phi(x')+[\Phi^{+}(x), \Phi^{-}(x')] \\ &= \Phi^{+}(x)\Phi^{+}(x')+\Phi^{-}(x)\Phi^{+}(x') \\ &\quad +\Phi^{-}(x')\Phi^{+}(x)+\Phi^{-}(x)\Phi^{-}(x'). \end{aligned} \tag{5.76}$$

We see that all the terms of this product which contain annihilation operators have these operators on the right; it directly follows that

$$\langle 0 | :\Phi(x)\Phi(x'): | 0 \rangle = 0. \tag{5.77}$$

Let us then define, for instance, the Lagrangian density of the neutral field as follows:

$$\mathcal{L} \equiv -\frac{1}{2} : (\Phi_{,\mu} \Phi_{,\mu} + m^2\Phi^2) : . \tag{5.78a}$$

We find the same field equations as previously, but the expressions of all the observables are given in terms of ordered products, for instance:

$$T_{\mu\nu} = :\Phi_{,\mu}\Phi_{,\nu} - \frac{1}{2}\delta_{\mu\nu}(\Phi_{,\sigma}\Phi_{,\sigma} + m^2\Phi^2): \tag{5.78b}$$

and its vacuum-expectation value is zero.

Note finally that the charge conjugate field $\Phi^{(c)}(x)$ [(3.113) and (3.116)] can be defined by

$$\Phi^{(c)}(x) = \tilde{\Phi}(x). \tag{5.79}$$

Field equations and commutation relations are invariant under this transformation, and it can be verified from formulae (5.70) that the charge conjugation exchanges $\mathcal{N}_{+}$ and $\mathcal{N}_{-}$.

Note that a phase factor, which we have arbitrarily chosen equal to +1, can be included in (5.79). The importance of this factor may be easily seen by investigating the transformation properties of the Hermitian fields $\Phi_1(x)$ and $\Phi_2(x)$ of formulae (5.50). Using (3.116) we readily see that

$$\Phi_1^{(c)}(x) = \Phi_1(x), \tag{5.80a}$$

$$\Phi_2^{(c)}(x) = -\Phi_2(x). \tag{5.80b}$$

The choice of the phase in (5.79) therefore obliges us to subject Φ_1 and Φ_2 to different transformations for the charge conjugation.

6. Commutation or anticommutation rules

We saw in Chap. II [formula (2.145)] that on a plane t = constant (or on a space-like surface σ) commutation or anticommutation rules are compatible, but only in a formal manner, with the field equations. The general method used in § 9 of the same chapter, which enabled us to compute the commutators of the wave functions for different times, can easily be extended to anticommutators: the result expressed by formula (2.101) remains unchanged.

But we also noted, in the specific example of a Hermitian scalar field, that this freedom of choice is in the last analysis only apparent and that the existence of the field equations determines the types of commutation rules (in other words the choice between commutators and anticommutators).†

We are now going to show that the structure of the commutation (or anticommutation) relations is itself responsible for this choice.

Consider first of all a Hermitian field quantized by anticommutators

$$[\Phi(x), \Phi(x')]_+ = i\Delta(x-x').$$

Exchange x and x': the anticommutator remains unchanged, whereas the odd function $\Delta(x-x')$ changes its sign. The scalar field cannot therefore be quantized by anticommutation relations.

Consider next a charged field which we again try to quantize by anticommutators

$$[\Phi(x), \tilde{\Phi}(x')]_+ = i\Delta(x-x').$$

Making $x = x'$, we get

$$[\Phi(x), \tilde{\Phi}(x)]_+ = 0.$$

It can be easily seen that the only solution of this relation is $\Phi(x) = 0$. We must therefore again choose commutators. This chain of reasoning will be extended to field quantities of arbitrary spin in § 11 of this chapter.‡

† Cf. p. 86 *et seq.*

‡ Consider the non-relativistic case: $\left\{i\frac{\partial}{\partial t}+\frac{1}{2m}\nabla^2\right\}\Phi(x) = 0$. The commutation relations $[\Phi(x), \Phi(x')]$ can be easily calculated by the method described on p. 172, and taking into account the fact that $\Pi(x) = i\tilde{\Phi}(x)$ (Lagrangian density: $\mathcal{L} = i\tilde{\Phi}\frac{\partial\Phi}{\partial t}-\frac{1}{2m}\nabla\tilde{\Phi}\nabla\Phi$), we see that their right side is

$$(2\pi)^{-3}\int e^{i\left[\boldsymbol{k}(\boldsymbol{x}-\boldsymbol{x}')-\frac{\boldsymbol{k}^2}{2m}(t-t')\right]}d^3\boldsymbol{k}.$$

We verify that the above argument no longer applies: such a field, whose field equation is not invariant under the Lorentz group, can be quantized either by commutators or anticommutators. This example brings out the close relations between the relativistic variance of the field and the statistics it obeys.

7. Field equations of first order for scalar and pseudo-scalar fields

Equation (5.1) can be easily replaced by a system of equations of first order. Indeed, let us set

$$\frac{\partial \Phi}{\partial x_\mu} = -m\Phi_\mu. \tag{5.81a}$$

The equation $(\Box - m^2)\Phi = 0$ is written

$$\frac{\partial \Phi_\mu}{\partial x_\mu} + m\Phi = 0. \tag{5.81b}$$

Equations (5.81*a*) and (5.81*b*) are the wave equations of first order of the scalar field.

In the case of a pseudo-scalar field, we must consider:

(a) the pseudo-scalar $\Phi_{\mu\nu\lambda\varrho}$ which is antisymmetric with respect to all its indices;

(b) the pseudo-vector $\Phi_{\sigma\tau\varrho}$ which is also antisymmetric with respect to all its indices.

Equations (5.81*a*) and (5.81*b*) are then to be replaced by the following ones:

$$\frac{\partial \Phi_{\mu\alpha\beta\gamma}}{\partial x_\mu} + m\Phi_{\alpha\beta\gamma} = 0, \tag{5.82a}$$

$$\frac{\partial \Phi_{\alpha\beta\gamma}}{\partial x_\mu} - \frac{\partial \Phi_{\beta\gamma\mu}}{\partial x_\alpha} + \frac{\partial \Phi_{\gamma\mu\alpha}}{\partial x_\beta} - \frac{\partial \Phi_{\mu\alpha\beta}}{\partial x_\gamma} + m\Phi_{\mu\alpha\beta\gamma} = 0. \tag{5.82b}$$

A detailed discussion of equations of first order for fields of spin 0 will be found in Chap. VI.

8. Vacuum-expectation values of certain bilinear expressions in Φ

It is very easy to obtain by a similar method a formula analogous to

$$\begin{aligned}\langle 0|\Phi(x)\tilde{\Phi}(x')|0\rangle &= \langle 0|\Phi^+(x)\tilde{\Phi}^-(x')|0\rangle \\ &= \langle 0|[\Phi^+(x), \tilde{\Phi}^-(x')]|0\rangle = -i\Delta^+(x-x').\end{aligned} \tag{5.83}$$

We shall see in Chap. IX, Part 2, the important role played by the "chronological product". In the case of a field of spin 0, it can be defined as follows:

$$P\{\Phi(x)\tilde{\Phi}(x')\} = \begin{cases}\Phi(x)\tilde{\Phi}(x') & \text{if } x_0 > x_0', \\ \tilde{\Phi}(x')\Phi(x) & \text{if } x_0 < x_0'.\end{cases} \tag{5.84a}$$

By referring to the properties of the functions $\Delta^+(x)$ and $\Delta^-(x)$ and to the definition of $\Delta_C(x)$, we find that

$$\langle 0|P\{\Phi(x)\tilde{\Phi}(x')\}|0\rangle = \begin{cases} i\Delta^+(x-x') & \text{if } x_0 > x_0' \\ -i\Delta^-(x-x') & \text{if } x_0 < x_0' \end{cases} \tag{5.84b}$$

$$\equiv \frac{1}{2}\Delta_C(x-x') = \lim_{\varepsilon=0} -\frac{i}{(2\pi)^4}\int d^4k\, \frac{e^{ik(x-x')}}{k^2+m^2-i\varepsilon}.$$

The left side of (5.84) will be referred to as the "propagator of a free particle of spin 0".

PART 2

GENERAL FORMALISM

9. Drawbacks of the Hamiltonian formalism

The method described in Chap. II, § 9, connects the determination of the commutation (or anticommutation) rules with the solution of the Hamilton equations. It requires therefore the calculation of the conjugate variables; we shall illustrate the difficulty of this calculation by considering eqn. (4.12*c*) which we shall derive from the Lagrangian density (4.36) which has the form

$$\mathcal{L} = -\tilde{Q}(x)\eta\beta_\mu Q,_\mu(x) - m\tilde{Q}(x)\eta Q(x). \tag{5.85}$$

The conjugate momenta are

$$\Pi(x) = -i\frac{\partial\mathcal{L}}{\partial Q_{,4}} = i\tilde{Q}(x)\eta\beta_4, \tag{5.86a}$$

$$\tilde{\Pi}(x) = 0. \tag{5.86b}$$

We note, first of all, that only $Q(x)$ and $\Pi(x)$ are canonical variables: indeed no conjugate variable corresponds to $\tilde{Q}(x)$.† The solution of the Hamilton equations must be expressed solely by means of the canonically independent variables $Q(x)$ and $\Pi(x)$ and the variables $\tilde{Q}(x)$ must therefore be eliminated. Now, the Hamiltonian density is written

$$\Lambda(x) = \Pi_A\dot{Q}_A - \mathcal{L}(\Pi_A, Q_A), \tag{5.87}$$

$\dot{Q}(x)$ is expressed in terms of $Q(x)$ from the equation

$$i\beta_4\frac{\partial Q(x)}{\partial t} = (\beta_j\partial_j + m)Q(x) \tag{5.88}$$

† Indeed, this difficulty is not a real one and a slight change in the form of (5.85) [cf. for instance (5.138)] would avoid it.

and $\tilde{Q}(x)$ must be expressed in terms of $\Pi(x)$ with the aid of the relation

$$\tilde{Q}(x)\eta\beta_4 = -i\Pi(x). \tag{5.89}$$

In simple examples (for instance the case of particles of spin $\frac{1}{2}$) the solution of these equations presents no difficulty, but this is no longer the case when one deals with particles of higher spins with β_μ's obeying more complicated algebraic relations.

In particular, by using this method one cannot state easily and *a priori* the commutation rules between field variables.

The question may now arise whether it is not possible to avoid resorting to the canonical formalism and to form the commutation rules directly from the action. We shall now see that such a method can be formulated.

10. General commutation rules

We propose therefore to solve the following problem:

Supposing the commutation rules for a field of simple type to be known (in all cases, we shall choose the scalar field), find the commutation rules for other fields by a change of the field variables.†

Let us carry out the following change of variables:

$$Q_A(x) = D_{AA'}(\partial)q_{A'}(x), \tag{5.90a}$$

where $D_{AA'}(\partial)$ is a polynomial (or an analytical function) in ∂_μ.

We clearly have

$$\tilde{Q}_A(x) = \{D_{AA'}(\partial)\}^*\tilde{q}_{A'}(x). \tag{5.90b}$$

Suppose now the commutation rules of the fields $q(x)$ to be known:

$$C_{AB}(x-x') = [q_A(x), \tilde{q}_B(x')] \tag{5.91}$$

and let us try to infer from them the commutation rules of the field $Q(x)$:

$$\mathcal{C}_{AB}(x-x') = [Q_A(x), \bar{Q}_B(x')] \equiv [Q_A(x), \tilde{Q}_{B'}(x')]\eta_{B'B}. \tag{5.92}$$

By virtue of the transformation (5.90), we have

$$\begin{aligned} \mathcal{C}_{AB}(x-x') &= D_{AA'}(\partial)C_{A'B''}(x-x')\{D_{B'B''}(\overleftarrow{\partial}')\}^*\eta_{B'B} \\ &= D_{AA'}(\partial)C_{A'B''}(x-x')\{D^T_{B''B'}(\overleftarrow{\partial}')\}^*\eta_{B'B}. \end{aligned} \tag{5.93}$$

Setting

$$\tilde{D}(\partial) = \{D^T(\partial)\}^*, \tag{5.94}$$

we finally obtain

$$\mathcal{C}(x-x') = D(\partial)C(x-x')\tilde{D}(\overleftarrow{\partial}')\eta. \tag{5.95}$$

† H. Umezawa and A. Visconti, *Nucl. Phys.*, **1**, 20 (1956).

On the other hand, let the action describing the field $Q(x)$ be:

$$\mathfrak{A}[Q] = \int \tilde{Q}(x)\eta\mathcal{F}(\partial)Q(x)\, d^4x; \tag{5.96}$$

the change of variables (5.90) expresses this action through the field $q(x)$:†

$$\begin{aligned}\mathfrak{A}[Q] &= \int (\{D_{A''A'}(\partial)\}^*\tilde{q}_{A'}(x))\eta_{A''A}\mathcal{F}_{AB}(\partial)\, D_{BB'}(\partial)q_{B'}(x)\, d^4x \\ &= \int \tilde{q}_{A'}(x)\{D_{A''A'}(-\partial)\}^*\eta_{A''A}\mathcal{F}_{AB}(\partial)\, D_{BB'}(\partial)q_{B'}(x)\, d^4x,\end{aligned} \tag{5.97}$$

where the second integral is inferred from the first one by an integration by parts.

Introducing the adjoint matrix of $D(\partial)$ [formula (5.94)], the action can finally be expressed in the form

$$\mathfrak{A}[Q] = \int \tilde{q}(x)\, \tilde{D}(-\partial)\eta F(\partial)\, D(\partial)q(x)\, d^4x. \tag{5.98}$$

Consequently, one gets the wave equations:

$$\left.\begin{aligned}\tilde{D}(-\partial)\eta\mathcal{F}(\partial)D(\partial)q(x) &= 0,\\ \tilde{q}(x)\tilde{D}(\underleftarrow{\partial})\eta\mathcal{F}(-\underleftarrow{\partial})D(-\underleftarrow{\partial}) &= 0,\end{aligned}\right\} \tag{5.99}$$

which are indeed mutually conjugate by virtue of formula (4.39).

We have now to determine $D(\partial)$ such that

$$\tilde{D}(-\partial)\eta\mathcal{F}(\partial)D(\partial) = (\Box - m^2)I, \tag{5.100a}$$

an expression which can also take the form

$$\mathcal{F}(\partial)D(\partial)\tilde{D}(-\partial)\eta = (\Box - m^2)I. \tag{5.100b}$$

We can therefore determine $\mathcal{D}(\partial)$ by the following equation:

$$D(\partial)\tilde{D}(-\partial)\eta = \mathcal{D}(\partial), \tag{5.101}$$

where $\mathcal{D}(\partial)$, a differential operator of $2S$th order for a field of spin S,‡ is defined by (4.21) or (4.23).

The action governing the field $q(x)$ is now

$$\mathfrak{A}[Q] = \sum_A \int \tilde{q}_A(x)(\Box - m^2)q_A(x)\, d^4x. \tag{5.102}$$

† The bracket in the first integral of (5.97) means that the differential operator $\{D_{A''A'}(\partial)\}^*$ only acts on $\tilde{q}_A(x)$.

‡ We ought to have assumed that $\mathcal{D}(\partial)$ is an analytical function of ∂ and shown, using the relativistic variance of the commutator, that it degenerates into a polynomial of degree $2S$. We did in fact prove this in Chap. IV, § 6, by investigating the structure of the commutation rule (5.103). It need hardly be said that the vicious circle is only an apparent one here.

It can be quantized by the same methods as were used for the charged scalar field; we readily find

$$[q_A(x), \tilde{q}_B(x')] = i\delta_{AB}\Delta(x-x').$$

Formula (5.95) is then written simply

$$[Q(x), \bar{Q}(x')] = iD(\partial)\tilde{D}(-\partial)\eta\Delta(x-x') = i\mathcal{D}(\partial)\Delta(x-x') \quad (5.103)$$

and completely solves the problem we set ourselves at the beginning of this section. We shall use this method to compute the commutators of the various fields we shall be encountering in the next chapter. In the case of the scalar field, we have

$$\mathcal{F}(\partial) = \Box - m^2.$$

Consequently, $\mathcal{D}(\partial) = 1$ and we are led to the commutation rules previously obtained.

It must be noted finally that the relation (5.103) is invariant under any proper Lorentz transformation:

$$x \to X = Lx, \qquad Q'(X) = \mathcal{J}(L)Q(x).$$

We have indeed

$$[Q'_A(X), \bar{Q}'_B(X')] = \mathcal{J}_{AA'}[Q_{A'}(x), \bar{Q}_{B'}(x')]\mathcal{J}^{-1}_{\bar{B}'B}$$
$$= i\{\mathcal{J}\mathcal{D}(\partial)\mathcal{J}^{-1}\}_{AB}\,\Delta(x-x'). \quad (5.104a)$$

Making use of formula (4.15)

$$\mathcal{J}\beta_\mu\partial_\mu\mathcal{J}^{-1} = L_{\nu\mu}\beta_\nu\partial_\mu = \beta_\nu\,\frac{\partial}{\partial X_\nu}$$

and of the Lorentz invariance of $\Delta(x)$,

$$\Delta(x) = \Delta(X),$$

we easily verify that

$$\mathcal{J}\mathcal{D}(\partial)\mathcal{J}^{-1}\Delta(x-x') = \mathcal{D}\left(\frac{\partial}{\partial X_\mu}\right)\Delta(X-X'), \quad (5.104b)$$

and that does in fact prove the invariance of the commutation rules.

These considerations may be extended to the full Lorentz group, when we take into account the remarks following formula (4.47); we shall verify in particular that if the time reversal is represented by a unitary transformation, we must take $\bar{Q}'(X) = \bar{Q}(x)\mathcal{J}^{-1}$, since $\Delta(x)$ is an odd function of time.

11. Anticommutation rules

Formulae (2.94) *et seq.* allow us to express (5.104) as follows:

$$[Q_A(x),\ \bar{Q}_B(x')] = -i\mathcal{D}_{AA'}(\partial)W^{(\Pi)}(t, t', \nabla)\ \delta_{A'B}\ \delta(\boldsymbol{x}-\boldsymbol{x}') \qquad (5.105)$$
$$= -i\mathcal{D}_{AA'}(\partial)W^{(\Pi)}(t, t', \nabla)[q_{A'}(\boldsymbol{x}, t), \Pi_B(\boldsymbol{x}', t)].$$

We have seen on the other hand (2.145) that on a plane t=constant, we can equally well choose anticommutation or commutation relations (bearing in mind all the reservations already expressed). The above formula therefore shows that (5.103) can be extended in the form

$$[Q_A(x), \bar{Q}_B(x')]_+ = i\mathcal{D}_{AB}(\partial)\ \Delta(x-x'), \qquad (5.106)$$

provided this extension is compatible with the structure of the function $\mathcal{D}(\partial)\Delta(x-x')$.

In order to investigate this point, consider an integer spin field and endeavour to quantize it by anticommutators. In Chap. IV, § 9, we saw that such a field can be represented by a set of variables with well-defined tensor variances.

Insert the expression (4.61*a*) of $Q(x)$ and $\bar{Q}(x)$ into (5.106); the spurs method as described in this section allows us to obtain anticommutation rules for each of the tensor components of the field quantities. We shall obtain, for instance:

$$[\varphi_A(x), \tilde{\varphi}_B(x')]_+ = id_{AB}(\partial)\Delta(x-x'),$$

where A and B stand for collections of indices. It is now clear that, taking into account the variance of the term $\beta_\sigma\partial_\sigma\Delta$ (p. 144), the tensor $d(\partial)$ is of even order, therefore its diagonal terms are necessarily even functions of ∂.

Adopting a method similar to the one used in § 6 of this chapter, let $A=B$ in the preceding anticommutator, then exchange x and x' and add the two commutators thus obtained; then by making $x=x'$, we obtain

$$[\varphi_A(x), \tilde{\varphi}_A(x)]_+ = 0,$$

which leads to $\varphi_A(x)=0$. We therefore conclude that integer spin fields must necessarily be quantized with the help of commutators. In § 14 we shall discuss the relations between spins and statistics.†

† Cf. also W. Pauli, *Phys. Rev.*, **58,** 716 (1940), who deals exclusively, as we have just been doing, with free fields. For a study of this question within the framework of axiomatic formalism, the reader is referred to Chap. XI, § 16.

12. Elementary solutions of the wave equations of arbitrary spin fields

The relation

$$\mathcal{F}(\partial)\mathcal{D}(\partial) = (\Box - m^2)I \tag{5.107a}$$

gives a method of obtaining solutions of the homogeneous equation

$$\mathcal{F}(\partial)G(x) = 0, \tag{5.107b}$$

and elementary solutions of the differential operator $\mathcal{F}(\partial)$.

Indeed, the function

$$G^{(e)}(x) \equiv \mathcal{D}(\partial)\Delta^{(e)}(x) \tag{5.108}$$

obeys the equation

$$\mathcal{F}(\partial)G^{(e)}(x) = B(x),$$

where the right side $B(x)$ stands for $(\Box - m^2)\Delta^{(e)}(x)$, $\Delta^{(e)}(x)$ being one of the functions $\Delta_{\text{ret}}(x)$, $\Delta_{\text{adv}}(x)$, ..., which we studied in § 2 of this chapter.

Let us set, for instance,

$$G(x) = \mathcal{D}(\partial)\,\Delta(x).$$

$G(x)$ is then a solution of the homogeneous equation (5.107*b*), whereas by setting

$$G^{\text{ret}}(x) = \mathcal{D}(\partial)\,\Delta_{\text{ret}}(x),$$

we obtain an elementary solution

$$\mathcal{F}(\partial)\,\Delta_{\text{ret}}(x) = -\delta(x-x'). \tag{5.109}$$

Note that *we generally do not have*

$$G^{\text{ret}}(x) = \theta_+(t)G(x)$$

except when $\mathcal{D}(\partial)$ is a constant (scalar field), or a differential polynomial of first order (field of spin $\frac{1}{2}$). This can be easily verified by using a calculation we shall be carrying out in Chap. VII, § 10, for more general Green functions.

13. Reduced operators and observables of a field

We are now going to investigate the various physical quantities associated with the free field $Q(x)$ and show that they can easily be expressed with the aid of the field $q(x)$ introduced in (5.90*a*). It is simpler to use, as in Chap. II, the momentum space. By a Fourier transformation, formula (5.90*a*) can be written in the form:

$$Q_A(k) = D_{AA'}(ik)q_{A'}(k). \tag{5.110}$$

Let P'_μ be then the displacement operator of the field $q(x)$:

$$k_\mu q_{A'}(k) = [q_{A'}(k), P'_\mu)]. \tag{5.111a}$$

Multiply the two sides of the preceding equation by $D_{AA'}(ik)$ and sum over A'; we obtain

$$k_\mu Q_A(k) = [Q_A(k), P_\mu]. \tag{5.111b}$$

If P_μ is the displacement operator of the field $Q(x)$, we also have

$$k_\mu Q_A(k) = [Q_A(k), P_\mu]. \tag{5.112}$$

Since these same relations hold for $\tilde{q}_A(k)$ and $\tilde{Q}_A(k)$, we infer that the operator $P'_\mu - P_\mu$ commutes with all the components of the field variable and its Hermitian conjugate. Since this operator is independent of x, it also commutes with all the derivatives, products and powers of the field variables: it therefore commutes with the conjugate momenta. Consequently, $P'_\mu - P_\mu$ is a c-number:

$$P'_\mu - P_\mu = C,$$

the existence of the vacuum for the system described by $Q(x)$ or by $q(x)$ is expressed as follows:

$$C = \langle 0 | P'_\mu - P_\mu | 0 \rangle = 0.$$

Thus the operator P_μ is the displacement operator of the field $q(x)$. It can be proved in similar fashion that e_{tot} is the total charge operator of this same field $q(x)$.

The analytical function $D(ik)$ must be determined from formula (5.101). Taking into account the fact that ∂_4 is purely imaginary, we find

$$D(ik)\,\tilde{D}(ik)\eta = \mathcal{D}(ik), \tag{5.113}$$

and this is moreover a condition concerning $\mathcal{D}(ik)$, namely that $\mathcal{D}(ik)\eta^{-1}$ is positive definite.

The field variables $Q(x)$ may now be expressed in terms of the reduced operators $q(x)$; we have only to consider (5.90) and the Fourier transforms of the q_A's as given in (5.68). We obtain

$$\begin{aligned} Q_A(x) &= D_{AA'}(\partial) q_{A'}(x) \\ &= (2\pi)^{-\frac{3}{2}} \int_{k_0=\omega} \frac{d^3\boldsymbol{k}}{\sqrt{2\omega}} \{D_{AA'}(ik) a^+_{A'}(\boldsymbol{k}) e^{ikx} + D_{AA'}(-ik) a^-_{A'}(\boldsymbol{k}) e^{-ikx}\} \\ &\equiv Q^+_A(x) + Q^-_A(x) \end{aligned} \tag{5.114a}$$

and

$$\begin{aligned} \tilde{Q}_A(x) &= \tilde{q}_{A'}(x)\, D^*_{AA'}(\overleftarrow{\partial}) \\ &= (2\pi)^{-\frac{3}{2}} \int_{k_0=\omega} \frac{d^3\boldsymbol{k}}{\sqrt{2\omega}} \{\tilde{a}^+_{A'}(\boldsymbol{k})\, \tilde{D}_{A'A}(-ik) e^{ikx} + \tilde{a}^-_{A'}(\boldsymbol{k}) \tilde{D}_{A'A}(ik) e^{-ikx}\} \\ &= \tilde{Q}^+_A(x) + \tilde{Q}^-_A(x). \end{aligned} \tag{5.114b}$$

The only non-zero commutators of the reduced variables are according to formulae (5.66) the following ones:

$$[a_A^+(\boldsymbol{k}), \tilde{a}_B^-(\boldsymbol{k}')] = \delta_{AB}\delta(\boldsymbol{k}-\boldsymbol{k}'), \tag{5.115a}$$

$$[\tilde{a}_A^+(\boldsymbol{k}), a_B^-(\boldsymbol{k}')] = \delta_{AB}\delta(\boldsymbol{k}-\boldsymbol{k}'). \tag{5.115b}$$

Finally, the definition of the vacuum state [cf. (3.141)] yields:

$$a^+(\boldsymbol{k})\,|\,0\rangle = 0, \qquad \tilde{a}^+(\boldsymbol{k})\,|\,0\rangle = 0. \tag{5.115c}$$

We now note that the remarks at the beginning of this section enable us to write observables of the field $Q(x)$ directly: they are identical with those of the field $q(x)$.

Consider the operator $\mathcal{N}$ [cf. (3.93)]; by virtue of (5.70) we have

$$\mathcal{N} = \int_{k_0=\omega} d^3\boldsymbol{k}[a_B^-(\boldsymbol{k})\tilde{a}_B^+(\boldsymbol{k}) - \tilde{a}_B^-(\boldsymbol{k})a_B^+(\boldsymbol{k})]$$

$$= \int_{k_0=\omega} d^3\boldsymbol{k}(\mathcal{N}_+(\boldsymbol{k}) - \mathcal{N}_-(\boldsymbol{k})). \tag{5.116a}$$

Similarly,

$$P_\mu = \int_{k_0=\omega} d^3\boldsymbol{k}k_\mu(\mathcal{N}_+(\boldsymbol{k}) + \mathcal{N}_-(\boldsymbol{k})). \tag{5.116b}$$

It will be noted that the observables thus defined have a zero expectation value in the vacuum state; their expression has therefore been obtained by the method described at the end of § 5 of this chapter for the elimination of the infinite value given by a careless calculation (cf. also the end of this section).

Taking up again the arguments used for the scalar field and those of Chap. III, § 11, we can conclude that the operators $a^+(\boldsymbol{k})$ and $\tilde{a}^+(\boldsymbol{k})$ respectively annihilate a particle of charge $-e$ and a particle of charge $+e$, whereas the operators $a^-(\boldsymbol{k})$ and $\tilde{a}^-(\boldsymbol{k})$ respectively create a particle of charge $+e$ and a particle of charge $-e$.

The name *annihilation and creation operators* is generally reserved for these reduced operators.

The above results may be generalized to anticommuting fields in the following manner. Note first of all that for such fields formula (5.113) must be extended. Consider indeed the anticommutator [cf. (5.103)]:

$$[\bar{Q}(x), Q(x')]_+$$
$$= i\mathcal{D}(\partial)\,\Delta(x-x')$$
$$= (2\pi)^{-3}\int_{k_0=\omega} \frac{d^3\boldsymbol{k}}{2\omega}\,[-\mathcal{D}(-ik)e^{-ik(x-x')} + \mathcal{D}(ik)e^{ik(x-x')}]. \tag{5.117}$$

Express $Q(x)$ and $\bar{Q}(x)$ with the aid of the reduced operators $\tilde{a}^{\pm}(\boldsymbol{k})$ and $a^{\pm}(\boldsymbol{k})$ [formulae (5.114)], obeying the relations (5.115) where the commutators are to be replaced by anticommutators

$$\left.\begin{aligned}[a_A^+(\boldsymbol{k}), \tilde{a}_{\bar{B}}^-(\boldsymbol{k}')]_+ &= \delta_{AB}\delta(\boldsymbol{k}-\boldsymbol{k}'),\\ [\tilde{a}_A^+(\boldsymbol{k}), a_{\bar{B}}^-(\boldsymbol{k}')]_+ &= \delta_{AB}\delta(\boldsymbol{k}-\boldsymbol{k}').\end{aligned}\right\} \qquad (5.118a)$$

It is then easy to see that the above substitution is only possible if the relation (5.113) is replaced by the two following ones:

$$D(\pm ik)\tilde{D}(\pm ik)\eta = \pm\mathcal{D}(\pm ik), \qquad k_0 = +\omega(\boldsymbol{k}). \qquad (5.118b)$$

It can then be verified that, in virtue of the preceding anticommutation relations, the expressions (5.116) of $\mathcal{M}$ and P_μ are still valid.†

The problem of the infinite vacuum-expectation values which we have assumed to be completely solved by writing the observables directly in the form (5.116) can also be solved with the aid of the ordered products, as we saw in the case of the scalar field (end of § 5). For integer spin fields, it is sufficient to apply the method established for the scalar field: for half-integer spin fields, one has merely to define the ordered product of two field variables by the following formula [in place of (5.76)]:

$$:Q_A(x)\bar{Q}_B(x'): = Q_A(x)\bar{Q}_B(x') - [Q_A^+(x), \bar{Q}_B(x')]_+; \qquad (5.119a)$$

we then have the essential property of the ordered products:

$$\langle 0 | :Q_A(x)\bar{Q}_B(x'): | 0\rangle = 0 \qquad (5.119b)$$

and the Lagrangian density can be written as in (5.78*a*) with the aid of ordered products. We shall come back to this point in the particular case of the Dirac field.

14. Commutation rules and statistics

We are now going to prove that if the field variables obey commutation rules, an indefinite number of indiscernible particles can be created in the same state (Bose–Einstein statistics), whereas if they obey anticommutation rules, only one particle can be created in each state (we then have the Pauli exclusion principle and the Fermi–Dirac statistics).

† The expression that would be obtained by replacing in P_μ inferred from the Lagrangian formalism [as given, for instance, by formula (5.140)] the field variables by the reduced variables would make P_0 a Hermitian operator, but with positive and negative eigenvalues. One gets indeed:

$$P_0 = \int d^3\boldsymbol{k}\,\omega(\boldsymbol{k})\left(\tilde{a}_{\bar{B}}^-(\boldsymbol{k})a_B^+(\boldsymbol{k}) - \tilde{a}_B^+(\boldsymbol{k})a_{\bar{B}}^-(\boldsymbol{k})\right),$$

an observable whose expectation value in the vacuum state is infinite. The sign of the second term of P_0 can be changed by making use of the anticommutation rules (5.118).

Indeed, in the first case the creation and annihilation operators may act n times on the vacuum in order to obtain a state with n particles. But if, on the contrary, we have anticommutators, the operator $a^-(\boldsymbol{k})$ acting twice on the vacuum gives zero:

$$a^-(\boldsymbol{k})a^-(\boldsymbol{k})\,|\,0\rangle = \frac{1}{2}\,[a^-(\boldsymbol{k}),\, a^-(\boldsymbol{k})]_+\,|\,0\rangle = 0.$$

It is therefore impossible to create two particles in the same state: the statistics agree with the Pauli exclusion principle.

Comparing this result with those obtained in § 12, we can finally conclude:

Any integer spin field must be quantized with the aid of commutators and the particles which it represents obey the Bose statistics.

Half-integer spin fields are then quantized by anticommutators and obey Fermi–Dirac statistics.

15. Sets of eigenvectors of the observables $\mathcal{N}, P_\mu$ and e_{tot}

We showed in Chap. III, § 11, that eigenvectors of some observables could be formed by making products of operators $Q_A^-(\boldsymbol{k})$ act on the vacuum.† The fact that the operators $\mathcal{N}$, P_μ and e_{tot} can be expressed by means of the reduced variables only proves that we can form eigenvectors of these three observables by making products of operators $a_A^-(\boldsymbol{k})$ and $\tilde{a}_A^-(\boldsymbol{k})$ (respectively creation operators of particles of charge $+e$ and $-e$) act on the vacuum.

Let us first consider, for simplicity's sake, the general formulae in the case of a Hermitian scalar field. The creation and annihilation operators of such a field which obey $[a^+(\boldsymbol{k}), a^-(\boldsymbol{k}')] = \delta(\boldsymbol{k}-\boldsymbol{k}')$ [cf. note ‡, p. 180] enable us to form the eigenvectors of $\mathcal{N} = \int d^3\boldsymbol{k}a^-(\boldsymbol{k})a^+(\boldsymbol{k})$ and $P\mu = \int d^3\boldsymbol{k}k_\mu a^-(\boldsymbol{k})a^+(\boldsymbol{k})$, where $k_0 = +\omega(\boldsymbol{k})$. By setting

$$\left.\begin{aligned} &|\,p\rangle \equiv a^-(\boldsymbol{p})\,|\,0\rangle, \qquad |\,p_1, p_2\rangle \equiv a^-(\boldsymbol{p}_1)a^-(\boldsymbol{p}_2)\,|\,0\rangle, \\ &|\,p_1, p_2, p_3\rangle \equiv a^-(\boldsymbol{p}_1)a^-(\boldsymbol{p}_2)a^-(\boldsymbol{p}_3)\,|\,0\rangle, \qquad \ldots, \end{aligned}\right\} \quad (5.120a)$$

† V. Fock, *Zeits. Phys.*, **75**, 622 (1932); M. Jean, *Thèse*, Masson, 1953; G. Bodiou, *Journal Phys.*, **15**, 39 (1954); A. S. Wightman and S. S. Schweber, *Phys. Rev.*, **98**, 812 (1955); J. G. Valatin, *Journal Phys.*, **12**, 134 (1951); D. Kastler, *Introduction à l'électrodynamique quantique*, Dunod, 1960. The space we are going to construct is called Fock space.

with

$$\left.\begin{array}{lll}\mathcal{N}|p\rangle = |p\rangle, & \mathcal{N}|p_1, p_2\rangle = 2|p_1, p_2\rangle, & \ldots; \\ P_\mu|p\rangle = p_\mu|p\rangle, & P_\mu|p_1, p_2\rangle = (p_{1\mu}+p_{2\mu})|p_1, p_2\rangle, & \ldots\end{array}\right\} \quad (5.120b)$$

these eigenvectors which are symmetric with respect to $p_1, \ldots, p_n$, obey the relations

$$\left.\begin{array}{c}\langle p'|p\rangle = \delta(\mathbf{p}'-\mathbf{p}), \\ \langle p_2', p_1'|p_1, p_2\rangle = \delta(\mathbf{p}_1'-\mathbf{p}_1)\delta(\mathbf{p}_2'-\mathbf{p}_2)+\delta(\mathbf{p}_1'-\mathbf{p}_2)\delta(\mathbf{p}_2'-\mathbf{p}_1); \\ \cdots\cdots\cdots\cdots\cdots\cdots\cdots\cdots\end{array}\right\} \quad (5.121)$$

in particular two vectors $|p_1, p_2, \ldots, p_n\rangle$ and $|p_1, \ldots, p_m\rangle$ are orthogonal provided that $m \neq n$ whatever the value of the total momenta $p_1+p_2+p_3+\ldots$ may be, since they are non-degenerate eigenvectors of $\mathcal{N}$. It may be shown that they form a complete system.

With the aid of this basis a state vector $|\Psi\rangle$ normalized to 1 can be formed; indeed, by writing

$$|\Psi\rangle = |0\rangle\, C_0 + \frac{1}{\sqrt{1!}}\int |p\rangle c(p)\, d^3\mathbf{p}$$

$$+\frac{1}{\sqrt{2!}}\int |p_1, p_2\rangle c(p_1, p_2)\, d^3\mathbf{p}_1 d^3\mathbf{p}_2 + \ldots \quad (5.122a)$$

with $c(p_1, p_2), \ldots, c(p_1, \ldots, p_n)$, symmetric functions of $p_1, p_2, \ldots, p_n$, the coefficients c obey the normalization condition

$$1 = \langle\Psi|\Psi\rangle = |C_0|^2 + \int |c(p)|^2\, d^3\mathbf{p} + \int |c(p_1, p_2)|^2\, d^3\mathbf{p}_1 d^3\mathbf{p}_2 + \ldots . \quad (5.122b)$$

This formula proves moreover that $|c(p)|^2, |c(p_1, p_2)|^2, \ldots$ are the densities of the transition probabilities from the states $\frac{1}{\sqrt{1!}}|p\rangle$, $\frac{1}{\sqrt{2!}}|p_1, p_2\rangle$, $\frac{1}{\sqrt{n!}}|p_1, \ldots, p_n\rangle$, to the state $|\Psi\rangle$.

To these formulae may be added the following one:

$$\begin{aligned} a^+(\mathbf{q})a^-(\mathbf{p}_1)\ldots a^-(\mathbf{p}_n)|0\rangle &= \delta(\mathbf{q}-\mathbf{p}_1)a^-(\mathbf{p}_2)\ldots a^-(\mathbf{p}_n)|0\rangle \\ &\quad + \delta(\mathbf{q}-\mathbf{p}_2)a^-(\mathbf{p}_1)a^-(\mathbf{p}_3)\ldots a^-(\mathbf{p}_n)|0\rangle + \ldots \end{aligned} \quad (5.123)$$

which can be proved recurrently and enables one to calculate the scalar products (5.121).

The extension of the above results to non-Hermitian integer spin fields (quantized by commutators) merely calls for a simple and direct generalization. We shall also content ourselves with a few simple formulae to enable the reader to write the extension of the above formulae to the case of arbitrary spin fields.

Define the basis of eigenvectors common to the operators e_{tot} and P_μ:

$$\left.\begin{aligned} |-e, A, p\rangle &= \tilde{a}_{\bar{A}}(\boldsymbol{p})|0\rangle, \\ |+e, A, p\rangle &= a_{\bar{A}}(\boldsymbol{p})|0\rangle, \end{aligned}\right\} \tag{5.124}$$

$$|-e, A_1, p_1; -e, A_2, p_2\rangle = \tilde{a}_{\bar{A}^1}(\boldsymbol{p}_1)\tilde{a}_{\bar{A}_2}(\boldsymbol{p}_2)|0\rangle, \ldots$$

and so on. For half-integer spin fields, the above vectors are antisymmetric in order to take into account the properties concerning the permutation of two particles (or antiparticles). Formulae analogous to (5.121) are obtained from the commutation or anticommutation rules (5.115) and (5.118). We have, for instance,

$$\left.\begin{aligned} &\langle p', A', -e|-e, A, p\rangle = \delta_{AA'}\delta(\boldsymbol{p}'-\boldsymbol{p}), \\ &\langle p', A', +e|-e, A, p\rangle = 0, \\ &\langle p_2', A_2', -e; p_1', A_1', -e|-e, A_1, p_1; -e, A_2, p_2\rangle \\ &= \delta_{A_1'A_1}\delta_{A_2'A_2}\delta(\boldsymbol{p}_2'-\boldsymbol{p}_2)\delta(\boldsymbol{p}_1'-\boldsymbol{p}_1)\mp\delta_{A_1'A_2}\delta_{A_2'A_1}\delta(\boldsymbol{p}_2'-\boldsymbol{p}_1)\delta(\boldsymbol{p}_1'-\boldsymbol{p}_2), \\ &\langle p_2', A_2', +e; p_1', A_1', -e|-e, A_1, p_1; +e, A_2, p_2\rangle \\ &= \delta_{A_1'A_1}\delta_{A_2'A_2}\delta(\boldsymbol{p}_1'-\boldsymbol{p}_1)\delta(\boldsymbol{p}_2'-\boldsymbol{p}_2), \\ &\ldots\ldots\ldots\ldots\ldots\ldots\ldots\ldots\ldots\ldots\ldots\ldots \end{aligned}\right\} \tag{5.125}$$

where the minus sign concerns the fermions. A formula analogous to (5.124) can be obtained, in which the coefficients $C_{A_1A_2A_3}\ldots(p_1, p_2, \ldots)$ are antisymmetric with respect to their arguments A and p when the field investigated is a fermion field. We obtain a formula similar to (5.123) with alternating plus and minus signs for the case of fermions.

16. Construction of a representation of the creation and annihilation operators

In order to obtain convenient formulae, we can imagine that the entire system is contained in a cube with side L and volume $V=L^3$ and that the operators obey periodical boundary conditions on the surface of this cube. Suppose for greater simplicity that the index A represents a single (spin) index varying from 1 to N.

The Fourier integrals expressing $Q(k)$ in terms of $Q(x)$ became Fourier series with

$$\boldsymbol{k}^{(l)} = \left(m_1^{(l)}\frac{2\pi}{L},\ m_2^{(l)}\frac{2\pi}{L},\ m_3^{(l)}\frac{2\pi}{L}\right), \tag{5.126a}$$

where the $m_j^{(l)}$'s take the values 0, 1, 2, . . . :

$$(2\pi)^{-3}\int d^3\boldsymbol{k} \rightarrow \frac{1}{V}\sum_k, \quad (2\pi)^{-3}\int d^3\boldsymbol{x} \rightarrow \frac{1}{V}\int_V d^3\boldsymbol{x}, \qquad (5.126b)$$

$$\frac{(2\pi)^3}{V}\,\delta(\boldsymbol{k}^{(l)}-\boldsymbol{k}^{(l')}) \rightarrow \delta_{\boldsymbol{k}^{(l)}\boldsymbol{k}^{(l')}}$$

and the latter formula is a consequence of the following equality:†

$$\frac{1}{V}\int_V d^3\boldsymbol{x}\, e^{i(\boldsymbol{k}-\boldsymbol{k}')\boldsymbol{x}} = \delta_{\boldsymbol{k}\boldsymbol{k}'}.$$

The vectors $k^{(l)}$ and the indices A thus form a denumerable set; they can be labelled by a single index r so that the operator $a_A(\boldsymbol{k}^{(l)})$ is to be replaced by the operator a_r.

We are now interested in building up sets of operators a_r and $\tilde{a}_s$ obeying

$$[a_r, \tilde{a}_s] \;= \delta_{rs}, \qquad (5.127a)$$

$$[a_r, \tilde{a}_s]_+ = \delta_{rs} \qquad (5.127b)$$

† The above results may be justified by appealing to the relations between Fourier series and integrals. For the sake of simplicity, let $f(x)$ be a periodical function of the one-dimensional variable x and let its period be L; its Fourier expansion is written

$$f(x) = \frac{1}{L}\sum_n \int_{-\frac{L}{2}}^{+\frac{L}{2}} f(\xi)e^{-in\frac{2\pi}{L}\xi}\,d\xi\, e^{in\frac{2\pi}{L}x} = \frac{1}{L}\sum_n \int_{-\frac{L}{2}}^{+\frac{L}{2}} f(\xi)\, e^{+ik^{(n)}(x-\xi)}\,d\xi$$

where as above $k^{(n)} = n\dfrac{2\pi}{L}$. By setting

$$\Delta k^{(n)} = k^{(n-1)} - k^{(n)} = \frac{2\pi}{L},$$

the above summation becomes an integral

$$f(x) = \frac{1}{2\pi}\int\!\!\int f(\xi)e^{ik(x-\xi)}\,d\xi\,dk,$$

whence the rule $V \rightarrow (2\pi)^3$ of (5.126). We are finally led to compute sums of the form $\sum_p f(k^{(p)})$ (for instance, a transition probability per unit of time) with the help of the formula

$$\sum_p f(k^{(p)}) = \sum_p f\left(p\frac{2\pi}{L}\right) = \frac{L}{2\pi}\sum_p f\left(p\frac{2\pi}{L}\right)\frac{2\pi}{L} = \int f(k)\varrho(k)\,dk,$$

where the density of states is $\varrho(k) = \dfrac{L}{2\pi}$. In the volume V this density function is $\varrho(k) = \left(\dfrac{L}{2\pi}\right)^3 = \dfrac{V}{(2\pi)^3}$.

respectively for the bosons and fermions by means of Kronecker products. Consider the case of bosons; denote by a the matrix with an infinite number of rows and columns:

$$a \equiv \begin{pmatrix} 0 & 1 & 0 & 0 & \dots \\ 0 & 0 & \sqrt{2} & 0 & \dots \\ 0 & 0 & 0 & \sqrt{3} & \dots \\ . & . & . & .. & \dots \end{pmatrix}. \tag{5.128}$$

Consider the following matrices:

$$a_r = I \otimes I \otimes \dots \otimes a \otimes \dots \otimes I, \tag{5.129a}$$

$$\tilde{a}_r = I \otimes I \otimes \dots \otimes \tilde{a} \otimes \dots \otimes I, \tag{5.129b}$$

where the matrices a and $\tilde{a}$ are the rth respective factors. The matrix elements (5.129) are expressed as follows:

$$(a_r)_{N_1, N_2, \dots, N_r, \dots; N_1', \dots, N_r', \dots} = \sqrt{N_r'}\,\delta_{N_r, N_r'-1} \prod_{l \neq r} \delta_{N_l, N_l'}, \tag{5.130a}$$

$$(\tilde{a}_r)_{N_1, N_2, \dots, N_r, \dots; N_1', \dots, N_r', \dots} = \sqrt{N_r}\,\delta_{N_r-1, N_r'} \prod_{l \neq r} \delta_{N_l, N_l'}, \tag{5.130b}$$

and

$$(\tilde{a}_r a_r)_{N_1, \dots, N_r, \dots; N_1', \dots, N_r', \dots} = N_r \prod_{l \neq r} \delta_{N_l, N_l'}. \tag{5.130c}$$

It is then easy to verify that they indeed obey the relations (5.127a) and are therefore a solution of the problem. It can be shown that any other irreducible representation of these matrices (let a_r' and $\tilde{a}_r'$ be the new matrices) is equivalent to the first one†

$$a_r' = S a_r S^{-1}, \quad \tilde{S} = S^{-1} \tag{5.131a}$$

provided that the vacuum state can be defined

$$\sum_{r=0}^{\infty} \tilde{a}_r a_r |0\rangle = 0. \tag{5.131b}$$

The representation of these same matrices for the fermions is a more delicate matter. Adopting Wigner's procedure, we can choose the matrices

$$\alpha = \begin{pmatrix} 0 & 1 \\ 0 & 0 \end{pmatrix}, \quad \tilde{\alpha} = \begin{pmatrix} 0 & 0 \\ 1 & 0 \end{pmatrix}, \quad \beta = \begin{pmatrix} 1 & 0 \\ 0 & -1 \end{pmatrix} \tag{5.132}$$

which obey the relations

$$[\alpha, \tilde{\alpha}] = I, \quad \beta = 1 - 2\tilde{\alpha}\alpha, \tag{5.133}$$

† J. von Neumann, *Compos. Math.*, **6**, 1 (1938); L. van Hove, *Physica*, **18**, 145 (1952); Gärding and Wightman, *Proc. Nat. Acad. Sc.*, **40**, 617 (1954).

and we define the matrices α_r and β_r:

$$\left.\begin{aligned}\alpha_r &= I\otimes I\otimes\ldots\otimes\alpha\otimes\ldots\otimes I,\\ \beta_r &= I\otimes I\otimes\ldots\otimes\beta\otimes\ldots\otimes I,\end{aligned}\right\}\qquad(5.134)$$

where the factors α and β occupy the rth place. It can then be verified that the matrices

$$a_r = \alpha_r \prod_{q=1}^{r-1} \beta_q, \qquad (5.135a)$$

$$\tilde{a}_r = \tilde{\alpha}_r \prod_{q=1}^{r-1} \beta_q, \qquad (5.135b)$$

do obey the commutation relations and that the uniqueness condition of the representation is still given by formula (5.131a).

Note in conclusion that all the calculations we shall be led to perform in the course of this work will be carried out in such a way as to avoid the need to resort to these explicit representations.

PART 3

GENERAL FORMALISM (Continued)

Before investigating them further, let us sum up the properties and formulae so far obtained concerning the field $(\beta_\mu\partial_\mu+m)Q(x) = 0$.

We have seen that by setting

$$\bar{Q}(x) = \tilde{Q}(x)\eta, \qquad (4.41)$$

with

$$\eta = \tilde{\eta}, \quad \tilde{\beta}_j = -\eta\beta_j\eta^{-1}, \quad \tilde{\beta}_4 = \eta\beta_4\eta^{-1}, \qquad (4.40)$$

we had

$$\bar{Q}(x)\{\beta_\mu\overleftarrow{\partial}_\mu - m\} = 0 \qquad (4.42)$$

and that the field equations could be derived from the action.†

$$\mathfrak{A} = -\int \bar{Q}(x)\{\beta_\mu\partial_\mu+m\}Q(x)\, d^4x.$$

We have also seen that under a proper Lorentz transformation (and with certain reservations, under a transformation of the full group), we had

$$Q'(x') = \mathfrak{J}Q(x), \qquad (4.13)$$

$$\bar{Q}'(x') = \bar{Q}(x)\mathfrak{J}^{-1}. \qquad (4.47)$$

† For the Hermitian character of the Lagrangian density, cf. note on p. 147.

We finally introduced the differential operator $\mathcal{D}(\partial)$ of $2S$th order for a field of spin S such that

$$-\{\beta_\mu\partial_\mu+m\}\mathcal{D}(\partial) = (\Box-m^2)I \tag{4.20}$$

and

$$\mathcal{D}(\partial) = m-\beta_\sigma\partial_\sigma-\frac{I}{m}\{\Box-\beta_{\sigma_1}\beta_{\sigma_2}\partial_{\sigma_1}\partial_{\sigma_2}\}+\ldots. \tag{4.23}$$

This operator enabled us:

(a) to obtain information concerning the algebra of the $\beta\mu$'s (Chap. IV, § 6);

(b) to write the elementary solutions of this wave equation (5.108);

(c) to write the commutation or anticommutation rules of the field variables (5.103) and (5.106).

We now intend to study some other characteristic properties of this field.

17. Observables of the field $(\beta_\mu\partial_\mu+m)Q(x) = 0$

Note first that the operators $Q^+(k)$ and $Q^-(k)$ defined by formulae (5.114) respectively obey equations

$$\left.\begin{aligned} i\beta_\mu k_\mu Q^+(k) &= -mQ^+(k),\\ i\beta_\mu k_\mu Q^-(k) &= +mQ^-(k),\\ (k_\mu k_\mu+m^2)Q^\pm(k) &= 0. \end{aligned}\right\} \tag{5.136}$$

The following relations can be easily verified making use of the expression of $\mathcal{D}(ik)$ obtained by Fourier transforming (4.23):

$$\left.\begin{aligned} &\frac{1}{2m}\mathcal{D}(ik)Q^+(k) = Q^+(k), && \frac{1}{2m}\mathcal{D}(ik)Q^-(k) = 0,\\ &\frac{1}{2m}\mathcal{D}(-ik)Q^+(k) = 0, && \frac{1}{2m}\mathcal{D}(-ik)Q^-(k) = Q^-(k). \end{aligned}\right\} \tag{5.137}$$

It is, moreover, from these properties that we shall determine in the case of the Dirac field the matrices $D(\pm ik)$ [cf. formula (6.1.92) *et seq.*].

In order to calculate the field observables $Q(x)$, we shall use the following Lagrangian density:

$$\mathcal{L} = -\frac{1}{2}(\bar{Q}\beta_\mu\partial_\mu Q-\bar{Q}\beta_\mu\overleftarrow{\partial}_\mu Q)-m\bar{Q}Q; \tag{5.138}$$

it differs from the one previously used [formula (4.36)] up to a four-dimensional divergence: it therefore leads to the same wave equations (Chap. I, end of § 7) and its Hermitian character allows us to obtain in addition observables which are Hermitian (particularly for the spin tensor).

We obtain first for the energy–momentum tensor:

$$T_{\mu\nu} = \frac{1}{2}(\bar{Q}\beta_\nu Q_{,\mu} - \bar{Q}_{,\mu}\beta_\nu Q)$$
$$= \bar{Q}(x)\beta_\nu Q_{,\mu}(x) - \frac{1}{2}\frac{\partial}{\partial x_\mu}\{\bar{Q}(x)\beta_\nu Q(x)\}, \tag{5.139}$$

making use of the wave equations. The expression of the total momentum operator can then be easily written

$$P_\mu = -i\int T_{\mu 4}\, d^3\boldsymbol{x} = -\frac{i}{2}\int (\bar{Q}\beta_4 Q_{,\mu} - \bar{Q}_{,\mu}\beta_4 Q)\, d^3\boldsymbol{x} \tag{5.140a}$$

or again in accordance with the current conservation

$$P_\mu = -i\int \bar{Q}\beta_4 Q_{,\mu}\, d^3\boldsymbol{x}. \tag{5.140b}$$

We verify by a partial integration that the components P_j are Hermitian; the anti-Hermitian character of P_4 can also be established by a very easy calculation; we have

$$\tilde{P}_4 = i\int \partial_4^* \tilde{Q}.\tilde{\beta}_4 \eta Q\, d^3\boldsymbol{x} = -i\int \partial_4 \tilde{Q}.\eta\beta_4\eta^{-1}\eta\, Q d^3\boldsymbol{x}$$
$$= -i\partial_4 \int \bar{Q}(x)\beta_4 Q(x)\, d^3\boldsymbol{x} + i\int \bar{Q}(x)\beta_4\partial_4 Q(x)\, d^3\boldsymbol{x}$$

and the first term is again zero by virtue of (4.45).

The angular momentum tensor splits into an orbital angular momentum tensor and a spin tensor; the orbital tensor (3.52*b*) can be easily written:

$$L_{\mu\nu} = -i\int \bar{Q}(x)\,\beta_4\{x_\nu\partial_\mu - x_\mu\partial_\nu\}\, Q(x)\, d^3\boldsymbol{x}. \tag{5.141}$$

An important remark must be made concerning the spin tensor (3.52*c*): in accordance with Note 2, Chap. II, § 5, we are treating the fields Q and $\bar{Q}$ as independent fields; a term with $\bar{Q}$ should therefore be added to formula (3.52*c*), as has in fact been done in all the preceding formulae.

But in the case of the spin tensor, it must also be borne in mind that this additive term must be calculated with $-J_{\mu\nu}$ instead of $+J_{\mu\nu}$, since by virtue of (4.47), under a Lorentz transformation, $\bar{Q}(x)$ is transformed into $\bar{Q}'(x') = \bar{Q}(x)\mathcal{J}^{-1}$.

We therefore have

$$\mathfrak{M}_{\mu\nu\lambda} = \frac{\partial \mathscr{L}}{\partial Q_{A,\lambda}} J_{\mu\nu} Q_A - \bar{Q}_A J_{\mu\nu} \frac{\partial \mathscr{L}}{\partial \bar{Q}_{A,\lambda}} = -\frac{1}{2} \bar{Q}[J_{\mu\nu}, \beta_\lambda]_+ Q, \quad (5.142)$$

as may be easily verified.

The expression of $\mathfrak{M}_{\mu\nu\lambda}$ thus obtained leads us to form the symmetrical energy–momentum tensor (3.55). We shall leave it to the reader to verify the Hermitian character of the corresponding observables. We shall apply these formulae to the Dirac field and then examine their properties in detail.

The expression of the current can be written down very easily, by virtue of formulae (3.91) and (3.94); we have†

$$J_\mu(x) = -ie\bar{Q}\beta_\mu Q, \quad (5.143)$$

an expression which leads to the total charge operator or in any case to the number operators for particles and antiparticles.

It can be easily proved that the first three components of the observable $J_\mu(x)$ are Hermitian and that the fourth is anti-Hermitian: we have, indeed,

$$\tilde{J}_\mu(x) = \{-ie\bar{Q}_A(x)(\beta_\mu)_{AB} Q_B(x)\}^\sim = ie\tilde{Q}_B(\beta_\mu)^*_{AB}\eta^*_{A'A} Q_{A'}$$
$$= ie\tilde{Q}_B(\eta\beta_\mu)^*_{A'B} Q_{A'}.$$

Now,

$$(\eta\beta_\mu)^*_{A'B} = (\eta\tilde{\beta}_\mu)_{BA'} = (\tilde{\beta}_\mu\eta)_{BA'};$$

taking (4.40*b*) into account, we verify that

$$\tilde{J}_k(x) = J_k(x), \qquad \tilde{J}_4(x) = -J_4(x).$$

The expressions of all these observables may be written without the least difficulty in terms of the reduced operators. We shall consider this problem in detail in connection with the Dirac field.

18. Charge conjugation

In Chap. III, § 10, we introduced the notion of charge conjugation. We are interested in defining a canonical transformation which introduces a field $Q^{(c)}(x)$ which obeys eqns. (3.114) and (3.115) if the field is charged, or similar formulae where e_{tot} is replaced by $\mathcal{N}$ if the field is neutral. In all cases, we shall say that the fields quantities $Q^{(c)}(x)$ describe *antiparticles*.

We represent the charge conjugation by the matrix $\mathscr{C}$ acting on the spin indices, such that

$$Q_A(x) \rightarrow Q^{(c)}_A(x) = \mathscr{C}_{AA'}\tilde{Q}_{A'}(x). \quad (5.144)$$

† Cf. also the invariant form under charge conjugation (5.162) with $e \rightarrow -e$.

We submit $Q^{(c)}(x)$ to obey the same equation as $Q(x)$:

$$\{\beta_\mu \, \partial_\mu + m\} Q^{(c)}(x) = 0; \tag{5.145}$$

then, denoting the spinor transposition by T, we multiply on the left by $(C^T)^{-1}$. It can then be easily seen that the preceding equation is equivalent to the following one:

$$\tilde{Q}(x)\{\mathcal{C}^T \beta_\mu^T (\mathcal{C}^T)^{-1} \overleftarrow{\partial}_\mu + m\} = 0. \tag{5.146}$$

Now, $\tilde{Q}(x)$ obeys the following equation:

$$\tilde{Q}(x)\{\tilde{\beta}_j \partial_j - \tilde{\beta}_4 \overleftarrow{\partial}_4 + m\} = 0, \tag{5.147}$$

by equating the last two equations and transposing the relations thus obtained, we finally obtain which relations $\mathcal{C}$ must obey:

$$\mathcal{C}^{-1} \beta_j \mathcal{C} = \beta_j^*, \qquad \mathcal{C}^{-1} \beta_4 \mathcal{C} = -\beta_4^*, \tag{5.148}$$

where the operation * denotes the complex conjugation (not to be confused with the Hermitian conjugation).

It is usual to introduce the matrix

$$\mathcal{C} = C\eta^*, \tag{5.149}$$

which leads to a more symmetric expression of formulae (5.148):

$$C^{-1} \beta_\mu C = -\beta_\mu^T. \tag{5.150}$$

This formula shows that $C^T C^{-1}$ commutes with all the β_μ's. If therefore the β_μ's are the matrices of an irreducible system, $C^T C^{-1}$ is necessarily a multiple of the identity (Schur lemma). This condition can in fact be required in all cases without contradicting (5.150).†

We shall now define the wave function $\overline{Q}^{(c)}(x)$:

$$\overline{Q}_B^{(c)} = \tilde{Q}_{B'}^{(c)} \eta_{B'B}^{(c)} = \mathcal{C}^*_{B'B''} Q_{B''} \eta_{B'B}^{(c)}, \tag{5.151}$$

where $\eta^{(c)}$ is a matrix.‡

In order to determine the conditions with regard to $\eta^{(c)}$, let us write that $\overline{Q}^{(c)}(x)$ obeys the same equation as $\overline{Q}(x)$, in other words

$$\mathcal{C}^*_{BB'} Q_{B'}(x) \eta_{BA'}^{(c)} \{-\beta_\mu \overleftarrow{\partial}_\mu + m\}_{A'A} = 0, \tag{5.152}$$

an equation which can also take the form:

$$\{-\beta_\mu \partial_\mu + m\}^T_{AA'} \eta_{A'B}^{(c)T} \mathcal{C}^*_{BB'} Q_{B'}(x) = 0.$$

† We are therefore free to introduce in the formula (5.144) a phase factor λ_c ($|\lambda_c| = 1$) which will define the "particle conjugation parity". If we further assume that by twice applying the particle conjugation operation to a given field, one gets back to the original field, then $\lambda_c = \pm 1$.

‡ Remarks similar to those made on η' [p. 148, note ‡] can be applied to $\eta^{(c)}$.

Introducing the matrix C and taking (5.150) into account, we get

$$C^{-1}\{\beta_\mu\partial_\mu+m\}C\eta^{(c)T}C^*\eta Q(x) = 0,$$

and this equation can be verified provided that

$$C\eta^{(c)T}C^*\eta = I, \tag{5.153}$$

a condition which $\eta^{(c)}$ must obey and which must be compatible with the relations (4.40) ensuring the Hermitian character of the action.†

The expressions of $\bar{Q}^{(c)}$ and $Q^{(c)}$ in terms of Q and $\bar{Q}$ can then be obtained:‡

$$\bar{Q}^{(c)}_A = \mathscr{C}^*_{BB'}Q_{B'}\eta^{(c)}_{BA} = (C^*\eta)_{BB'}Q_{B'}\eta^{(c)}_{BA} = (\eta^{(c)T}C^*\eta)_{AB'}Q_{B'} = (Q^TC^{-1})_A, \tag{5.154a}$$

$$Q^{(c)}_A = \mathscr{C}_{AA'}\tilde{Q}_{A'} = (C\eta^*)_{AA'}\tilde{Q}_{A'} = C_{AA'}\bar{Q}_{A''}\eta^{*T}_{A'A''} = (C\bar{Q}^T)_{A'}, \tag{5.154b}$$

as can be easily verified.

The invariance of the commutation rules under a charge conjugation will provide us with a further condition for C. Writing down the commutation or anticommutation rules for the fields $Q^{(c)}$:

$$[Q^{(c)}_A(x), \bar{Q}^{(c)}_B(x')]_\pm = \pm C_{AA'}C^{-1}_{\bar{B}B'}[Q_{B'}(x'), \bar{Q}_{A'}(x)]_\pm .$$

Consequently, we have

$$\begin{aligned}[Q^{(c)}_A(x), \bar{Q}^{(c)}_B(x')] &= \pm iC_{AA'}\mathscr{D}^T_{A'B'}(\partial')C^{T-1}_{B'B}\,\Delta(x'-x)\\ &= \mp iC_{AA'}\mathscr{D}^T_{A'B'}(-\partial)C^{T-1}_{B'B}\,\Delta(x-x'),\end{aligned} \tag{5.155a}$$

an expression which must be identical to

$$i\mathscr{D}_{AB}(\partial)\,\Delta(x-x'). \tag{5.155b}$$

Consider now the explicit form (4.23) of $\mathscr{D}(\partial)$; by identifying the two terms in m, we obtain the condition

$$C^T = \mp C \tag{5.156}$$

where we must choose the minus sign for fermion fields and the plus sign for boson fields. By taking a matrix C which obeys (5.156), the identity of the other terms of the expressions (5.155*a*) and (5.155*b*) can be easily proved.

This is a particularly important result: it shows in fact that a theory enabling particles and antiparticles to be defined and containing a conjugation matrix C can only be quantized in agreement with the relation (5.156). The charge conjugation thus helps determine the type of commu-

† Note that we can have an arbitrary matrix in the right side of (5.153), a matrix commuting with the four β_μ's (and therefore a multiple of the identity when the β_μ's form an irreducible set of matrices, as is the case in the Dirac equation).

‡ For the scalar field: $\Phi^{(c)}(x) = \tilde{\Phi}(x)$, cf. formula (5.79) *et seq.*

tation rules to be adopted. In particular, for the Dirac field, C can be easily determined and we can verify that $C^T = -C$: this field must therefore be quantized by anticommutators.†

Taking up an argument already used, we shall infer from the fact that field equations and commutation rules remain invariant under the charge conjugation, that this is a canonical transformation in the space $\mathcal{R}$. A unitary operator Γ known as the "charge parity operator" can then be defined such that

$$Q^{(c)}(x) = \Gamma Q(x)\Gamma^{-1}. \tag{5.157}$$

We also assume that

$$\Gamma|0\rangle = |0\rangle. \tag{5.158}$$

Let us finally investigate the variance of the expectation values of observables. We shall suppose that the state vector Ψ is independent of t, in other words we shall adopt the Heisenberg picture, which amounts to considering the field under investigation as a free field.

Take the total charge operator; by virtue of formula (3.1), we have

$$\begin{aligned}\langle e_{\text{tot}}\rangle^{(c)} &= \langle \Psi|\Gamma\left(-e\int \bar{Q}(x)\beta_4 Q(x)\, d^3\boldsymbol{x}\right)\Gamma^{-1}|\Psi\rangle \\ &= -e\langle\Psi|\int \bar{Q}^{(c)}(x)\beta_4 Q^{(c)}(x)\, d^3\boldsymbol{x}\,|\Psi\rangle. \end{aligned} \tag{5.159}$$

Reverse formulae (5.154):

$$Q_B(x) = C_{BA}\bar{Q}_A^{(c)}(x), \qquad \bar{Q}_B(x) = C^{-1}_{BA}Q_A^{(c)}(x) \tag{5.160}$$

and insert them into the expression of $\langle e_{\text{tot}}\rangle^{(c)}$; we obtain by using (5.156):

$$\begin{aligned}\langle e_{\text{tot}}\rangle^{(c)} &= -e\langle\Psi|\int C^{-1}_{AA'}Q_{A'}(x)(\beta_4)_{AB}C_{BB'}\bar{Q}_{B'}(x)\, d^3\boldsymbol{x}\,|\Psi\rangle \\ &= \pm e\langle\Psi|\int Q_A(x)(C^{-1}\beta_4 C)_{AB}\bar{Q}_B(x)\, d^3\boldsymbol{x}\,|\Psi\rangle,\end{aligned}$$

where the plus and minus signs respectively concern fermion fields and boson fields. On the other hand, apart from an additive infinite term, we have‡

$$Q_A(x)\bar{Q}_B(x) \simeq \mp\bar{Q}_B(x)Q_A(x),$$

† This point was investigated by S. Watanabe, *Phys. Rev.*, **84**, 1008 (1951). Cf. also the remarks at the end of this chapter.

‡ It is perhaps worth noting that we only obtain $J_\mu^{(c)} = -J_\mu$ in a quantized theory.

where the minus and plus signs are respectively related to the fermions and bosons. Thus, making use of (5.150), we finally obtain apart from an infinite additive term:

$$\langle e_{\text{tot}}\rangle^{(c)} \simeq +\langle \Psi \,|\, e \int \bar{Q}(x)\beta_4 Q(x)\, d^3\boldsymbol{x} \,|\, \Psi\rangle = -\langle e_{\text{tot}}\rangle, \qquad (5.161)$$

the charge conjugation therefore changes the sign of the charge by replacing particles by antiparticles. We obtain an expression which no longer includes any infinite term by defining the current as follows:

$$J_\mu(x) = -\frac{ie}{2}\left(\bar{Q}(x)\beta_\mu Q(x) - \bar{Q}^{(c)}\beta_\mu Q^{(c)}(x)\right), \qquad (5.162a)$$

in place of the expression (5.143). This current changes its sign under a charge conjugation: moreover, in view of the fact that the fields $Q(x)$ and $Q^{(c)}(x)$ obey the same equations and the same commutation rules, we infer by splitting the field $Q^{(c)}(x)$ into reduced variables, that

$$\langle 0 \,|\, J_\mu(x) \,|\, 0\rangle = 0. \qquad (5.162b)$$

The current $J_\mu(x)$ can also take the form:

$$\begin{aligned} J_\mu(x) &= -\frac{ie}{2}\left(\bar{Q}_A(x)(\beta_\mu)_{AB} Q_B(x) \pm Q_B(x)(\beta_\mu^T)_{BA}\bar{Q}_A(x)\right) \\ &= -\frac{ie}{2}\,[\bar{Q}_A(x), Q_B(x)]_\pm (\beta_\mu)_{AB}, \end{aligned} \qquad (5.163)$$

where anticommutator and commutator correspond respectively to bosons and fermions; and it can be shown by a simple calculation that the definitions (5.162) and (5.163) of the current are equivalent.

As we have already noted [cf. (5.119)] the difficulties involved in computing the vacuum-expectation values can be avoided by using ordered products: an expression identical to (5.163) is thus obtained.

Repeating a method described on p. 115, we can, moreover, easily prove that if we set

$$P_\mu^{(c)} = \Gamma P_\mu \Gamma^{-1}, \qquad e_{\text{tot}}^{(c)} = \Gamma e_{\text{tot}} \Gamma^{-1},$$

we have

$$P_\mu^{(c)} = P_\mu, \qquad e_{\text{tot}}^{(c)} = -e_{\text{tot}}$$

by virtue of the canonical equations. We shall prove it for instance for the total charge observable, thus confirming the relations (5.161).

Writing (3.99*a*) as transformed by the operator Γ, namely

$$e\Gamma Q_A(x)\Gamma^{-1} = [e_{\text{tot}}^{(c)}, \Gamma Q_A(x)\Gamma^{-1}],$$

then making use of (5.157), (5.154) and (3.93*b*), we express the above relation in the form

$$C_{AA'}[e_{\text{tot}}^{(c)}+e_{\text{tot}}, \bar{Q}_{A'}(x)] = 0,$$

which leads to the equality stated between $e_{\text{tot}}^{(c)}$ and e_{tot}. The relation $P_\mu^{(c)} = P_\mu$ shows that Γ commutes with P_4: it is therefore a constant of motion in the Heisenberg picture (free fields). This operator defines "good quantum numbers" and makes it possible to state selection rules.

A close connection must finally be noted between charge conjugation and the Lorentz group: it is expressed in the first place by the simple fact that the charge conjugation is reflected by an operation in the spin space. In the second place, the charge conjugation must be an invariant operation under the Lorentz group; we must therefore have both

$$Q^{(c)\prime}(x') = \mathcal{J}Q^{(c)}(x), \quad \text{and} \quad Q^{(c)\prime}(x') = C\bar{Q}'(x').$$

With the aid of formulae (5.154*b*) and (4.47) the last two formulae can be expressed in terms of $\bar{Q}(x)$, so that we find between $\mathcal{J}$ and C the compatibility relation

$$C = \mathcal{J}C\mathcal{J}^T. \tag{5.164a}$$

We can easily verify that under proper Lorentz transformations, it is equivalent to

$$-J_{\mu\nu}^T = C^{-1}J_{\mu\nu}C. \tag{5.164b}$$

19. Space symmetry

We introduced space symmetries in Chap. III, § 7; the existence of a field equation assumed to be invariant (cf. also Chap. IV, § 16) will enable us to obtain a number of properties of the matrix σ.

Let us write that the equation of the field variable $Q^{(\sigma)}(x')$ has the same form as the equation from which we started. Expressing the variables x' in terms of the variables x and multiplying on the left the equation thus obtained by σ^{-1}, we get

$$\{-\sigma^{-1}\beta_j\sigma\,\partial_j+\sigma^{-1}\beta_4\sigma\,\partial_4+m\}Q(x) = 0. \tag{5.165}$$

In order that this equation be identical to the equation of $Q(x)$, it is sufficient that

$$\sigma^{-1}\beta_j\sigma = -\beta_j, \qquad \sigma^{-1}\beta_4\sigma = \beta_4 \tag{5.166a}$$

or

$$[\beta_j, \sigma]_+ = 0, \qquad [\beta_4, \sigma]_- = 0. \tag{5.166b}$$

We easily infer from the above equations that

$$[\sigma^2, \beta_\mu] = 0, \tag{5.167}$$

from which we can conclude that, when the β_μ's form a set of irreducible matrices (for instance, in the Dirac equation), σ^2 is a multiple of the identity operator. Note also that σ must obey (5.164*a*) expressing the invariance of the charge conjugation under the Lorentz transformations.†

It is also easy to verify that the commutation rules are invariant under space symmetries. We first obtain, starting from (4.47):

$$\bar{Q}^{(\sigma)}(x') = \bar{Q}(x)\sigma^{-1}. \tag{5.168}$$

Denoting by x and x' the transformed points, the commutation rules are written

$$\begin{aligned}[Q_A^{(\sigma)}(x), \bar{Q}_B^{(\sigma)}(x')]_\pm &= \sigma_{AA'}[Q_{A'}(-\boldsymbol{x}, x_0), \bar{Q}_{B'}(-\boldsymbol{x}', x_0')]_\pm \sigma^{-1}_{B'B} \\ &= i\{\sigma\mathcal{D}(-\nabla, \partial_4)\sigma^{-1}\}_{AB}\,\Delta(-\boldsymbol{x}+\boldsymbol{x}', x_0-x_0').\end{aligned}$$

Making use of the fact that $\Delta(x-x')$ is an odd function in $\boldsymbol{x}-\boldsymbol{x}'$, the preceding expression takes the form

$$[Q^{(\sigma)}(x), \bar{Q}^{(\sigma)}(x')]_\pm = i\sigma\mathcal{D}(-\nabla, \partial_4)\sigma^{-1}\,\Delta(x-x'). \tag{5.169}$$

The right side of (5.169) is equal to $\mathcal{D}(\partial)\,\Delta(x-x')$; indeed, a term such as $\beta_\mu\partial_\mu$ has been transformed by a space symmetry into $\{-\beta_j\partial_j+\beta_4\partial_4\}$ and we have

$$\sigma\{-\beta_j\partial_j+\beta_4\partial_4\}\sigma^{-1} = \beta_\mu\partial_\mu. \tag{5.170}$$

Referring to the expression (4.23) of $\mathcal{D}(\partial)$, we easily infer the desired equality. The above considerations therefore enable us to regard a space symmetry as a canonical operation of the space $\mathcal{R}$ and to introduce the unitary operator Π given by (3.78).

20. Time reversal

Proceeding as in the above section, we can easily represent the time reversal

$$\boldsymbol{x} \to \boldsymbol{x}' = \boldsymbol{x}, \qquad x_0 \to x''_0 = -x_0$$

by a linear transformation $\tau: Q^{(\tau)}(x') = \tau Q(x)$. We find that

$$[\beta_j, \tau]_- = 0, \qquad [\beta_4, \tau]_+ = 0 \tag{5.171}$$

and we must have [cf. remarks following formula (4.47)]:

$$\bar{Q}^{(\tau)}(x') = -\bar{Q}(x)\tau^{-1}. \tag{5.172}$$

But when we come to study the canonical equations [formula (5.182) *et seq.*], we shall see that such a linear representation does not generate a

† There still remains, as in the previous section, an arbitrary phase factor which specifies the "intrinsic parity" of the field (cf. eqn. (6.1.14) for the case of Dirac field).

canonical transformation in the space $\mathcal{R}$. We shall therefore express a time reversal by an antilinear transformation, as Wigner has suggested.†

Let us therefore represent the time reversal by the transformation

$$(Q_A^{(t)}(x') = T_{AA'}\tilde{Q}_{A'}(x), \tag{5.173}$$

where T is a matrix in the space of the A's (spin space) and the Hermitian conjugation is to be taken with respect to the space $\mathcal{R}$.

Since the field equation must be invariant, we obtain as in the preceding section:

$$\{T^{-1}\beta_j T\partial_j - T^{-1}\beta_4 T\partial_4 + m\}_{AA'}\tilde{Q}_{A'}(x) = 0,$$

and by comparing this equation with the equation of $\tilde{Q}(x)$:

$$\{\beta_j^*\partial_j - \beta_4^*\partial_4 + m\}_{AA'}\tilde{Q}_{A'}(x) = 0,$$

we finally have:

$$T^{-1}\beta_\mu T = \beta_\mu^*. \tag{5.174}$$

It will be noted that the operation denoted by the symbol * in the above formula is simply the complex conjugation (and not the Hermitian conjugation in the space of the indices A) and that TT^* which commutes with the four β_μ's can be made equal to a multiple of the identity matrix.

We now define

$$\bar{Q}^{(t)}(x) = \tilde{Q}^{(t)}(x)\eta^{(t)}$$

and propose to determine $\eta^{(t)}$ by writing that $\bar{Q}^{(t)}(x)$ obeys the same equation as $\bar{Q}(x)$:

$$\bar{Q}^{(t)}(x)\{\beta_\mu \overleftarrow{\partial}_\mu - m\} = 0.$$

A few simple transformations enable us to express this equation in the form

$$\tilde{Q}(x)T^T\eta^{(t)*}T^{-1}\{\beta_\mu \overleftarrow{\partial}_\mu - m\} = 0.$$

We can therefore take

$$T^T\eta^{(t)*}T^{-1} = \alpha\eta, \tag{5.175}$$

where α is a real or complex number.‡

Going back to the definition of $\bar{Q}^{(t)}(x)$, we get

$$\begin{aligned}\bar{Q}_A^{(t)}(x) &= \tilde{Q}_{A'}^{(t)}(x)\eta_{A'A}^{(t)} = T^*_{A'A''}Q_{A''}(x)\eta^{(t)}_{A'A}\\ &= (\tilde{T}\eta^{(t)})_{A''A}Q_{A''}(x) = \alpha^*(\eta T)^*_{A''A}Q_{A''}(x)\\ &= \alpha^*(\tilde{Q}_{A''}(x)\eta_{A''A'})^{\sim}T^*_{A'A} = \alpha^*(\bar{Q}_{A'}(x))^{\sim}T^*_{A'A}.\end{aligned}$$

By making T [cf. the remark following (5.174)] obey

$$\alpha^*TT^* = I, \tag{5.176}$$

we finally obtain

$$\bar{Q}_A^{(t)}(x') = \bar{Q}_{\tilde{A}'}(x)T^{-1}_{A'A}. \tag{5.177}$$

† E. P. Wigner, *Nachr. Akad. Wiss., Göttingen*, 546 (1932).

‡ We verify that $\alpha = -1$ for a field of spin $\frac{1}{2}$ (6.1.16*b*) and 1 for a scalar field.

Let us now investigate the commutation or anticommutation rules according to the spin†‡

$$\begin{aligned}[Q_A^{(t)}(x), \bar{Q}_B^{(t)}(y)]_\pm &= T_{AA'}[\tilde{Q}_{A'}(\boldsymbol{x}, -x_0), \bar{Q}_{\tilde{B}'}(\boldsymbol{y}, -y_0)]_\pm T^{-1}_{\bar{B}'B} \\ &= T_{AA'}[\bar{Q}_{B'}(\boldsymbol{y}, -y_0), Q_{A'}(\boldsymbol{x}, -x_0)]^{\sim}_{\pm} T^{-1}_{\bar{B}'B} \\ &= \pm T_{AA'}[Q_{A'}(\boldsymbol{x}, -x_0), \bar{Q}_{B'}(\boldsymbol{y}, -y_0)]^{\sim}_{\pm} T^{-1}_{\bar{B}'B} \\ &= \mp i T_{AA'} \mathcal{D}^*_{A'B'}(\nabla, -\partial_4) T^{-1}_{\bar{B}'B} \Delta^*(\boldsymbol{x}-\boldsymbol{y}, -(x_0-y_0)) \\ &= \pm i \mathcal{D}_{AB}(\partial) \Delta(x-y), \end{aligned} \tag{5.178}$$

where we have taken into account the two well-known properties of $\Delta(x)$:

$$\Delta^*(x) = \Delta(x) \quad \text{and} \quad \Delta(\boldsymbol{x}, -x_0) = -\Delta(x).$$

It is obvious that (5.178) cannot be obeyed by writing that $Q^{(t)}(x)$ and $\bar{Q}^{(t)}(x)$ are canonically transformed variables of $Q(x)$ and $\bar{Q}(x)$. But we shall prove that if, following Wigner, we set:

$$\left.\begin{aligned} Q_A^{(t)}(x) &= \mathcal{T}\tilde{Q}_A(x)\mathcal{T}^{-1}, \\ \bar{Q}_A^{(t)}(x) &= \mathcal{T}\bar{Q}_A^{\sim}(x)\mathcal{T}^{-1}, \end{aligned}\right\} \tag{5.179}$$

where $\mathcal{T}$ is an antiunitary operator,|| eqn. (5.178) can be identically ful-

† The Hermitian conjugation is performed in $\mathcal{R}$.

‡ The former consideration applied to the scalar charged field gives

$$\Phi^{(t)}(x') = \tilde{\Phi}(\boldsymbol{x}, -t) = \mathcal{T}\tilde{\Phi}(x)\mathcal{T}^{-1}.$$

Taking Hermitian conjugates of both sides, one also gets

$$\Phi(\boldsymbol{x}, -t) = \mathcal{T}\Phi(x)\mathcal{T}^{-1},$$

and finally, expanding Φ in terms of annihilation and creation operators, one verifies the two formulae:

$$\begin{aligned} a^+(\boldsymbol{k}) &= \mathcal{T}a^+(-\boldsymbol{k})\mathcal{T}^{-1}, \\ a^-(\boldsymbol{k}) &= \mathcal{T}a^-(-\boldsymbol{k})\mathcal{T}^{-1}. \end{aligned}$$

|| An antiunitary operator U obeys the following conditions:

(1) Let α be any complex number; the image $|v\rangle$ of the vector $\alpha|u\rangle$ is

$$|v\rangle = U\alpha|u\rangle = \alpha^* U|u\rangle;$$

(2) If $|v\rangle$ and $|v'\rangle$ are the images by U of the vectors $|u\rangle$ and $|u'\rangle$, we have

$$\langle v'|v\rangle = \langle u'|u\rangle^* = \langle u|u'\rangle.$$

In order to build such operators, we begin by defining the conjugation operator K such that $K^2=1$ and

$$\langle u'|K|u\rangle = \langle u|K|u'\rangle.$$

Let A denote a linear operator; by setting

$$|w\rangle = K|u\rangle, \qquad |w'\rangle = K|u'\rangle,$$

we obtain

$$\langle w'|w\rangle = \langle u'|u\rangle^*$$

$$\langle u'|KA|u\rangle = \langle w'|A|u\rangle^*, \qquad \langle u'|AK|u\rangle = \langle w|\tilde{A}|u'\rangle^*.$$

(*Continued on p.* 209)

filled. Indeed, we obtain

$$\begin{aligned}[Q_A^{(t)}(x), \bar{Q}_B^{(t)}(y)]_{\pm} &= \mathcal{T}[\tilde{Q}_A(x), \tilde{\bar{Q}}_B(y)]_{\pm}\mathcal{T}^{-1} \\ &= \mathcal{T}[\bar{Q}_B(y), Q_A(x)]_{\pm}^{\sim}\mathcal{T}^{-1} \\ &= \pm\mathcal{T}(-i\mathcal{D}_{AB}^{*}(\partial)\Delta^{*}(x-y))\mathcal{T}^{-1} \\ &= \pm i\mathcal{D}_{AB}(\partial)\Delta(x-y). \end{aligned} \tag{5.180}$$

Unlike the charge conjugation, the time reversal provides no information as to the type of commutation (or anticommutation) rules to be adopted.

It can be easily shown that under a time reversal, as defined above, the total charge of the field does not change its sign. This can be proved from formula (3.1*b*); indeed,†

$$\begin{aligned}\langle e_{\text{tot}}\rangle^{(t)} &= \langle\Psi|\mathcal{T}\left(-e\int\bar{Q}(x')\beta_4 Q(x')\,d^3\boldsymbol{x}'\right)\mathcal{T}^{-1}|\Psi\rangle \\ &= -e\langle\Psi|\int\bar{Q}^{(t)\sim}(x')\mathcal{T}^{-1}\beta_4\mathcal{T}\tilde{Q}^{(t)}(x')\,d^3\boldsymbol{x}'|\Psi\rangle,\end{aligned}$$

since

$$\mathcal{T}\beta_4 = \beta_4^{*}\mathcal{T},$$

and in accordance with (5.174), we have

$$\langle e_{\text{tot}}\rangle^{(t)} = -e\langle\Psi|\int\bar{Q}_B^{(t)\sim}(x')T_{\bar{B}B'}^{-1}(\beta_4)_{B'B''}T_{B''A}\tilde{Q}_A^{(t)}(x')\,d^3\boldsymbol{x}'|\Psi\rangle.$$

An antiunitary operator U can then be written $U = VK$, where V is unitary; indeed.

$$\begin{aligned}\langle v'|v\rangle = \langle v'|VK|u\rangle &= \langle w|V^{-1}|v'\rangle = \langle w|K|u'\rangle^{*} \\ &= \langle w|w'\rangle = \langle u|u'\rangle = \langle u'|u\rangle^{*}.\end{aligned}$$

We can also prove by virtue of the above relations that if

$$A = U^{-1}\tilde{B}U,$$

then

$$\tilde{A} = U^{-1}BU.$$

Indeed,

$$\begin{aligned}\langle u'|A|u\rangle &= \langle u'|KV^{-1}\tilde{B}VK|u\rangle = \langle u'|KV^{-1}\tilde{B}V|w\rangle \\ &= \langle w'|V^{-1}\tilde{B}V|w\rangle^{*} = \langle w|V^{-1}BV|w'\rangle = \langle w|V^{-1}BVK|u'\rangle \\ &= \langle u|KV^{-1}BVK|u'\rangle^{*} \equiv \langle u|\tilde{A}|u'\rangle^{*}.\end{aligned}$$

† We must take the Hermitian conjugate of the total charge operator which is one of the factors of the expression $\langle e_{\text{tot}}\rangle^{(t)}$, but this operator is identical to its adjoint. Because of the antiunitary character of $\mathcal{T}$, the formulae expressing the transformed variables are generally inferred by reading the original formulae from right to left, changing i into $-i$, and labelling the field variables with the index (t) (cf. Chap. VII, § 6d).

Inverting formulae (5.173) and (5.177):

$$\bar{Q}_A^{(t)\sim}(x')T_{AB}^{-1} = \alpha^*\bar{Q}_B(x), \quad T_{BA}\tilde{Q}_A^{(t)}(x') = \frac{1}{\alpha^*}Q_B(x), \qquad (5.181a)$$

we finally obtain

$$\langle e_{\text{tot}}\rangle^{(t)} = -e\langle\Psi|\int \bar{Q}(x)\beta_4 Q(x)\, d^3\boldsymbol{x}|\Psi\rangle = \langle e_{\text{tot}}\rangle. \qquad (5.181b)$$

We can easily prove, furthermore, by repeating a method described on pp. 115 and 205, that if we set

$$P_\mu^{(t)} = \mathcal{T}\tilde{P}_\mu\mathcal{T}^{-1}, \quad e_{\text{tot}}^{(t)} = \mathcal{T}e_{\text{tot}}\mathcal{T}^{-1}, \qquad (5.182)$$

we have

$$P_j^{(t)} = -P_j, \quad P_4^{(t)} = P_4, \quad e_{\text{tot}}^{(t)} = e_{\text{tot}}.$$

The last of these relations confirms (5.181); the first two show that

$$\boldsymbol{P}^{(t)} = -\boldsymbol{P},$$

and that the sign of the energy is conserved.

The operator $\mathcal{T}$ is a constant of motion in the Heisenberg picture (it anticommutes with P_4, but commutes with $P_0 \equiv -iP_4$); moreover, by introducing this operator we can satisfy the correspondence principle since P_μ has indeed this variance under a time reversal in classical mechanics, the energy of the field remaining positive.

If we look back over all the transformations investigated so far, we see that we have been led to introduce four matrices η, C, σ and T given by formulae (4.40), (5.149), (5.166) and (5.173) (and which we shall write explicitly in the case of the Dirac field):

$$\left.\begin{array}{c} \eta\beta_j\eta^{-1} = -\tilde{\beta}_j, \quad \eta\beta_4\eta^{-1} = \tilde{\beta}_4, \quad \tilde{\eta} = \eta, \\ C^{-1}\beta_\mu C = -\beta_\mu^T, \\ \sigma^{-1}\beta_j\sigma = -\beta_j, \quad \sigma^{-1}\beta_4\sigma = \beta_4, \\ T^{-1}\beta_\mu T = \beta_\mu^*. \end{array}\right\} \qquad (5.183)$$

These matrices are obviously not independent; we easily verify that we have

$$[\beta_\mu, \sigma TC^*\eta]_+ = 0, \qquad (5.184a)$$

and

$$[TC^*\eta, \beta_j] = 0, \qquad [TC\eta^*, \beta_4]_+ = 0. \qquad (5.184b)$$

From these last formulae and the property of the matrix τ in (5.171) we see that we could set

$$TC^*\eta = \alpha\tau, \qquad (5.185)$$

where α is a phase factor.

Note that a time reversal can also be defined by using a unitary operator instead of the antiunitary $\mathcal{T}$.† But the definitions are then no longer intrinsic; the field variables $Q_A(x)$ must be considered as matrices in a complete orthogonal set of vectors spanning the space $\mathcal{R}$, a set which is arbitrary but chosen once for all. This set can be built from the eigenvectors of an arbitrary field observable, for instance the energy–momentum.

$Q_A(x)$ is then an infinite matrix with elements $\langle a | Q_A(x) | b \rangle$ and we can define the matrix $Q_A^*(x)$ with elements $\langle a | Q_A(x) | b \rangle^*$ in the basis we choose. The operator $Q_A^*(x)$ whose matrix elements would be $\langle a | Q(x | b \rangle^*$ (where a and b are vectors of any basis) is not linear, but antilinear.

We can now represent the *time reversal* with the help of the matrix T [which *a priori* may be different from the one used in (5.173)] of the spin space

$$Q_A^{(t)}(x') = T_{AA'} Q_{A'}^*(x). \tag{5.186a}$$

It is easy to see that by making $Q^{(t)}(x')$ obey the same wave equation as $Q(x)$, the matrix T obeys (5.174). We verify that

$$\bar{Q}_A^{(t)}(x') = \bar{Q}_{A'}^*(x) T_{\bar{A}'A}^{-1}, \tag{5.186b}$$

and that the commutation rules are unchanged. This transformation is therefore a canonical transformation and is represented in the space $\mathcal{R}$ by a linear operator; we can also see that the sign of the charge is unchanged.

In short, we have just introduced three operations; the time reversal represented by the operator $\mathcal{T}$, the exchange between particles and antiparticles represented by the operator Γ and the space-symmetry represented by the operator Π, these three operators being defined in the space $\mathcal{R}$. Each of these operators commutes with P_0; so does their product $\mathcal{T}\Gamma\Pi$ (the factors can moreover be permuted). This property, commonly known as the "TCP theorem", allows of an important generalization. It can be proved, in fact, that it holds for any local and Lorentz invariant theory. It leads to interesting selection rules for the decays of elementary particles (cf. Chap. VII, § 8d).

It will be remembered, finally, that one of the symmetry operations, the charge conjugation, provided us with valuable information concerning the choice of the type of commutation rules. In Chap. II, § 13, we had already studied the consequences of the Schwinger–Feynman principle with regard to the choice of the commutation rules. In § 6 of this chapter, we showed that the scalar or pseudo-scalar field could not be quantized by

† S. Watanabe, *Phys. Rev.*, **84**, 1008 (1951).

anticommutators; in § 11 we extended this result to any integer spin field. In a note in § 13, we pointed out that for a half-integer spin field only the anticommutation rules lead to a positive definite P_0. Finally, in § 19, we showed that for any field where particles and antiparticles are defined, the conjugation matrix obeys the characteristic relation (5.156). These remarks, together with the investigation of statistics spin relations in § 14, represent a fairly complete study of this question as far as free fields are concerned; a more general approach will be adopted in Chaps. VII, § 8d, and XI, § 16.

CHAPTER VI

Quantization of linear field equations with constant coefficients. Examples

To ILLUSTRATE the general theories presented in the last chapter, we intend to examine here three fields of particular importance in their applications; the Dirac field (spin $\frac{1}{2}$), the Maxwell field (photons) and the vector field.

PART 1

THE DIRAC FIELD (FREE ELECTRONS)

We shall divide this study into two parts: in the first part, we shall investigate the non-quantized Dirac equation (the wave function will consequently be a column matrix whose elements are functions of x) while in the second part we shall study its quantization.

A. DIRAC EQUATION

1. Algebra of the Dirac matrices

Let us write down the Dirac equation which describes particles of spin $\frac{1}{2}$ in its general form as given in Chap. IV:

$$\{\gamma_\mu \partial_\mu + m\}\psi(x) = 0. \tag{6.1.1}$$

We saw in that chapter that the algebra of the γ_μ matrices had to be determined from the relations [cf. (4.25)]:†

$$[\gamma_\mu, \gamma_\nu]_+ = 2\,\delta_{\mu\nu}. \tag{6.1.2}$$

† The algebra of the γ matrices was investigated by W. Pauli, *Ann. Inst. Henri Poincaré*, **6**, 109 (1936); cf. also D. Kastler, *Introduction à l'électrodynamique quantique*, Dunod, 1960.

We propose to begin by investigating certain properties of these matrices. New matrices can be formed in the linear space in which the γ_μ's are defined, by carrying out products of these matrices. Now, the square of any γ_μ matrix is the unit matrix; on the other hand, to change the order of the factors in a product of γ_μ where all the factors are different, is equivalent, by virtue of (6.1.2), to multiplying this product by ± 1. We can, therefore, in the last analysis, only consider $2^4 = 16$ new matrices which are independent:

$$\left.\begin{array}{lll} I, & & \\ \gamma_1, \gamma_2, \gamma_3, \gamma_4 & \equiv \gamma_\mu, & \\ i\gamma_1\gamma_2, i\gamma_1\gamma_3, i\gamma_1\gamma_4, i\gamma_2\gamma_3, i\gamma_2\gamma_4, i\gamma_3\gamma_4 & \equiv \gamma_{\mu\nu} & (\mu \neq \nu), \\ i\gamma_1\gamma_2\gamma_3, i\gamma_1\gamma_2\gamma_4, i\gamma_2\gamma_3\gamma_4, i\gamma_3\gamma_1\gamma_4 & \equiv \gamma_{\mu\nu\varrho} & (\mu \neq \nu \neq \varrho). \\ \gamma_1\gamma_2\gamma_3\gamma_4 & \equiv \gamma_5 & \end{array}\right\} \tag{6.1.3}$$

It will be noted that the matrices $\gamma_{\mu\nu}$ and $\gamma_{\mu\nu\varrho}$ are antisymmetric† with respect to their indices and that the matrix γ_5 can also be written

$$\gamma_5 = \frac{1}{4!}\varepsilon_{\lambda\mu\nu\varrho}\gamma_\lambda\gamma_\mu\gamma_\nu\gamma_\varrho, \tag{6.1.4a}$$

where $\varepsilon_{\lambda\mu\nu\varrho}$ is equal to $+1$ if λ, μ, ν and ϱ are an even permutation of the numbers 1, 2, 3, 4; to -1 if these indices are an odd permutation; and to 0 if two of the indices are equal.‡

The same matrix γ_5 anticommutes with all the γ_μ's; we also have

$$\gamma_\mu\gamma_\nu\gamma_5 = -\frac{1}{2}\varepsilon_{\mu\nu\lambda\varrho}\gamma_\lambda\gamma_\varrho, \quad \mu \neq \nu. \tag{6.1.4b}$$

In a Hermitian representation|| of the γ_μ's, it is easy to verify that all the matrices in table (6.1.3) are Hermitian (this was our reason for introducing the factors i). It can also be verified that the square of each of these matrices is the unit matrix; for instance, we have

$$(\gamma_{\mu\nu})^2 = -\gamma_\mu\gamma_\nu\gamma_\mu\gamma_\nu = I,$$
$$(\gamma_5)^2 = \gamma_1\gamma_2\gamma_3\gamma_4\gamma_1\gamma_2\gamma_3\gamma_4 = I.$$

On the other hand, by virtue of formula (6.1.2) and the well-known relation:

$$\text{Sp}\,(AB) = \text{Sp}\,(BA)$$

† Note that $\gamma_5\gamma_\mu\gamma_\nu$ and $\gamma_5\gamma_\mu$ can replace $\gamma_\mu\gamma_\nu$ and $\gamma_{\mu\nu\varrho}$ in table (6.1.3).

‡ A permutation is even or odd according as the number of transpositions necessary to obtain it is itself even or odd.

|| We shall write such a representation in explicit form [cf. (6.1.30)].

we verify that the spurs of all the matrices in table (6.1.3) (except Sp I) are zero; for instance, if we consider Sp γ_1, we have by virtue of (6.1.2)

$$\text{Sp } \gamma_1 = \text{Sp } \gamma_1\gamma_2\gamma_2 = -\text{Sp } \gamma_2\gamma_1\gamma_2,$$

then

$$\text{Sp } (\gamma_1\gamma_2)\gamma_2 = -\text{ Sp } \gamma_2(\gamma_1\gamma_2) = -\text{ Sp } \gamma_1\gamma_2\gamma_2.$$

It is then easy to prove that the matrices in table (6.1.3) are linearly independent; consider indeed the linear combination

$$C_1 + C_\mu\gamma_\mu + C_{12}\gamma_{12} + C_{13}\gamma_{13} + \ldots + C_{1234}\gamma_5.$$

Suppose this expression to be zero and multiply it by one of the sixteen matrices in table (6.1.3), for instance γ_{12}:

$$C_1\gamma_{12} + C_\mu\gamma_{12}\gamma_\mu C_{12} + \ldots = 0.$$

Taking the spur of this expression, we get

$$C_{12} = 0.$$

The existence of these sixteen linearly independent quantities thus allows us to represent the γ_μ's by 4×4 matrices.

On the other hand, starting from the Schur lemma, we can prove the following theorem.†

If γ_μ and γ_μ' are two sets of 4×4 matrices which obey (6.1.2), a non-singular matrix S can be found such that

$$\gamma'_\mu = S\gamma_\mu S^{-1}. \tag{6.1.5}$$

Now consider the matrices $\tilde{\gamma}_\mu$, γ_μ^T and γ_μ^*: they all obey (6.1.2); they are therefore connected with the matrices γ_μ by an expression of the form (6.1.5), by virtue of the preceding theorem. As a further application of the Schur lemma, we can also prove that any 4×4 matrix commuting with all the γ_μ's is a multiple of the unit matrix. Assuming that the matrix A commutes with all the γ_μ's, the matrix $A-\lambda I$, where λ is a c-number, also commutes. Let us then choose λ such that det $|A-\lambda I| = 0$; according

† Let Σ and Σ' be two sets of matrices A and A' operating in two vector spaces E and E' which are respectively n- and n'-dimensional.

Suppose these sets of matrices to be irreducible, i.e. such that the only two subspaces invariant under each of the matrices are the spaces 0 (null vector), E and E'. Let there be a matrix M with n columns and n' rows such that to any matrix $A \in \Sigma$ there corresponds a matrix $A' \in \Sigma'$ with:

$$AM = MA',$$

and vice versa. The matrix M is either zero (all its elements are zeros), or regular, in which case $n=n'$. In order to prove (6.1.5), we explicitly build up S by expressing it as bilinear combinations of the sixteen matrices (6.1.3) in the original set and in the primed set (cf. W. Pauli, *op. cit.*, footnote, p. 214).

to the lemma in question, since $A-\lambda I$ is a singular matrix it must therefore be the zero matrix; consequently,

$$A = \lambda I.$$

2. Dirac equation and Lorentz invariance

The Dirac equation therefore belongs to the general type of equations investigated in Chap. IV: the γ_μ matrices and the column matrix $\psi(x)$ are defined in a four-dimensional linear space.

We have seen (4.40) that if a matrix η is introduced such that

$$\tilde{\gamma}_j = -\eta\gamma_j\eta^{-1}, \quad \tilde{\gamma}_4 = \eta\gamma_4\eta^{-1} \tag{6.1.6}$$

and if the row matrix $\bar{\psi}(x)$ (4.41) is defined

$$\bar{\psi}(x) = \tilde{\psi}(x)\eta \tag{6.1.7}$$

or

$$\bar{\psi}_\alpha(x) = \tilde{\psi}_\mu(x)\eta_{\mu\alpha},$$

the latter wave function obeys the equation

$$\bar{\psi}(x)\{\gamma_\mu \overleftarrow{\partial}_\mu - m\} = 0. \tag{6.1.8}$$

When the γ_μ's are Hermitian [standard representation (6.1.30)], η is identical to γ_4.

Consider now the proper Lorentz transformations: it is easy to see, taking (6.1.2) into account, that the relations (4.16) and (4.17) can be satisfied by taking

$$J_{\mu\nu} = \frac{1}{4}[\gamma_\mu, \gamma_\nu] \tag{6.1.9a}$$

or again

$$J_{\mu\nu} = \frac{1}{2}\gamma_\mu\gamma_\nu \quad (\mu \neq \nu), \tag{6.1.9b}$$

from which it follows for the spin operator $\mathbf{S}$ (3.62):

$$S_k = -\frac{i}{2}\gamma_l\gamma_m \quad \text{(where } k, l, m \text{ is a cyclic permutation of 1, 2, 3).} \tag{6.1.10}$$

We therefore set

$$\mathcal{J} = I + \frac{1}{8}\varepsilon_{\mu\nu}[\gamma_\mu, \gamma_\nu] \tag{6.1.11}$$

and formulae (4.13) and (4.47) give the transformation laws of ψ and $\bar{\psi}$:

$$\left.\begin{aligned} \psi(x) &\to \psi'(x') = \mathcal{J}\psi(x), \\ \bar{\psi}(x) &\to \bar{\psi}'(x') = \bar{\psi}(x)\mathcal{J}^{-1} \end{aligned}\right\}. \tag{6.1.12}$$

Furthermore, the compatibility equation (5.164*b*) is found to be satisfied.

Let us now turn to the full Lorentz group; we shall first consider a space symmetry; following the formulae (5.166), one has†

$$\sigma = \lambda\gamma_4, \tag{6.1.13}$$

where λ is an indeterminate *c*-number. It is then possible to determine it in such a way that the wave function is transformed according to the two-valued representation of the Lorentz group, in other words that

$$\sigma^2 = \pm I. \tag{6.1.14}$$

The number λ can therefore take one of the four values ± 1, $\pm i$; the values $\lambda = \pm i$ must be chosen in order that (5.164*a*) be satisfied.‡

Let us turn now to the time reversal: since it belongs to the Lorentz group, we may (at least for the time being) represent it by the linear transformation τ; referring to formulae (5.172), we find that

$$\tau = \lambda'\gamma_1\gamma_2\gamma_3, \tag{6.1.15a}$$

where λ' is a certain *c*-number (which is chosen of modulus 1). The operator $\gamma_1\gamma_2\gamma_3$ is sometimes referred to as the Racah operator. It must be noted that the transformation of $\bar{\psi}$ is

$$\bar{\psi}(x) \to \bar{\psi}^{(\tau)}(x') = -\bar{\psi}(x)\tau^{-1} \tag{6.1.15b}$$

and a justification of the minus sign will be found in the remarks following formula (4.47).

We now suppose the time reversal to be represented by the matrix T (5.174); we have seen (5.174) that $TT^* = \alpha I$, where α is a constant which, as may be easily proved, is independent of the choice of the representation of the γ_μ's and is equal to $-I$. By choosing the standard representation (6.1.30), formula (5.174) leads to

$$[\gamma_1, T]_+ = 0, \quad [\gamma_3, T]_+ = 0, \quad [\gamma_2, T]_- = 0, \quad [\gamma_4, T]_- = 0.$$

† The proof of the uniqueness of the expression (6.1.9) of $J_{\mu\nu}$ can be easily given. Suppose $J'_{\mu\nu}$ to be another solution of eqns. (4.16) and (4.17); from (4.16) it follows that

$$[J_{\mu\nu} - J'_{\mu\nu}, \gamma_\lambda] = 0.$$

Thus $J_{\mu\nu} - J'_{\mu\nu} = CI$, C being a scalar.

On the other hand, with the standard representation of γ's and $\eta = \gamma_4$, the matrix $\mathcal{J}$ is unimodular: $\det \mathcal{J} = 1$, therefore $\det J_{\mu\nu} = \det J'_{\mu\nu} = 0$, and $C = 0$.

‡ This formula is written:

$$C = \sigma C \sigma^T$$

or

$$C = \lambda^2 \gamma_4 C \gamma_4^T$$

and making use of formula (5.150) we indeed find the expected value of λ.

We can therefore set

$$T = \lambda''\gamma_1\gamma_3, \quad |\lambda''| = I; \tag{6.1.16a}$$

we also have [cf. (6.1.32)]:

$$TT^* = T^*T = -1. \tag{6.1.16b}$$

Finally, it is easy to see that the total symmetry operation ($\boldsymbol{x} \to -\boldsymbol{x}$, $x_0 \to -x_0$) is represented by the operator

$$\lambda'''\gamma_5, \tag{6.1.17}$$

where λ''' is a number of modulus one.

3. Transformation laws of certain bilinear tensor expressions

The results of the above section will now enable us to investigate the relativistic transformation laws of certain expressions, namely products of the following factors: $\bar{\psi}$ is the first factor, one of the sixteen matrices in table (6.1.3) is the second, and ψ is the third.

Take for instance the expression $\bar{\psi}\psi$; one has

$$\bar{\psi}(x)\psi(x) \to \bar{\psi}'(x')\psi'(x') = \bar{\psi}(x)\mathcal{J}^{-1}\mathcal{J}\psi(x) = \bar{\psi}(x)\psi(x).$$

Similarly, if we consider $\bar{\psi}\gamma_\mu\psi$, we obtain

$$\bar{\psi}\gamma_\mu\psi \to \bar{\psi}'(x')\gamma_\mu\psi'(x') = \bar{\psi}(x)\mathcal{J}^{-1}\gamma_\mu\mathcal{J}\psi(x) = A_{\mu\nu}\bar{\psi}\gamma_\nu\psi.$$

The first column of the table at the end of this section was formed in this way.

Let us now consider the space symmetry and time reversal; we shall take as an example the expression $\bar{\psi}\gamma_5\psi$; for a space symmetry we have

$$\begin{aligned}\bar{\psi}\gamma_5\psi \to \bar{\psi}^{(\sigma)}(x')\gamma_5\psi^{(\sigma)}(x') &= \bar{\psi}(x)\sigma^{-1}\gamma_5\sigma\psi(x)\\ &= \bar{\psi}(x)\gamma_4\gamma_5\gamma_4\psi(x) = -\bar{\psi}(x)\gamma_5\psi(x),\end{aligned}$$

while for a time reversal:

$$\begin{aligned}\bar{\psi}\gamma_5\psi \to \bar{\psi}^{(\tau)}(x')\gamma_5\psi^{(\tau)}(x') &= -\bar{\psi}(x)\tau^{-1}\gamma_5\tau\psi\\ &= -\bar{\psi}(x)\gamma_3\gamma_2\gamma_1\gamma_5\gamma_1\gamma_2\gamma_3\psi(x)\\ &= +\bar{\psi}(x)\gamma_5\psi(x);\end{aligned}$$

$\bar{\psi}\gamma_5\psi$ is said to be a pseudo-scalar with respect to the Lorentz group. The results of the third and fourth columns of the table will be verified in this way.

The determination of the transformation laws of the expressions given in the table below is particularly important, since in order to obtain the Lagrangian density corresponding to the interaction of two fields we are led to form a scalar under the Lorentz transformations by means of the given fields.

Bilinear expressions in ψ and $\bar{\psi}$

	Proper Lorentz transformations	Space symmetry	Designation in the ortho-chronous Lorentz group	Time reversal considered as a linear operation
$J = \bar{\psi}\psi \ldots\ldots$	$J' = J$	$J^{(\sigma)} = +J$	Scalar	$J^{(\tau)} = -J$
$J_5 = \bar{\psi}\gamma_5\psi \ \ldots$	$J'_5 = J_5$	$J_5^{(\sigma)} = -J_5$	Pseudo-scalar	$J_5^{(\tau)} = +J_5$
$J_\mu = \bar{\psi}\gamma_\mu\psi \ldots$	$J'_\mu = L_{\mu\nu}J_\nu$	$\begin{Bmatrix} J_k^{(\sigma)} = -J_k \\ J_4^{(\sigma)} = +J_4 \end{Bmatrix}$	Vector	$\begin{cases} J_k^{(\tau)} = -J_k \\ J_4^{(\tau)} = +J_4 \end{cases}$
$J_{5\mu} = \bar{\psi}\gamma_5\gamma_\mu\psi \ldots$	$J'_{5\mu} = L_{\mu\nu}J_{5\nu}$	$\begin{Bmatrix} J_{5k}^{(\sigma)} = +J_k \\ J_{54}^{(\sigma)} = -J_{54} \end{Bmatrix}$	Pseudo-vector	$\begin{cases} J_{5k}^{(\tau)} = +J_{5k} \\ J_{54}^{(\tau)} = -J_{54} \end{cases}$
$J_{\mu\nu} = \bar{\psi}\gamma_\mu\gamma_\nu\psi \ldots$	$J'_{\mu\nu} = L_{\mu\lambda}L_{\nu\varrho}J_{\lambda\varrho}$	$\begin{Bmatrix} J_{kl}^{(\sigma)} = +J_{kl} \\ J_{k4}^{(\sigma)} = -J_{k4} \\ J_{44}^{(\sigma)} = +J_{44} \end{Bmatrix}$	Tensor	$\begin{cases} J_{kl}^{(\tau)} = -J_{kl} \\ J_{k4}^{(\tau)} = +J_{k4} \\ J_{44}^{(\tau)} = -J_{44} \end{cases}$
$J_{5\mu\nu} = \bar{\psi}\gamma_5\gamma_\mu\gamma_\nu\psi$	$J'_{5\mu\nu} = L_{\mu\lambda}L_{\nu\varrho}J_{5\lambda\varrho}$	$\begin{Bmatrix} J_{5kl}^{(\sigma)} = -J_{5kl} \\ J_{5k4}^{(\sigma)} = +J_{5k4} \\ J_{544}^{(\sigma)} = -J_{544} \end{Bmatrix}$	Pseudo-tensor	$\begin{cases} J_{5kl}^{(\tau)} = +J_{5kl} \\ J_{5k4}^{(\tau)} = -J_{5k4} \\ J_{544}^{(\tau)} = +J_{544} \end{cases}$

4. The representations of the γ_μ matrices

The linearization of the expression: $p^2 - m^2$, in other words the solution of the equation in σ:

$$\mathbf{p}^2 - m^2 = (\boldsymbol{\sigma}\cdot\mathbf{p} + m)(\boldsymbol{\sigma}\cdot\mathbf{p} - m)$$

introduces three operators σ_1, σ_2 and σ_3 which have to obey the following anticommutation relations:

$$[\sigma_i, \sigma_j]_+ = 2\delta_{ij}. \tag{6.1.18}$$

They have the following unitary representations:

$$\sigma_1 = \begin{pmatrix} 0 & 1 \\ 1 & 0 \end{pmatrix}, \quad \sigma_2 = \begin{pmatrix} 0 & -i \\ i & 0 \end{pmatrix}, \quad \sigma_3 = \begin{pmatrix} 1 & 0 \\ 0 & -1 \end{pmatrix}; \tag{6.1.19}$$

σ_1, σ_2 and σ_3 are the Pauli matrices. We also obtain

$$\sigma_i\sigma_j = i\sigma_k, \tag{6.1.20}$$

provided that i, j, k form a cyclic permutation of the numbers 1, 2, 3. If we then define the matrices:

$$J_k = \frac{\sigma_k}{2}, \tag{6.1.21}$$

we verify that J_1, J_2 and J_3 can be considered as the generating operators of the group SO(3) of space rotations†

$$\mathcal{R} = I + i\boldsymbol{\varepsilon}.\boldsymbol{J} \qquad (6.1.22a)$$

since, as may be easily verified,

$$[J_i, J_j] = iJ_k, \qquad (6.1.23)$$

provided that i, j, k form a cyclic permutation of the numbers 1, 2, 3.

Formulae (6.1.22*a*) can be written in a more symmetrical manner by defining the skew-symmetrical rotation tensor: $\varepsilon_{ij} = \varepsilon_k$ (i, j, k, a cyclic permutation of the numbers 1, 2, 3, therefore $\varepsilon_1 = \varepsilon_{23}$, $\varepsilon_2 = \varepsilon_{31}$, $\varepsilon_3 = \varepsilon_{12}$) and similarly, $J_{ij} = J_k$; we then have

$$\mathcal{R} = I + \frac{1}{2}\varepsilon_{ij}J_{ij}. \qquad (6.1.22b)$$

We therefore obtain a representation SU(2) of the space rotations in spinor space.

A peculiarity of this representation is worth noting, however: a rotation through an angle θ about the space-axis Oz induces the following unitary transformation of the spinor space:

$$\mathcal{R}_\theta = e^{iJ_3\theta} = e^{\frac{i}{2}\sigma_3\theta}. \qquad (6.1.24)$$

Making use of $\sigma_3^2 = I$, we verify that the above expression can be written in the form‡

$$\mathcal{R}_\theta = \cos\frac{\theta}{2} + i\sigma_3 \sin\frac{\theta}{2}. \qquad (6.1.25)$$

Under this rotation a particular spinor $|u\rangle$ (a vector of the representation space under consideration) is transformed as follows:

$$|u\rangle \to |u'\rangle = \mathcal{R}_\theta |u\rangle, \qquad (6.1.26a)$$

in particular if $\theta = 2\pi$:

$$|u'\rangle = -|u\rangle. \qquad (6.1.26b)$$

Thus a rotation through the angle 2π about an axis of the three-dimensional space changes the orientation of a spinor whereas a rotation through the angle 4π restores it: the spinor representation of the group of space rotations is not therefore a faithful representation.

† By space rotations, we mean the rotations in the three-dimensional Euclidean space.

‡ A rotation through an angle θ about the axis of the unitary vector $\boldsymbol{u}$ is given by:

$$\mathcal{R}_{\theta,\boldsymbol{u}} = \cos\frac{\theta}{2} + i\boldsymbol{\sigma u} \sin\frac{\theta}{2}.$$

Let us now return to the γ matrices; it is convenient to introduce the matrices $\alpha_1, \alpha_2, \alpha_3, \beta$ defined as follows:

$$\alpha_j \equiv i\beta\gamma_j = -i\gamma_j\beta, \quad \beta \equiv \gamma_4, \tag{6.1.27a}$$

whence

$$\gamma_j = i\alpha_j\gamma_4 = -i\gamma_4\alpha_j, \quad \gamma_4 \equiv \beta. \tag{6.1.27b}$$

It is then easy to verify, starting from the anticommutation relations of the γ's , that we have

$$[\alpha_j, \alpha_k]_+ = 2\delta_{jk}, \quad [\alpha_j, \beta]_+ = 0. \tag{6.1.28}$$

By adding to the three Pauli matrices the unit 2×2 matrix, we can give the following representation of the matrices α_j and β:

$$\alpha_j = \begin{pmatrix} 0 & \sigma_j \\ \sigma_j & 0 \end{pmatrix}, \quad \beta = \begin{pmatrix} 1 & 0 \\ 0 & -1 \end{pmatrix} \tag{6.1.29}$$

and formulae (6.1.27*b*) enable us to write the γ-matrices explicitly:

$$\gamma_j = i\begin{pmatrix} 0 & -\sigma_j \\ \sigma_j & 0 \end{pmatrix}, \quad \gamma_4 \equiv \beta = \begin{pmatrix} 1 & 0 \\ 0 & -1 \end{pmatrix}; \tag{6.1.30}$$

this representation will be called the "standard" representation: its matrices are Hermitian and unitary.

The following expressions will be found helpful in the applications:

$$\left.\begin{aligned} &\eta \equiv \gamma_4, \\ &\gamma_j\gamma_k = i\begin{pmatrix} \sigma_l & 0 \\ 0 & \sigma_l \end{pmatrix} \quad (j, k, l, \text{cyclic permutation of } 1, 2, 3 \text{ and } j \neq k), \\ &\gamma_j\gamma_k\gamma_l = \begin{pmatrix} 0 & 1 \\ -1 & 0 \end{pmatrix} \quad (j \neq k \neq l \text{ and form a cyclic permutation of } 1, 2, 3), \\ &\gamma_5 = -\begin{pmatrix} 0 & 1 \\ 1 & 0 \end{pmatrix}. \end{aligned}\right\} \tag{6.1.31}$$

We also obtain

$$\left.\begin{aligned} \gamma_1 = -\gamma_1^* = -\gamma_1^T, \quad \gamma_2 = \gamma_2^* = \gamma_2^T, \\ \gamma_3 = -\gamma_3^* = -\gamma_3^T, \quad \gamma_4 = \gamma_4^* = \gamma_4^T. \end{aligned}\right\} \tag{6.1.32}$$

A further representation of the γ_μ's, namely

$$\gamma_j = i\begin{pmatrix} 0 & \sigma_j \\ -\sigma_j & 0 \end{pmatrix}, \quad \gamma_4 = \begin{pmatrix} 0 & 1 \\ 1 & 0 \end{pmatrix}, \tag{6.1.33}$$

is such that the non-zero elements of all the matrices γ_μ are arranged in the same way.

Form the generating operators of the proper Lorentz transformations:

$$\left.\begin{aligned} J_{jk} &= \frac{1}{2}\gamma_j\gamma_k = \frac{i}{2}\begin{pmatrix}\sigma_l & 0\\ 0 & \sigma_l\end{pmatrix} \quad (j, k, l, \text{cyclic permutation of } 1, 2, 3),\\ J_{k4} &= \frac{1}{2}\gamma_k\gamma_4 = \frac{i}{2}\begin{pmatrix}+\sigma_k & 0\\ 0 & -\sigma_k\end{pmatrix}, \end{aligned}\right\} \tag{6.1.34}$$

the operator $\mathcal{J}$ (6.1.11) of the proper Lorentz transformations therefore takes the form:

$$\mathcal{J} = 1 + \sum_{j,k,l}^{(P)}\left\{i\varepsilon_{jk}\begin{pmatrix}\sigma_l & 0\\ 0 & \sigma_l\end{pmatrix} + i\varepsilon_{l4}\begin{pmatrix}\sigma_l & 0\\ 0 & -\sigma_l\end{pmatrix}\right\}, \tag{6.1.35}$$

where the sign $\Sigma^{(P)}$ denotes a summation over j, k, l such that the indices j, k, l form a cyclic permutation of the numbers 1, 2, 3 and where only the ε_{jk}'s and $\varepsilon_{j'k'}$'s of the first term of the summation (6.1.35) with j, k different from j', k' are taken into account.

Denote by $\psi_{\mathrm{I}}(x)$ the column matrix formed by $\psi_1(x), \psi_2(x)$ and by $\psi_{\mathrm{II}}(x)$ the column matrix formed by $\psi_3(x), \psi_4(x)$; since the matrices in the expression (6.1.35) of the matrix $\mathcal{J}$ are all diagonal, it follows that ψ_{I} and ψ_{II} are transformed independently under a Lorentz transformation. Indeed, we obtain

$$\psi'_I(x') = \{I + \textstyle\sum^{(P)}(i\varepsilon_{jk}\sigma_l + i\varepsilon_{l4}\sigma_l)\}\psi_I(x), \tag{6.1.36a}$$

$$\psi'_{II}x') = \{I + \textstyle\sum^{(P)}(i\varepsilon_{jk}\sigma_l - i\varepsilon_{l4}\sigma_l)\}\psi_{II}(x). \tag{6.1.36b}$$

$\psi_{\mathrm{II}}(x)$ is called a contravariant spinor. We shall not enter into details such as the definition of covariant spinors or conjugate complex spinors, etc., which can be found in books or articles devoted to spinor calculus.† It is sufficient to note that this last representation of the γ's brings out clearly the fact that ψ is formed by two spinors. Note in conclusion that the spin operator S (6.1.10) is given, in all these representations of the γ's, by the matrix

$$S_l = \frac{1}{2}\begin{pmatrix}\sigma_l & 0\\ 0 & \sigma_l\end{pmatrix}. \tag{6.1.37}$$

5. Monochromatic plane waves

The Dirac equation can be written with the aid of the matrices α and β; it is sufficient to multiply eqn. (6.1.1) on the left by $i\beta$; we obtain

$$\left\{\frac{\partial}{\partial t} + \boldsymbol{\alpha}.\nabla + im\beta\right\}\psi(x) = 0 \tag{6.1.38}$$

† E. Cartan, *Leçons sur la théorie des spineurs*, Hermann, 1938; W. L. Bade and H. Jehle, *Rev. Mod. Phys.*, **25**, 714 (1953), where bibliographical references will be found.

or

$$i\frac{\partial\psi(x)}{\partial t} = \{-i\boldsymbol{\alpha}.\nabla + m\beta\}\psi(x) \equiv H_0(\nabla)\psi(x). \tag{6.1.39}$$

$H_0(\nabla)$ is the Dirac Hamiltonian in wave mechanics.

We are interested in a monochromatic plane wave solution of (6.1.39):

$$\psi(x) = A(\boldsymbol{p})e^{ipx}, \tag{6.1.40}$$

where $A(\boldsymbol{p})$ is a column matrix the elements of which are functions of $\boldsymbol{p}$. Inserting (6.1.40) into (6.1.39) we obtain

$$p_0 A(\boldsymbol{p}) = (\boldsymbol{\alpha}.\boldsymbol{p} + m\beta)A(\boldsymbol{p}) = H_0(\boldsymbol{p})A(\boldsymbol{p}), \tag{6.1.41}$$

a homogeneous system of four linear equations with four variables. Multiply both sides of this system by $\boldsymbol{\alpha}\cdot\boldsymbol{p}+m\beta$; we get

$$(p_0^2 - \boldsymbol{p}^2 - m^2)A(\boldsymbol{p}) = 0, \tag{6.1.42a}$$

which is the compatibility relation of the four homogeneous equations. It is easily seen that there exist both positive and negative energy solutions:

$$p_0 = \pm\sqrt{\boldsymbol{p}^2 + m^2} = \pm\omega(\boldsymbol{p}). \tag{6.1.42b}$$

The solutions can therefore be classified according to the sign of the eigenvalues of $H_0(\boldsymbol{p})$. It is helpful to form a second operator the eigenvalues of which completely define this classification. Form with the help of the S spin defined in (6.1.10) the helicity matrix† (Chap. IV, § 4):

$$\Sigma(\boldsymbol{p}) = \frac{\boldsymbol{S}.\boldsymbol{p}}{|\boldsymbol{p}|}. \tag{6.1.43}$$

† It is often helpful to describe the helicity with the aid of Stokes parameters. Introduce the components ψ_{I} and ψ_{II} of the wave functions as on p. 223 and note that

$$2\langle\psi, S_3\psi\rangle = i\langle\psi_I, \sigma_3\psi_I\rangle + i\langle\psi_{II}, \sigma_3\psi_{II}\rangle = i(s_1^I + s_1^{II})$$
$$2\langle\psi, S_1\psi\rangle = i\langle\psi_I, \sigma_1\psi_I\rangle + i\langle\psi_{II}, \sigma_1\psi_{II}\rangle = i(s_2^I + s_2^{II})$$
$$2\langle\psi, S_2\psi\rangle = i\langle\psi_I, \sigma_2\psi_I\rangle + i\langle\psi_{II}, \sigma_2\psi_{II}\rangle = i(s_3^I + s_3^{II}).$$

$\boldsymbol{S}$ is the spin (6.1.37). The numbers s_1, s_2 and s_3, whose definition is apparent, form with the scalar:

$$\langle\psi, \psi\rangle = |\psi_I|^2 + |\psi_{II}|^2 = s_0$$

the four components of the Stokes vector. These parameters describe the polarization of the particle of spin $\frac{1}{2}$ and we shall meet them again when we investigate the polarization of the photons (cf. W. McMaster, *Rev. Mod. Phys.*, 33, 8 (1961), for a more complete discussion).

It is easy to see that†

$$[H_0(\boldsymbol{p}), \Sigma(\boldsymbol{p})]_- = 0, \quad \Sigma^2 = \frac{1}{4}. \tag{6.1.44}$$

The first relation shows that we can consider eigenvalue solutions common to H_0 and Σ; the second relation shows that the eigenvalues of Σ have modulus $\frac{1}{2}$. This is in fact the expected result, since $+\frac{1}{2}$ and $-\frac{1}{2}$ are the possible values of the spin along the direction of the unitary propagation vector of the plane wave. The solutions can therefore be classified according to the sign of the energy and that of the polarization in the direction of the momentum.

The actual computation of a solution is an easy matter; adopting the standard representation of the γ's, we have to solve the following linear system:

$$\begin{aligned}
(p_0-m)A_1 \qquad\qquad\qquad - \quad p_3A_3 \quad -(p_1-ip_2)A_4 &= 0,\\
(P_0-m)\,A_2 \; -(p_1-ip_2)A_3 \; + \quad p_3A_4 \qquad &= 0,\\
- \quad p_3A_1 \quad -(p_1-ip_2)A_2 \; +(p_0+m)\,A_3 \qquad\qquad &= 0,\\
-(p_1+ip_2)A_1 + \quad p_3A_2 \qquad\qquad\qquad +(p_0+m)A_4 &= 0.
\end{aligned}$$

Its determinant is zero provided that the energy–momentum relation (6.1.42*b*) is satisfied. It will also be verified that its minors of first order are zero: the solutions are therefore double and we can choose the first two components of each solution arbitrarily. In all we obtain four linearly independent solutions, $A^{(1)}(\boldsymbol{p})$, $A^{(2)}(\boldsymbol{p})$, $A^{(3)}(\boldsymbol{p})$, $A^{(4)}(\boldsymbol{p})$, which we express in the form $\sqrt{\dfrac{\omega+m}{2\omega}}\,A^{(\varrho)\prime}(\boldsymbol{p})$ with

† The proof of (6.1.44) is simple. We first note that

$$[\Sigma(p),\ \alpha_j p_j+m\gamma_4] = [\Sigma(p),\ \beta](i\gamma_j p_j+m)+\beta[\Sigma(p),\ i\gamma_j p_j+m];$$

on the other hand, taking the formula (6.1.10) into account, we may express $\Sigma(p)$ as

$$\Sigma(p) = -\frac{i}{4}\,\varepsilon_{klm}\gamma_l\gamma_m\,\frac{p_k}{|\boldsymbol{p}|}.$$

Finally, the anticommutation properties of the γ-matrices show that

$$[\Sigma(p),\ \gamma_4] = -\frac{i}{4}\,\varepsilon_{klm}\,\frac{p_k}{p}\,[\gamma_l\gamma_m,\ \gamma_4] = 0$$

and

$$[\Sigma(p),\ \gamma_j]p_j = -i\,\frac{p_kp_j}{|p|}\,\varepsilon_{klj}\gamma_l.$$

Again this expression is zero since the factors p_kp_j and ε_{klj} are respectively symmetric and antisymmetric in k and j.

$$\left.\begin{array}{c|cccc}
\mu \backslash \varrho & 1. & 2. & 3. & 4. \\
1 & 1 & 0 & -\dfrac{p_3}{\omega+m} & -\dfrac{p_1-ip_2}{\omega+m} \\
2 & 0 & 1 & -\dfrac{p_1+ip_2}{\omega+m} & \dfrac{p_3}{\omega+m} \\
3 & \dfrac{p_3}{\omega+m} & \dfrac{p_1-ip_2}{\omega+m} & 1 & 0 \\
4 & \dfrac{p_1+ip_2}{\omega+m} & -\dfrac{p_3}{\omega+m} & 0 & 1 \\
 & \multicolumn{2}{c}{\underbrace{\qquad\qquad\qquad}_{p_0 = +\omega}} & \multicolumn{2}{c}{\underbrace{\qquad\qquad\qquad}_{p_0 = -\omega}}
\end{array}\right\} A_\mu^{(\varrho)\prime}(\boldsymbol{p}); \tag{6.1.45}$$

the normalization volume of $\psi(x)$ has been chosen equal to 1.

When the direction of propagation ($p_1 = p_2 = 0$) is chosen as the axis x_3, we easily verify that the first and third columns correspond to the eigenvalue $\frac{1}{2}$ of Σ and the two remaining columns to $-\frac{1}{2}$: consequently, the spin is in the direction of propagation for the first two solutions and in the opposite direction for the last two.

We also obtain the orthogonality relations

$$\sum_\mu A_\mu^{(\varrho)*}(\boldsymbol{p}) A_\mu^{(\varrho')}(\boldsymbol{p}) = \delta_{\varrho\varrho'}, \tag{6.1.46a}$$

$$\sum_\varrho A_\mu^{(\varrho)*}(\boldsymbol{p}) A_{\mu'}^{(\varrho)}(\boldsymbol{p}) = \delta_{\mu\mu'}. \tag{6.1.46b}$$

The first relation expresses the fact that the solutions are orthonormal; the second one is a closure relation.

Let us furthermore remark that from (6.1.45) and (6.1.30) we obtain the well-known formula of wave mechanics:

$$\tilde{A}(\boldsymbol{p})\boldsymbol{\alpha}A(\boldsymbol{p}) = i\tilde{A}(\boldsymbol{p})\beta\boldsymbol{\gamma}A(\boldsymbol{p}) = \boldsymbol{v}, \tag{6.1.47}$$

making use of $\boldsymbol{p} = |p_0|\,\boldsymbol{v}$ [cf. formula (1.33)].

We have already noted that the Dirac spinors given by table (6.1.45) are pure helicity states only for the case $p_1 = p_2 = 0$. Although this is all that is needed in most of the applications, it is worth while to write down simultaneous eigenvalues of $H_0(\boldsymbol{p})$ and $\Sigma(\boldsymbol{p})$. Let us begin with a few comments: we note that the matrix (6.1.45) can be expressed (up to the factor

$\sqrt{\dfrac{\omega+m}{2\omega}}\Big)$ by means of 2×2 matrices as follows:

$$M(\boldsymbol{p}) = \sqrt{\frac{\omega+m}{2}}\begin{pmatrix} I & \dfrac{-\sigma \boldsymbol{p}}{\omega+m} \\ \dfrac{\sigma \boldsymbol{p}}{\omega+m} & I \end{pmatrix} \tag{6.1.48}$$

that the former matrix is unitary and that one gets the solutions $A^{(\varrho)}(\boldsymbol{p})$ by taking the products of M with the spinors:

$$\begin{pmatrix}1\\0\\0\\0\end{pmatrix} \quad \begin{pmatrix}0\\1\\0\\0\end{pmatrix} \quad \begin{pmatrix}0\\0\\1\\0\end{pmatrix} \quad \begin{pmatrix}0\\0\\0\\1\end{pmatrix}.$$

We also note that:

$$\Sigma(\boldsymbol{p}) = \boldsymbol{S}\boldsymbol{n} = \frac{1}{2}\begin{pmatrix}\sigma_l & 0\\ 0 & \sigma_l\end{pmatrix} n_l = \frac{1}{2}\begin{pmatrix} n_3 & n_1-in_2 & 0 & 0\\ n_1+in_2 & -n_3 & 0 & 0\\ 0 & 0 & n_3 & n_1-in_2\\ 0 & 0 & n_1+in_2 & -n_3\end{pmatrix} \tag{6.1.49}$$

where $\boldsymbol{n} = \boldsymbol{p}/P$, with $P = |\boldsymbol{p}|$.

On the other hand, since $(\sigma\boldsymbol{n})^2 = I$, the projection matrices:

$$\frac{I+\sigma\boldsymbol{n}}{2} = \frac{1}{2}\begin{pmatrix}1+n_3 & n_1-in_2\\ n_1+in_2 & 1-n_3\end{pmatrix}; \quad \frac{I-\sigma\boldsymbol{n}}{2} = \frac{1}{2}\begin{pmatrix}1-n_3 & -n_1+in_2\\ -n_1-in_2 & 1+n_3\end{pmatrix}$$

are such that:

$$\sigma\boldsymbol{n}\frac{I+\sigma\boldsymbol{n}}{2} = \frac{I+\sigma\boldsymbol{n}}{2}, \quad \sigma\boldsymbol{n}\frac{I-\sigma\boldsymbol{n}}{2} = -\frac{I-\sigma\boldsymbol{n}}{2}.$$

In particular, the following spinors:

$$x^{(1)} = c\begin{pmatrix}1+n_3\\ n_1+in_2\\ 0\\ 0\end{pmatrix}, \quad x^{(2)} = c\begin{pmatrix}-n_1+in_2\\ 1+n_3\\ 0\\ 0\end{pmatrix}, \quad x^{(3)} = c\begin{pmatrix}0\\ 0\\ 1+n_3\\ n_1+in_2\end{pmatrix},$$

$$x^{(4)} = c\begin{pmatrix}0\\ 0\\ -n_1+in_2\\ 1+n_3\end{pmatrix} \tag{6.1.50}$$

with $c = (2+2n_3)^{-1/2}$ form an orthonormal system of Dirac spinors and are degenerate eigenvectors of $\Sigma(\boldsymbol{p})$:

$$\Sigma(\boldsymbol{p})x^{(l)} = s^{(e)}x^{(l)}$$

with $s^{(l)} = 1/2$ for $\varrho = 1,3$ and $s^{(e)} = -1/2$ for $\varrho = 2,4$.

We finally note that since $M(\boldsymbol{p})$ is a unitary matrix, the spinors $M(\boldsymbol{p})x^{(l)}$ are still orthonormal and that†

$$\Sigma(\boldsymbol{p})M(\boldsymbol{p})x^{(l)} = M(\boldsymbol{p})\Sigma(\boldsymbol{p})x^{(l)} = s^{(e)}M(\boldsymbol{p})x^{(l)},$$

i.e. the $M(\boldsymbol{p})x^{(l)}$ spinors remain eigenvectors of $\Sigma(\boldsymbol{p})$.

Considering the four spinors:

$$Mx^{(1)} = C\begin{pmatrix} 1+n_3 \\ n_1+in_2 \\ \dfrac{P}{\omega+m}(s+n_3) \\ \dfrac{P}{\omega+m}(n_1+in_2) \end{pmatrix}, \qquad Mx^{(2)} = C\begin{pmatrix} -n_1+in_2 \\ 1+n_3 \\ \dfrac{-P}{\omega+m}(-n_1+in_2) \\ \dfrac{-P}{\omega+m}(1+n_3) \end{pmatrix}$$

$$Mx^{(3)} = C\begin{pmatrix} \dfrac{-P}{\omega+m}(1+n_3) \\ \dfrac{-P}{\omega+m}(n_1+in_2) \\ 1+n_3 \\ n_1+in_2 \end{pmatrix}, \qquad Mx^{(4)} = C\begin{pmatrix} \dfrac{P}{\omega+m}(-n+in_2) \\ \dfrac{P}{\omega+m}(1+n_3) \\ -n_1+in_2 \\ s+n_3 \end{pmatrix} \tag{6.1.51}$$

with $P = |\boldsymbol{p}|$, $C = \sqrt{\dfrac{\omega+m}{4\omega(1+n_3)}}$, we note that they form an orthonormal system and are linear combinations of the $A^{(\varrho)}(\boldsymbol{p})$, the first two being linear combinations of $A^{(1)}$ and $A^{(2)}$, the last two being linear combinations of $A^{(3)}$ and $A^{(4)}$. Therefore, the first one corresponds to $s = 1/2$, $p_0 = \omega$; the second to $s = -1/2$, $p_0 = \omega$; the third to $s=1/2$, $p_0=-\omega$ and the fourth to $s=-1/2$, $p_0=-\omega$. The above considerations imply that $1+n_3 \neq 0$; if $1+n_3 = 0$, then it follows that $n_1 = n_2 = 0$ and table 6.1.45 gives the required solution.

6. Feynman notation

Feynman introduced a special notation, which is often used and is worth knowing.

With the metric

$$g_{\mu\nu} = 0 \quad \text{if} \quad \mu \neq \nu, \quad g_{00} = +1; \quad g_{jj} = -1,$$

† The matrices $M(\boldsymbol{p})$ and $\Sigma(\boldsymbol{p})$ commute, as can be easily verified.

the scalar product of two four-vectors a and b is given by

$$a.b = a_0 b_0 - \boldsymbol{a}.\boldsymbol{b}.$$

Taking the following representation of the γ's:

$$\gamma_0 = \begin{pmatrix} \mathrm{I} & 0 \\ 0 & -\mathrm{I} \end{pmatrix}, \quad \gamma_j = \begin{pmatrix} 0 & \sigma_j \\ -\sigma_j & 0 \end{pmatrix},$$

it is easy to verify that these matrices obey the anticommutation relations

$$[\gamma_\mu, \gamma_\nu] = 2g_{\mu\nu}.$$

With the four-vector a, Feynman associated

$$\hat{a} = \gamma_0 a_0 - \gamma_j a_j \equiv \gamma_0 a_0 - \boldsymbol{\gamma}.\boldsymbol{a} \tag{6.1.52}$$

and we can easily verify the following relations

$$\begin{aligned} \hat{a}\hat{b} &= 2ab - \hat{b}\hat{a}, \quad \hat{a}\hat{a} = aa = a^2, \\ \gamma_\mu \hat{a} \gamma_\mu &= -2\hat{a}, \quad \gamma_\mu \hat{a}\hat{b}\gamma_\mu = 2(\hat{a}\hat{b} + \hat{b}\hat{a}) = 4ab, \\ \gamma_\mu \hat{a}\hat{b}\hat{c}\gamma_\mu &= -2\hat{c}\hat{b}\hat{a}, \quad \gamma_\mu \hat{a} = \hat{a}\gamma_\mu + 2a_\mu. \end{aligned}$$

We also define the four-dimensional operator

$$\nabla = \left(\frac{\partial}{\partial x_0}, -\nabla\right), \tag{6.1.53a}$$

so that the Dirac equation takes the form

$$(i\nabla - m)\psi(x) = 0, \tag{6.1.53b}$$

with

$$\hat{\nabla} = \gamma_0 \frac{\partial}{\partial x_0} + \gamma_j \frac{\partial}{\partial x_j}$$

in agreement with (6.1.52).

7. Charge conjugation, Majorana spinors†

The charge conjugation formalism can easily be applied to the spinors case: the existence of the C matrix is ensured by the fact that the γ_μ^T's obey the same anticommutation relations as the γ_μ's (6.1.2). In the standard representation of the γ's, we easily verify, in virtue of (6.1.32), that the C matrix defined by (5.144) obeys

$$\left.\begin{aligned} [\mathcal{C}, \gamma_1]_+ = [\mathcal{C}, \gamma_3]_+ &= [\mathcal{C}, \gamma_4]_+ = 0, \\ [\mathcal{C}, \gamma_2] &= 0. \end{aligned}\right\} \tag{6.1.54}$$

† E. Majorana, *Nuovo Cimento*, **14**, 171 (1937).

We can therefore take

$$\mathscr{C} = \alpha\gamma_2 \qquad (6.1.55a)$$

and in agreement with (5.149):

$$C = \alpha\gamma_2\beta; \quad C^{-1}\gamma_\mu C = -\gamma_\mu^T \qquad (6.1.55b)$$

where $\alpha^*\alpha = 1$ in order that C may be a unity matrix. We shall take $\alpha = i$ such that $c^2 = I$.

It will also be verified that the expressions (6.1.31) of η, (6.1.15*a*) of τ, (6.1.16) of T and (6.1.55*b*) of C do obey (5.181).

We can form from ψ and $\psi^{(c)}$ new wave functions which obey the Dirac equation and are invariant under charge conjugation:

$$\psi^{(1)}(x) = \frac{1}{\sqrt{2}}(\psi(x) + \psi^{(c)}(x)), \quad \psi^{(2)}(x) = \frac{1}{i\sqrt{2}}(\psi(x) - \psi^{(c)}(x)). \qquad (6.1.56)$$

Any spinor equal to its charge conjugate represents a field whose particles and antiparticles are identical: they are called Majorana particles.

Suppose now that we consider the spinors $\chi(x)$, which we shall refer to as pseudo-spinors; they are transformed under a space symmetry as follows:

$$\chi(x) \to \chi^{(\sigma)}(x') = \pm\gamma_4\chi(x) \qquad (6.1.57)$$

(instead of $\chi^{(\sigma)} = \pm i\gamma_4\chi$). Using calculations similar to those which led to (5.164), we can easily see that we have

$$\chi^{(c)\,(\sigma)}(x') = \mp\gamma_4\chi^{(c)}(x). \qquad (6.1.58)$$

The Majorana spinors and pseudo-spinors were introduced for the investigation of the β-decay; for a detailed discussion of this question, we refer readers to the book quoted in the footnote below.†

It may be noted finally that the well-known problem in wave mechanics of waves with negative frequencies which represent states with negative energies does not exist in quantum field theory: the energy operator $-iP_4$ is indeed positive definite as we shall see later on [formula (6.1.108)].

It must be added that when a particle state is described by means of a negative frequency

$$\psi(x) = A(\boldsymbol{k})e^{i\boldsymbol{k}.\boldsymbol{x}-i\omega t},$$

the antiparticle

$$\psi^{(c)}(x) = C\bar{A}^T(\boldsymbol{k})e^{-i\boldsymbol{k}.\boldsymbol{x}+i\omega t}$$

does in fact correspond to a positive frequency.

† Kai Siegbahn, *Beta and Gamma Spectroscopy*, North Holland, 1955, where bibliographical indications will be found.

8. Summations over the spins and energies

In quantum field theory, as in wave mechanics, we often have occasion to consider squares of matrix elements of particular operators of the spin space, built therefore with the aid of the γ matrices. For instance, let O be such an operator (Hermitian matrix); we have to compute expressions of the form

$$|\tilde{A}^{(\varrho)}(\boldsymbol{p})OA^{(\beta)}(\boldsymbol{k})|^2 = \left|\sum_{\nu,\nu'} A_{\nu}^{(\varrho)*}(\boldsymbol{p})O_{\nu\nu'}A_{\nu'}^{(\beta)}(\boldsymbol{k})\right|^2. \qquad (6.1.59)$$

We often have to take the expectation values over all the final spin states, which amounts to multiplying by the factor $\frac{1}{2}$ then carrying out a summation over ϱ ($\varrho=1, 2$ or $\varrho=3, 4$), or alternatively over the spin states and the two signs of final frequencies ($\varrho=1, 2, 3, 4$). The first case corresponds to an experiment in which the polarization of the particles is not observed.

Summations of these kinds involve considerable simplifications in the calculations. Suppose, for instance, that we sum over the ϱ's, ϱ running from 1 to 4, which would correspond to a summation over the two energy signs and the spins. Formula (5.1.59) takes the form

$$\sum_{\varrho} |\tilde{A}^{(\varrho)}(\boldsymbol{p})OA^{(\beta)}(\boldsymbol{k})|^2$$
$$= \sum_{\varrho=1}^{4} \sum_{\substack{\mu,\mu'\\ \nu,\nu'}} A_{\mu}^{(\beta)*}(\boldsymbol{k})O_{\mu\mu'}A_{\mu'}^{(\varrho)}(\boldsymbol{p})A_{\nu}^{(\varrho)*}(\boldsymbol{p})O_{\nu\nu'}A_{\nu'}^{(\beta)}(\boldsymbol{k}), \qquad (6.1.60)$$

where the energy $\boldsymbol{p}$ of the intermediate state is constant.

This is a particular case of more general expressions which we intend to investigate in this section. Consider, for instance, the following summation:

$$\sum_{\varrho} (\tilde{A}^{(\alpha)}(\boldsymbol{k}_1)O^{(1)}A^{(\varrho)}(\boldsymbol{p}))\,(\tilde{A}^{(\varrho)}(\boldsymbol{p})O^{(2)}A^{(\beta)}(\boldsymbol{k})); \qquad (6.1.60a)$$

making use of (6.1.46*b*), we can carry out the summation over ϱ; this formula is written

$$A^{(\alpha)}(\boldsymbol{k}_1)O^{(1)}O^{(2)}A^{(\beta)}(\boldsymbol{k}_2) \qquad (6.1.60b)$$

and the dependence of (6.1.60) with respect to $\boldsymbol{p}$ has disappeared.

Similarly,

$$\sum_{\varrho} \tilde{A}^{(\varrho)}(\boldsymbol{p})OA^{(\varrho)}(\boldsymbol{p}) = \sum_{\mu} O_{\mu\mu} = \mathrm{Sp}O \qquad (6.1.61)$$

and the summation is reduced to the computation of a spur.

We also have to perform other types of summation in which we only sum over the ϱ's which correspond to a single sign of the frequency ω, and therefore over the two values of the spin. These summations can be simplified by introducing projection operators on the spaces of the wave functions with positive and negative frequencies. Consider indeed the wave functions

$$\psi(x) = A(\boldsymbol{p}, \pm\omega)e^{i(\boldsymbol{p}.\boldsymbol{x}\mp\omega x_0)};$$

the operators

$$\lambda^{\pm}(\boldsymbol{p}) = \frac{\pm(\boldsymbol{\alpha}.\boldsymbol{p}+\beta m)+\omega}{2\omega} \tag{6.1.62a}$$

are the required projection operators, since

$$\lambda^{+}(\boldsymbol{p})A(\boldsymbol{p}, \omega) = A(\boldsymbol{p}, \omega), \quad \lambda^{+}(\boldsymbol{p})A(\boldsymbol{p}, -\omega) = 0,$$
$$\lambda^{-}(\boldsymbol{p})A(\boldsymbol{p}, \omega) = 0, \quad \lambda^{-}(\boldsymbol{p})A(\boldsymbol{p}, -\omega) = A(\boldsymbol{p}, -\omega).$$

It will be noted that these operators can also be written

$$\lambda^{\pm} = \mp\frac{i\gamma_u p_u - m}{2\omega}\beta, \tag{6.1.62b}$$

where p_0 is equal to $\pm\omega$ according to the superscript of λ. Using the same theorems as before, we easily verify that the summation

$$\sum_{\varrho,\, p_0>0}\left(\tilde{A}^{(\alpha)}(\boldsymbol{k}_1)O^{(1)}A^{(\varrho)}(\boldsymbol{p})\right)\left(\tilde{A}^{(\varrho)}(\boldsymbol{p})O^{(2)}A^{(\beta)}(\boldsymbol{k}_2)\right) \tag{6.1.63a}$$

takes the form

$$\tilde{A}^{(\alpha)}(\boldsymbol{k}_1)O^{(1)}\lambda^{+}(\boldsymbol{p})O^{(2)}A^{(\beta)}(\boldsymbol{k}_2) \tag{6.1.63b}$$

and that we also have

$$\sum_{\varrho,\, p_0>0}\tilde{A}^{(\varrho)}(\boldsymbol{p})OA^{(\varrho)}(\boldsymbol{p}) = \mathrm{Sp}\,\lambda^{+}(\boldsymbol{p})O. \tag{6.1.64}$$

We are therefore finally led to compute spurs of products of γ matrices. It can be easily proved that the spur of a product of an odd number of γ matrices is zero. Indeed, taking into account the fact that γ_5 anticommutes with the four γ_μ's, we can write

$$\gamma_5\gamma_{\mu_1}\ldots\gamma_{\mu_n}\gamma_5 = (-1)^n\gamma_{\mu_1}\ldots\gamma_{\mu_n}, \tag{6.1.65}$$

from which it follows, by virtue of formula $\mathrm{Sp}\,\{AB\} = \mathrm{Sp}\,\{BA\}$, that

$$\mathrm{Sp}\,\{\gamma_5(\gamma_{\mu_1}\ldots\gamma_{\mu_n}\gamma_5)\} = \mathrm{Sp}\{(\gamma_{\mu_1}\ldots\gamma_{\mu_n}\gamma_5)\gamma_5\} = \mathrm{Sp}\{\gamma_{\mu_1}\ldots\gamma_{\mu_n}\}.$$

Returning to (6.1.65), we get

$$(-1)^n\mathrm{Sp}\{\gamma_{\mu_1}\ldots\gamma_{\mu_n}\} = \mathrm{Sp}\{\gamma_{\mu_1}\ldots{}_{\mu_n}\}. \tag{6.1.66}$$

Consequently, the spur of a product of an odd number of γ matrices is zero.

If n is even the anticommutation relations can always be used to reduce a product of n factors to an expression with $n-2$ factors; we thus obtain

$$\left.\begin{aligned} &\text{(a)}\quad \mathrm{Sp}\{\gamma_\mu\gamma_\nu\} = 4\delta_{\mu\nu},\\ &\text{(b)}\quad \mathrm{Sp}\{\gamma_\mu\gamma_\nu\gamma_\varrho\gamma_\sigma\} = 4\delta_{\mu\sigma}\delta_{\nu\varrho}-4\delta_{\mu\varrho}\delta_{\nu\sigma}+4\delta_{\mu\nu}\delta_{\varrho\sigma},\\ &\text{(c)}\quad \mathrm{Sp}\{\gamma_\mu a_\mu\} = 0,\\ &\text{(d)}\quad \mathrm{Sp}\{\gamma_\mu a_\mu\gamma_\nu b_\nu\} = 4a_\mu b_\mu; \end{aligned}\right\} \tag{6.1.67}$$

in formulae (c) and (d), a and b are four-vectors. We also have

$$\left.\begin{aligned} &\mathrm{Sp}\{\gamma_5\gamma_\mu\} = 0, \qquad \mathrm{Sp}\{\gamma_5\gamma_\mu\gamma_\nu\} = 0,\\ &\mathrm{Sp}\{\gamma_5\gamma_\mu\gamma_\nu\gamma_\lambda = 0, \qquad \mathrm{Sp}\{\gamma_\mu\gamma_\nu\gamma_\varrho\gamma_\sigma\gamma_5\} = 4\varepsilon_{\mu\nu\varrho\sigma}. \end{aligned}\right\} \tag{6.1.68}$$

9. Foldy–Wouthuysen transformation†

This is a canonical transformation which diagonalizes the Dirac Hamiltonian (in wave mechanics). We shall investigate the simple case of a free particle. The matrix

$$U(\boldsymbol{k}, \omega) = \frac{\boldsymbol{\alpha}.\boldsymbol{k}+\beta(m+\omega)}{\sqrt{2\omega(\omega+m)}}\beta = \frac{-i\gamma_j p_j+m+\omega}{\sqrt{2\omega(\omega+m)}} \tag{6.1.69a}$$

is a unitary matrix, as may be easily verified, the inverse of which is

$$U^{-1}(\boldsymbol{k}, \omega) = \beta\frac{\boldsymbol{\alpha}.\boldsymbol{k}+\beta(m+\omega)}{\sqrt{2\omega(\omega+m)}}, \tag{6.1.69b}$$

and it is equally easy to verify that

$$U^{-1}(\boldsymbol{k}, \omega)(\boldsymbol{\alpha}.\boldsymbol{k}+m\beta)U(\boldsymbol{k}, \omega) \equiv U^{-1}H_0U = \beta\omega. \tag{6.1.70}$$

Consequently, in the standard representation of β, the eigenfunctions of H_0 which correspond respectively to $\pm\omega$ are

$$\begin{pmatrix}1\\0\\0\\0\end{pmatrix}, \begin{pmatrix}0\\1\\0\\0\end{pmatrix}; \begin{pmatrix}0\\0\\1\\0\end{pmatrix}, \begin{pmatrix}0\\0\\0\\1\end{pmatrix}. \tag{6.1.71}$$

The transform of certain remarkable matrices are worth giving:

$$\left.\begin{aligned} \boldsymbol{\alpha} \to \boldsymbol{\alpha}' &= \boldsymbol{\alpha}+\frac{\beta\boldsymbol{p}}{\omega}-\frac{(\boldsymbol{\alpha}.\boldsymbol{p})\boldsymbol{p}}{\omega(\omega+m)},\\ \beta \to \beta' &= \frac{-\boldsymbol{\alpha}.\boldsymbol{p}+m\beta}{\omega},\\ \boldsymbol{\sigma} \to \boldsymbol{\sigma}' &= \boldsymbol{\sigma}+\frac{i\beta(\boldsymbol{\alpha}\wedge\boldsymbol{p})}{\omega}-\frac{\boldsymbol{p}\wedge(\boldsymbol{\sigma}\wedge\boldsymbol{p})}{\omega(\omega+m)}. \end{aligned}\right\} \tag{6.1.72}$$

† L. L. Foldy and S. A. Wouthuysen, *Phys. Rev.*, **78**, 29 (1950).

We verify that β' and σ' have non-zero matrix elements only for wave functions with positive frequencies (they are called even operators) while α has non-zero matrix elements for wave functions with negative frequencies (they are called odd operators). This method is helpful whenever we have to investigate transitions between states with frequencies of the same sign; we shall also make use of it to build solutions of the Dirac equation (6.1.100).

B. QUANTIZATION OF THE DIRAC EQUATION†

10. The elementary solutions of the Dirac equation

Elementary solutions of the Dirac equation can be built, as indicated in (5.108) with the help of the differential operator $\mathcal{D}(\partial)$ defined by (4.20):

$$-\{\gamma_\mu\partial_\mu+m\}\mathcal{D}(\partial) = \square - m^2. \tag{6.1.73a}$$

The identification of the two sides gives the anticommutation relations of the γ_μ's [cf. (4.25)] and the expression of $\mathcal{D}(\partial)$:

$$\mathcal{D}(\partial) = -\gamma_\mu\partial_\mu+m. \tag{6.1.73b}$$

Returning therefore to formula (5.108), we can define the various functions $G(x)$, which are denoted by $S(x)$ in the case of the Dirac equation as follows:

$$\left.\begin{aligned} S(x) &= \{\gamma_\mu\partial_\mu-m\}\,\Delta(x),\\ S^{\pm}(x) &= \{\gamma_\mu\partial_\mu-m\}\,\Delta^{\pm}(x),\\ S^{\substack{\text{ret}\\ \text{adv}}}(x) &= \{\gamma_\mu\partial_\mu-m\}\,\Delta^{\substack{\text{ret}\\ \text{adv}}}(x),\\ \bar{S}(x) &= \{\gamma_\mu\partial_\mu-m\}\,\bar{\Delta}(x),\\ S^{(C)}(x) &= \{\gamma_\mu\partial_\mu-m\}\,\Delta_C(x), \end{aligned}\right\} \tag{6.1.74}$$

the functions $S(x)$ are obviously 4×4 matrices the elements of which are functions of x (distributions, to be more precise).

Since the differential operator $\mathcal{D}(\partial)$ is of first order, these functions obey the following relations:

$$\left.\begin{aligned} S^{\substack{\text{ret}\\ \text{adv}}}(x) &= \mp\theta\pm(t)\{\gamma_\mu\partial_\mu-m\}\,\Delta(x),\\ \bar{S}(x) &= -\frac{1}{2}\,\varepsilon(t)\{\gamma_\mu\partial_\mu-m\}\Delta(x). \end{aligned}\right\} \tag{6.1.75}$$

† The quantization of the field of spin $\frac{1}{2}$ was investigated by P. Jordan and E. Wigner, *Zeits. Phys.*, **47**, 631 (1928); V. Fock, *Zeits. Phys.*, **75**, 622 (1932).

Let us prove for instance the first relation; we have

$$-\{\gamma_\mu\partial_\mu - m\}\theta_+(t)\Delta(x) = -(\gamma_\mu\partial_\mu\theta_+(t))\Delta(x) - \theta_+(t)\{\gamma_\mu\partial_\mu - m\}\Delta(x)$$
$$= i\gamma_4\,\delta(t)\Delta(x) - \theta_+(t)\{\gamma_\mu\partial_\mu - m\}\Delta(x)$$

and the first term of the last line is zero since $\Delta(\boldsymbol{x},0) = 0$. The functions S (distributions) and their representations have been collected in a table at the end of this section. We see that $S(x)$ and $S_\pm(x)$ are both solutions of the Dirac equation, but that the other S functions are Green functions (elementary solutions).

Note finally that the function $S(x)$ is a solution of the following Cauchy problem: given the solution of the Dirac equation over a spacelike surface σ, find this solution for a point x outside this surface.† We propose to prove that we have, indeed:

$$\psi(x) = \int_\sigma S(x-x')\gamma_\mu\psi(x')\,d\sigma'_\mu. \tag{6.1.76a}$$

Let us first prove that the above integral is a functional independent of σ. By forming

$$f_\mu(x') = S(x-x')\gamma_\mu\psi(x'),$$

we see by virtue of (1.133) that

$$\frac{\delta\psi[\sigma|x]}{\delta\sigma(x')} = \frac{\partial f_\mu(x')}{\partial x'_\mu} = 0,$$

since

$$\frac{\partial}{\partial x'_\mu} f_\mu(x') = \left(\frac{\partial}{\partial x'_\mu} S(x-x')\right)\gamma_\mu\psi(x') + S(x-x')\gamma_\mu\frac{\partial}{\partial x'_\mu}\psi(x')$$
$$= S(x-x')\left\{\gamma_\mu\overleftarrow{\frac{\partial}{\partial x'_\mu}} - m\right\}\psi(x') = 0.$$

Taking the plane t = constant for σ, we then obtain

$$\psi(x) = \lim_{t=t'} -i\int\{\gamma_\mu\partial_\mu - m\}\Delta(x-x')\beta\psi(x')\,d^3\boldsymbol{x}'$$

† We may reach the same conclusion by noting that for $x_0 = 0$, $S(x)$ reduces to $i\gamma_4\delta(x)$. Indeed, we may write:

$$S(x) = -\{\gamma_\mu\partial_\mu - m\}\frac{1}{(2\pi)^3}\int_{k_0=\omega} e^{i\boldsymbol{kx}}\frac{\sin\omega x_0}{\omega}\,d^3\boldsymbol{k} = -\frac{1}{(2\pi)^3}\{\gamma_j\partial_j - m\}$$
$$\int_{k_0=\omega} e^{i\boldsymbol{kx}}\frac{\sin\omega x_0}{\omega}\,d^3\boldsymbol{k} + \frac{i\gamma_4}{(2\pi)^3}\frac{\partial}{\partial x_0}\int_{k_0=0} e^{i\boldsymbol{kx}}\frac{\sin\omega x_0}{\omega}\,d^3\boldsymbol{k}$$

for $x_0 = 0$: the first term is 0, the second one becomes $i\gamma_4\delta(\boldsymbol{x})$.

and passing to the limit:

$$\psi(x) = \lim_{t=t'} -i \int \frac{\partial \Delta(x-x')}{\partial x_4} \psi(x')\, d^3\boldsymbol{x}' = \psi(x).$$

Finally, by virtue of the definitions (6.1.74), one has:

$$\{\gamma\partial + m\} S(x) = 0, \qquad \{\gamma\partial + m\} S^{\text{ret}}_{\text{adv}}(x) = -\delta(x), \qquad \dots \quad (6.1.76b)$$

TABLE OF THE FUNCTIONS $S(x)$:

$$S^{(\mathscr{C})}(x) = \{\gamma\partial - m\} \Delta^{(\mathscr{C})}(x) = -\frac{1}{(2\pi)^4} \int_{(\mathscr{C})} d^4k e^{ikx} \frac{i\gamma k - m}{k_\mu k_\mu + m^2}.$$

$$S(x) = -\frac{i}{(2\pi)^3} \int d^4k e^{ikx} (i\gamma k - m)\, \varepsilon(k)\, \delta(k^2 + m^2)$$

$$S^{\pm}(x) = \mp \frac{i}{(2\pi)^3} \int d^4k e^{ikx} (i\gamma k - m) \theta_{\pm}(k)\, \delta(k^2 + m^2)$$

$$\overline{S}(x) = \frac{P}{(2\pi)^4} \int d^4k e^{ikx} \frac{i\gamma k - m}{k_\mu k_u + m^2}$$

$$S^{(1)}(x) = \frac{1}{(2\pi)^3} \int d^4k e^{ikx} (i\gamma k - m)\, \delta(k^2 + m^2) = i(S^+(x) - S^-(x))$$

$$S^{(c)}(x) = \frac{2}{(2\pi)^3} \int d^4k e^{ikx} (i\gamma k - m)\, \delta_-(k^2 + m^2)$$

$$= \lim_{\varepsilon=0} -\frac{2i}{(2\pi)^4} \int d^4k e^{ikx} \frac{i\gamma k - m}{k^2 + m^2 - i\varepsilon} = \begin{cases} -2iS^-(x) & \text{if } x_0 < 0 \\ +2iS^+(x) & \text{if } x_0 > 0. \end{cases}$$

11. Anticommutation rules and Hamiltonian

The anticommutation rules are directly obtained from formula (5.106); we get

$$[\psi_\alpha(x), \bar\psi_\beta(x')]_+ = -i\{\gamma_\mu\partial_\mu - m\}_{\alpha\beta} \Delta(x-x') = -iS_{\alpha\beta}(x-x'). \quad (6.1.77)$$

It is interesting, however, if only as an example of an application of the Hamiltonian theory formulated in Chap. II, to write the Hamiltonian of the Dirac field explicitly and to infer from it the anticommutation rules.

We can start from the Lagrangian density:

$$\mathscr{L} = -\bar\psi(x) \{\gamma_\mu\partial_\mu + m\} \psi(x), \quad (6.1.78)$$

from which we infer the conjugate momenta:

$$\Pi(x) = i\bar\psi(x)\beta = i\tilde\psi(x), \quad \tilde\Pi(x) = 0; \quad (6.1.79)$$

the variables $\tilde{\psi}(x)$ and $\tilde{\Pi}(x)$ cannot therefore be considered as canonical variables. Then we form the Hamiltonian with the help of the Dirac equation; we obtain

$$\mathcal{H}(x) = \Pi_\mu(x)\dot{\psi}_\mu(x) = \Pi(x)\{-\alpha.\nabla - im\beta\}\psi(x) \qquad (6.1.80)$$
$$= \tilde{\psi}(x)\{-i\alpha.\nabla + m\beta\}\psi(x).$$

It will be noted that all the non-canonical variables have been eliminated.†

It is then easy to infer the Hamilton equations for the canonical variables; we get

$$\frac{\partial\psi(x)}{\partial t} = -i\beta\{\gamma.\nabla + m\}\psi(x), \quad \frac{\partial\Pi(x)}{\partial t} = -i\Pi(x)\beta\{\gamma.\nabla - m\}. \quad (6.1.81)$$

One can then prove (6.1.77) by the method described in Chap. II, § 9. It is sufficient indeed to solve the Dirac equation formally by means of the evolution operator of wave mechanics (Chap. IX, § 18):

$$\psi(\boldsymbol{x}, t) = e^{-i(t-t')\{\beta\gamma.\nabla + m\beta\}}\,\psi(\boldsymbol{x}, t'); \qquad (6.1.82)$$

carrying this expression into the left side of (6.1.77):

$$[\psi_\alpha(x), \tilde{\psi}_\beta(x')]_+ = -i\{e^{-i(t-t')\{\beta\gamma.\nabla + m\beta\}}\}_{\alpha\nu}[\psi_\nu(\boldsymbol{x}, t'), \Pi_\mu(\boldsymbol{x}', t')]_+\beta_{\mu\beta} \qquad (6.1.83)$$

and using the methods described in Chap. IX, § 3, this expression leads us once again to (6.1.77) [cf. in particular (9.175*a*)].

We shall finally arrive at the anticommutation rules (6.1.77) by a direct method.

Form

$$[\psi_\alpha(x), \tilde{\psi}_\beta(x')]_+ = -i[\psi_\alpha(x), \Pi_\mu(x')]_+\beta_{\mu\beta} \qquad (6.1.84)$$

and note incidentally that, given the invariance of the theory under the translations, the above anticommutator is a function of $x - x'$ (and not of x and x' separately); let $\mathcal{C}_{\alpha\beta}(x - x')$ therefore be the expression of (6.1.84).

† It will also be noted that denoting the Hamiltonian of the Dirac theory in wave mechanics by

$$H_0^{(\mathrm{cl})} = -i\alpha.\nabla + m\beta,$$

we get

$$H = \int \mathcal{H}(x)\, d^3\boldsymbol{x} = \int \tilde{\psi}(x) H_0^{(\mathrm{cl})}\psi(x)\, d^3\boldsymbol{x},$$

an expression which represents the expectation value of the energy operator in wave mechanics. In the same way, formula (6.1.86) gives the mean value of the momentum operator $-i\partial_\mu$ of wave mechanics, (6.1.87) the mean value of the angular momentum operator $-i(x_\nu\partial_\mu - x_\mu\partial_\nu)$ and (6.1.88*b*) the mean value of the spin operator $i/2\gamma_j\gamma_k$.

When $x_0 = x_0'$ according to the canonical anticommutation rules, $\mathscr{C}(\boldsymbol{x}-\boldsymbol{x}', 0) = \beta\delta(\boldsymbol{x}-\boldsymbol{x}')$; we can therefore say that $\mathscr{C}(x)$ is a solution of the Dirac equation which, when $t = 0$, becomes equal to $\beta\delta(\boldsymbol{x})$. Taking as a surface σ in (6.1.76) the equitemporal plane $t' = 0$, we obtain

$$\mathscr{C}(x) = -i \int_{t'=0} S(x-x')\, \delta(\boldsymbol{x}')\, d^3\boldsymbol{x}' = -iS(x)$$

and find once more (6.1.77).†

12. The Dirac field observables

The expressions of these observables can be obtained by a straightforward application of the formulae in the last chapter (§ 17).

The energy–momentum tensor can be readily written by virtue of (5.139):

$$T_{\mu\nu} = \bar{\psi}(x)\gamma_\nu \frac{\partial\psi(x)}{\partial x_\mu} \tag{6.1.85}$$

and the energy–momentum vector by virtue of (5.140*b*):

$$P_\mu = -i \int \bar{\psi}(x)\beta \frac{\partial\psi(x)}{\partial x_\mu}\, d^3\boldsymbol{x}. \tag{6.1.86}$$

The orbital angular momentum tensor (5.141) and the spin-tensor (5.142) are expressed as follows:

$$L_{\mu\nu} = -i \int \tilde{\psi}(x)\{x_\nu\partial_\mu - x_\mu\partial_\nu\}\psi(x)\, d^3\boldsymbol{x}, \tag{6.1.87}$$

$$\mathfrak{M}_{\mu\nu} = \frac{i}{4} \int \bar{\psi}(x)[\gamma_\mu\gamma_\nu, \beta]_+\psi(x)\, d^3\boldsymbol{x}. \tag{6.1.88a}$$

It will be noted that the anti-Hermitian components of $\mathfrak{M}_{\mu\nu}$ are zero since $[\gamma_j\gamma_4, \gamma_4]_+ = 0$ and that the Hermitian components are

$$\mathfrak{M}_{jk} = \frac{i}{2} \int \bar{\psi}(x)\gamma_j\gamma_k\beta\psi(x)\, d^3\boldsymbol{x}. \tag{6.1.88b}$$

We can finally write the symmetrical energy–momentum tensor (3.55):

$$\Theta_{\mu\nu} = \frac{1}{2}\bar{\psi}\gamma_\nu \frac{\partial\psi(x)}{\partial x_\mu} + \frac{1}{2}\bar{\psi}(x)\gamma_\mu \frac{\partial\psi(x)}{\partial x_\nu}. \tag{6.1.89}$$

The expression of the current is obtained by a direct application of formula (5.143):

$$J_\mu(x) = -ie\bar{\psi}(x)\gamma_\mu\psi(x). \tag{6.1.90a}$$

† This is a direct consequence of the footnote on p. 235.

Finally, two further expressions of $J_\mu(x)$, with zero vacuum-expectation values [cf. (8.118)], can be obtained from (5.162) and (5.163):

$$J_\mu(x) = -\frac{ie}{2}\,[\bar{\psi}_\alpha(x), \psi_\beta(x)](\gamma_\mu)_{\alpha\beta}, \tag{6.1.90b}$$

$$J_\mu(x) = -\frac{ie}{2}\,(\bar{\psi}(x)\gamma_\mu\psi(x) - \bar{\psi}^{(c)}(x)\gamma_\mu\psi^{(c)}(x)). \tag{6.1.90c}$$

It is convenient for the theory of the magnetic moment of the electron to express the current as a sum of two currents by the Gordon method. Taking into account the equations obeyed by $\psi(x)$ and $\bar{\psi}(x)$, it is easy to see that

$$J_\nu(x) = J_\nu^{(\mathrm{orb})}(x) + J_\nu^{(\mathrm{spin})}(x) \tag{6.1.91a}$$

with

$$J_\nu^{(\mathrm{orb})}(x) = +\frac{ie}{2m}\left(\bar{\psi}(x)\frac{\partial\psi(x)}{\partial x_\nu} - \frac{\partial\bar{\psi}(x)}{\partial x_\nu}\psi(x)\right), \tag{6.1.91b}$$

$$J_\nu^{(\mathrm{spin})}(x) = -\frac{ie}{4m}\frac{\partial}{\partial x_\mu}\{\bar{\psi}(x)[\gamma_\mu, \gamma_\nu]\psi(x)\}. \tag{6.1.91c}$$

$J_\mu^{(\mathrm{orb})}$ is called the orbital current while $J_\mu^{(\mathrm{spin})}$ is called the spin current. The reasons for these definitions will be discussed later on (Chap. X, § 12).

All the considerations relating to the charge conjugation (Chap. V, § 18) can be applied to the Dirac field; we refer readers to formulae (5.160) *et seq.* and to the remarks which follow them.

13. Creation and annihilation operators. Particle properties

According to formulae (5.119), the introduction of the creation and annihilation operators calls for a computation of the matrices $D(\pm ik)$ which have to obey the following equations:

$$D(\pm ik)\tilde{D}(\pm ik)\beta = \pm\mathcal{D}(\pm ik) = \pm(\mp i\gamma_\mu k_\mu + m), \qquad k_0 = \omega \tag{6.1.92a}$$

or

$$D(\pm ik)\tilde{D}(\pm ik) = (\boldsymbol{\alpha}.\boldsymbol{k} \pm m\beta + \omega). \tag{6.1.92b}$$

To determine these matrices we shall use the properties of $(1/2m)\mathcal{D}(\pm ik)$ expressed by formulae (5.137), namely that $(1/2m)\mathcal{D}(ik)$ and $(1/2m)\mathcal{D}(-ik)$ are respectively projection operators for the wave functions with positive and negative frequencies.

Consider therefore the orthogonal solutions of the equations [cf. (5.136)]:

$$\left.\begin{aligned} (\ \ i\gamma_\varrho k_\varrho + m)_{\alpha\mu} u_\mu^{(r)}(\boldsymbol{k}) &= 0 \quad (6.1.93a)\\ (-i\gamma_\varrho k_\varrho + m)_{\alpha\mu} v_\mu^{(r)}(\boldsymbol{k}) &= 0 \quad (6.1.93b)\end{aligned}\right\} k_0 = +\omega(\boldsymbol{k});$$

since the sign of the energy is fixed, r can only take two values (+ and −) corresponding to the eigenvalues $\pm\frac{1}{2}$ of the helicity matrix $\Sigma(\boldsymbol{p})$ as given by (6.1.43).

By virtue of formulae (5.137), we can write

$$\mathcal{D}_{\alpha\beta}(ik) = 2m \sum_r u_\alpha^{(r)}(\boldsymbol{k})\bar{u}_\beta^{(r)}(\boldsymbol{k}), \tag{6.1.94a}$$

$$\mathcal{D}_{\alpha\beta}(-ik) = 2m \sum_r v_\alpha^{(r)}(\boldsymbol{k})\bar{v}_\beta^{(r)}(\boldsymbol{k}), \tag{6.1.94b}$$

provided we normalize appropriately $u^{(r)}$ and $v^{(r)}$, as we shall do in (6.1.96). We can then choose, in agreement with (6.1.92):

$$D_{\alpha r}(i\boldsymbol{k}, -\omega) = \sqrt{2m}\, u_\alpha^{(r)}(k), \quad D_{\alpha r}(i\boldsymbol{k}, +\omega) = \sqrt{2m}\, v_\alpha^{(r)}(-k). \tag{6.1.95}$$

In other terms, the matrix $D(i\boldsymbol{k}, -\omega)$ is a matrix with only its two first columns different from 0 whereas the matrix $D(i\boldsymbol{k}, \omega)$ has only its two last columns different from 0, the index corresponding to these columns being related to the helicity.

We now normalize these functions as follows:

$$\tilde{u}^{(r)}(\boldsymbol{k})u^{(r')}(\boldsymbol{k}) = \frac{\omega}{m}\,\delta_{rr'}, \qquad \tilde{v}^{(r)}(\boldsymbol{k})v^{(r')}(\boldsymbol{k}) = \frac{\omega}{m}\,\delta_{rr'};\dagger \tag{6.1.96}$$

then, by identical calculations to those which led to formula (4.50), we verify that this norm leads to

$$\left.\begin{aligned} \bar{u}^{(r)}(\boldsymbol{k})u^{(r')}(\boldsymbol{k}) &= \delta_{rr'}, \\ \bar{v}^{(r)}(\boldsymbol{k})v^{(r')}(\boldsymbol{k}) &= -\delta_{rr'}, \end{aligned}\right\} \tag{6.1.98a}$$

$$\begin{aligned} \bar{u}^{(r)}(\boldsymbol{k})v^{(r')}(\boldsymbol{k}) &= \bar{v}^{(r)}(\boldsymbol{k})u^{(r')}(\boldsymbol{k}) \\ &= \bar{v}^{(r)}(\boldsymbol{k})\gamma_4 u^{(r')}(-\boldsymbol{k}) = \bar{u}^{(r)}(\boldsymbol{k})\gamma_4 v^{(r')}(-\boldsymbol{k}) = 0, \end{aligned} \tag{6.1.98b}$$

formulae which justify the relations (6.1.94). A straightforward application of formula (5.114) then enables us to write $\psi(x)$ in the form

$$\begin{aligned} \psi_\alpha(x) &= (2\pi)^{-\frac{3}{2}} \int_{k_0=\omega} d^3\boldsymbol{k} \sqrt{\frac{m}{\omega(\boldsymbol{k})}} \\ &\quad \times \left[\sum_r u_\alpha^{(r)}(\boldsymbol{k})\, a_r^+(\boldsymbol{k})\, e^{ikx} + v_\alpha^{(r)}(\boldsymbol{k})\, a_r^-(\boldsymbol{k})\, e^{-ikx} \right] \\ &\equiv \psi_\alpha^+(x) + \psi_\alpha^-(x), \end{aligned} \tag{6.1.99a}$$

$$\begin{aligned} \bar{\psi}_\alpha(x) &= (2\pi)^{-\frac{3}{2}} \int d^3\boldsymbol{k} \sqrt{\frac{m}{\omega(\boldsymbol{k})}} \\ &\quad \times \left[\sum_r \bar{v}_\alpha^{(r)}(\boldsymbol{k})\tilde{a}_r^+(\boldsymbol{k})e^{ikx} + \bar{u}_\alpha^{(r)}(\boldsymbol{k})\tilde{a}_r^-(\boldsymbol{k})e^{-ikx} \right] \\ &\equiv \bar{\psi}_\alpha^+(x) + \bar{\psi}_\alpha^-(x). \end{aligned} \tag{6.1.99b}$$

† We no longer have

$$\tilde{u}^{(r)}(\boldsymbol{k})v^{(r')}(\boldsymbol{k}) = \tilde{v}^{(r)}(\boldsymbol{k})u^{(r')}(\boldsymbol{k}) = 0. \tag{6.1.97}$$

The operator $a_r^+(k)[\tilde{a}_r^+(k)]$ annihilates a negatron [positron] represented by the wave function (of wave mechanics) $u^{(r)}(\boldsymbol{k})[\bar{v}^{(r)}(\boldsymbol{k})]$, whereas $a_r^-(\boldsymbol{k})[\tilde{a}_r^-(\boldsymbol{k})]$ creates a positron [negatron], represented by the wave function $v^{(r)}(\boldsymbol{k})[\bar{u}(\boldsymbol{k})]$.†

We shall now give another expression of the matrices $D(\pm ik)$ in terms of the matrix $U(\boldsymbol{k}, \omega)$ of the Foldy–Wouthuysen transformation. Consider the first of the formulae (6.1.92*b*); it can also take the form

$$D(ik)\tilde{D}(ik) = \omega U(\boldsymbol{k}, \omega)\beta U^{-1}(\boldsymbol{k}, \omega)+\omega$$

$$= \sqrt{2\omega}\, U(\boldsymbol{k}, \omega)\, \frac{1+\beta}{2}\, \sqrt{2\omega}\, \frac{1+\beta}{2}\, \tilde{U}(\boldsymbol{k}, \omega), \qquad (6.1.100)$$

† Creation and annihilation operators can be introduced without reference to the general theory of Chap. V. The spinors $u(\boldsymbol{k})$ and $v(\boldsymbol{k})$ being solutions of eqns. (6.1.93), we define the operator $a_r^{\pm}(\boldsymbol{k})$, $\tilde{a}_r^{\pm}(\boldsymbol{k})$ by the formulae (6.1.99) and we determine their anticommutation relations in such a way that the anticommutators of ψ and $\bar{\psi}$ be given by (6.1.77). The right side of (6.1.77) takes the form

$$[\psi_\alpha(x), \bar{\psi}_\beta(x')]_+ = (2\pi)^{-3} \int_{\substack{k_0=\omega \\ k'_0=\omega'}} d^3\boldsymbol{k}\, d^3\boldsymbol{k}' \frac{m}{\sqrt{\omega\omega'}} \Big\{ [a_r^+(\boldsymbol{k}), a_s^+(\boldsymbol{k}')]_+ u_\alpha^{(2)}(\boldsymbol{k})$$

$$\bar{v}_\beta^{(1)}(\boldsymbol{k}')e^{ikx+ik'x'} + [a_r^+(\boldsymbol{k}), \tilde{a}_s^-(\boldsymbol{k}')]_+ u_\alpha^{(2)}(\boldsymbol{k})\bar{u}_\beta^{(1)}(\boldsymbol{k}')e^{ikx-ik'x'}$$

$$+[a_r^-(\boldsymbol{k}), \tilde{a}_s^+(\boldsymbol{k}')]_+ v_\alpha^{(2)}(\boldsymbol{k})\bar{v}_\beta^{(1)}(\boldsymbol{k}')e^{-ikx+ik'x'}$$

$$+[a_r^-(\boldsymbol{k}), \tilde{a}_s^-(\boldsymbol{k}')] v_{+\alpha}^{(2)}(\boldsymbol{k})\bar{u}_\beta^{(1)}(\boldsymbol{k}')e^{-ikx-ik'x'} \Big\}$$

while the left side can be written as

$$-iS(x-x') = -i\{(\gamma\partial-m\}(\Delta^+(x-x')+\Delta^-(x-x'))$$

$$= (2\pi)^{-3} \int_{\substack{k_0=\omega \\ k'_0=\omega'}} d^3\boldsymbol{k}\, d^3\boldsymbol{k}'\, \delta(\boldsymbol{k}-\boldsymbol{k}')$$

$$\left(\frac{-i\gamma k+m}{2\omega} e^{ikx-ik'x'} - \frac{i\gamma k+m}{2\omega} e^{-ikx+ik'x'}\right).$$

Identification of both sides gives terms analogous to the following one:

$$\frac{m}{\omega}[a_r^-(\boldsymbol{k}), \tilde{a}_s^-(\boldsymbol{k}')]_+ u_\alpha^{(r)}(\boldsymbol{k})\bar{u}_\beta^{(s)}(\boldsymbol{k}') = \left(\frac{-i\gamma k+m}{2\omega}\right)_{\alpha\beta} \delta(\boldsymbol{k}-\boldsymbol{k}').$$

If we now multiply to the left by $\bar{u}_\alpha^{(r')}(\boldsymbol{k})$ and sum over α, we get one of the anticommutation relations:

$$[a_{r'}^+(\boldsymbol{k}), \tilde{a}_s^-(\boldsymbol{k}')]_+ = \delta(\boldsymbol{k}-\boldsymbol{k}')\, \delta_{r's};$$

all the other relations are obtained in the same way.

whence

$$D(ik) = \sqrt{2\omega}\, U(\boldsymbol{k}, \omega)\frac{1+\beta}{2} = -\frac{-i\gamma k+m}{\sqrt{\omega+m}}\frac{1+\beta}{2}, \qquad k_0 = \omega. \quad (6.1.101)$$

Similarly, by taking the second of the formulae (6.1.92), we should verify that†

$$D(-ik) = \sqrt{2\omega}\, U(-\boldsymbol{k}, \omega)\frac{1-\beta}{2} = \frac{i\gamma k+m}{\sqrt{\omega+m}}\frac{1-\beta}{2}, \qquad k_0 = \omega. \quad (6.1.102)$$

With the standard representation (6.1.30) of the γ_μ's and in the special case $k_1 = k_2 = 0$, we easily verify that the solutions $u^{(r)}(\boldsymbol{k})$ and $v^{(r)}(\boldsymbol{k})$ are given in terms of the $A^{(r)'}(\boldsymbol{k})$ in table (6.1.45) as follows: $u(\boldsymbol{k})$ is given by one of the first two columns of the table, whereas $v(\boldsymbol{k})$ is given by one of the last two with $-\boldsymbol{k}$ as momentum.†

As a last comment let us show that the spinor $u^{(r)}(\boldsymbol{p})$ being chosen, we may take as $v^{(r)}(\boldsymbol{p})$ the spinor $C\bar{u}^{(r)T}(\boldsymbol{p})$. Indeed, one can easily prove that on the one hand, this spinor is a solution of the eqn. $(-i\gamma k+m)v^{(r)} = 0$. On the other hand, from the eqn. $\Sigma(\boldsymbol{p})u^{(r)}(\boldsymbol{p}) = s^{(r)}u^{(r)}(\boldsymbol{p})$, we then infer,

$$\bar{u}^{(r)}(\boldsymbol{p})\Sigma(\boldsymbol{p}) = s^{(r)}u^{(r)}(\boldsymbol{p}),$$

taking the transpose of the former equation and introducing the charge conjugation matrix C;

$$\Sigma(\boldsymbol{p})C\bar{u}^{(r)T}(\boldsymbol{p}) = s^{(r)}C\bar{u}^{(r)}(\boldsymbol{p}),$$

i.e. $C\bar{u}^{(r)T}$ is an eigenvector of the helicity matrix. Thus, up to a phase factor, it can be identified with $v^{(r)}(\boldsymbol{p})$.

Let us now study the behaviour of the Dirac field under space symmetry and charge conjugation.

It can be easily seen that the wave functions $u^{(r)}$ and $v^{(r)}$ behave differently under a space symmetry: to be more precise, they have opposite parities. Indeed, the space symmetry operation is defined by the matrix (6.1.13): $i\beta$ transforms the spinor $u^{(r)}(\boldsymbol{k})$ into the spinor:

$$u^{(r)(\sigma)}(\boldsymbol{k}') = i\beta u^{(r)}(\boldsymbol{k}).$$

† The matrix $A\frac{I+\beta}{2}$ is inferred from the matrix A by replacing its last two columns by 0 while the matrix $A\frac{I-\beta}{2}$ is inferred from the matrix A by replacing its first two columns by zeros. This is in agreement with what we said about the structure of the matrices $D(\pm ik)$ on p. 240.

Now, $u^{(r)(\sigma)}(\boldsymbol{k}')$ is a solution of the Dirac equation $(i\gamma k'+m)w = 0$; on the other hand, its helicity is given by the matrix $\Sigma(\boldsymbol{k}')$:

$$\Sigma(\boldsymbol{k}')u^{(r)(\sigma)}(\boldsymbol{k}') = i\beta\Sigma(-\boldsymbol{k})u^{(r)}(\boldsymbol{k}) = -s^{(r)}i\beta u^{(r)}(\boldsymbol{k}) = -s^{(r)}u^{(r)(\sigma)}(\boldsymbol{k}')$$

and is $-s^{(r)}$. But the spinor $u^{(r)}(-\boldsymbol{k})$ has the same behaviour as $u^{(r)(\sigma)}(\boldsymbol{k}')$, hence we may identify these spinors and take:

$$u^{(r)(\sigma)}(\boldsymbol{k}') = u^{(r)}(-\boldsymbol{k}), \tag{6.1.103a}$$

$$v^{(r)(\sigma)}(\boldsymbol{k}) = -v^{(-r)}(-\boldsymbol{k}). \tag{6.1.103b}$$

Incidentally, we have also been able to verify that $u^{(r)}(\sigma)$, $v^{(r)}(\sigma)$ correspond to the projections of the spin on the direction of propagation, with opposite signs (in agreement with the pseudo-vector character of the spin).

Introducing the expressions

$$\psi^{(\sigma)}(x) = \Pi\psi(x)\Pi^{-1} \tag{6.1.104}$$

[cf. (3.78)], we see that

$$\left.\begin{aligned} \Pi a_r^+(\boldsymbol{k})\Pi^{-1} &= a_{-r}^+(-\boldsymbol{k}),\\ \Pi a_r^-(\boldsymbol{k})\Pi^{-1} &= -a_{-r}^-(-\boldsymbol{k}),\\ &\ldots\ldots\ldots\ldots\ldots\ldots \end{aligned}\right\} \tag{6.1.105}$$

Coming now to charge conjugation, in agreement with the remark on p. 230, we first note that we may take

$$\left.\begin{aligned} u^{(r)(c)}(\boldsymbol{k}) &= C\bar{u}^{(r)T}(\boldsymbol{k}) \equiv v^{(r)}(\boldsymbol{k}),\\ v^{(r)(c)}(\boldsymbol{k}) &= C\bar{v}^{(r)T}(\boldsymbol{k}) \equiv u^{(r)}(\boldsymbol{k}) \end{aligned}\right\} \tag{6.1.106}$$

up to a phase which has been fixed arbitrarily.

By the same considerations which lead us to formulae (6.1.104) and (6.1.105) we may see

$$\left.\begin{aligned} \Gamma^{-1}a_r^+(\boldsymbol{k})\Gamma &= \tilde{a}_r^+(\boldsymbol{k}),\\ \Gamma^{-1}a_r^-(\boldsymbol{k})\Gamma &= \tilde{a}_r^-(\boldsymbol{k}). \end{aligned}\right\} \tag{6.1.107}$$

These results are particularly important for the theory of the positronium.

Let us now investigate the expressions of P_μ (6.1.86) and of the total charge operator e_{tot} (6.1.90). Making use of formulae (6.1.98), we can write

$$\frac{e_{\text{tot}}}{e} \equiv \mathcal{N} = -\int_{k_0=\omega} d^3\boldsymbol{k}[\tilde{a}_r^-(\boldsymbol{k})a_r^+(\boldsymbol{k})-a_r^-(\boldsymbol{k})\tilde{a}_r^+(\boldsymbol{k})]$$

$$= -\int_{k_0=\omega} d^3\boldsymbol{k}(\mathcal{N}_-(\boldsymbol{k})-\mathcal{N}_+(\boldsymbol{k})), \tag{6.1.108a}$$

$$P_\mu = \int_{k_0=\omega} d^3\boldsymbol{k}k_\mu(\mathcal{N}_+(\boldsymbol{k})+\mathcal{N}_-(\boldsymbol{k})), \tag{6.1.108b}$$

in agreement with formulae (5.116) and apart from constant infinite terms. The elimination of these infinite terms (already carried out in the preceding formulae) can be performed by the more elegant method of ordered products described above (5.119).†

Let us note in conclusion that the charge conjugation leaves all the Dirac field observables invariant, except the current and the charge which charge their signs under this operation.

14. Expectation values in the vacuum state of certain bilinear expressions in $\psi(x)$

Using calculations similar to those in Chap. V, § 8, we first obtain

$$\langle 0 \,|\, \psi_\alpha(x)\bar{\psi}_\beta(x') \,|\, 0 \rangle = \langle 0 \,|\, [\psi_\alpha^+(x), \bar{\psi}_\beta^-(x')]_+ \,|\, 0 \rangle;$$

we then express the right side of the above formulae with the aid of formulae (6.1.99):

$$(2\pi)^{-3} \int_{\substack{k_0=\omega \\ k'_0=\omega'}} d^3\boldsymbol{k}\, d^3\boldsymbol{k}' \frac{m}{\sqrt{\omega(\boldsymbol{k})\omega(\boldsymbol{k}')}} u_\alpha^{(r)}(\boldsymbol{k})\bar{u}_\beta^{(r')}(\boldsymbol{k}')[a_r^+(\boldsymbol{k}), \tilde{a}_{r'}^-(\boldsymbol{k}')]_+ e^{ikx}e^{-ik'x'}$$

$$= (2\pi)^{-3} \int_{k_0=\omega} \frac{d^3\boldsymbol{k}}{2\omega} (-i\gamma_\mu k_\mu + m)_{\alpha\beta} e^{ik(x-x')}.$$

By virtue of formulae (5.118), (6.1.94*a*) and (6.1.73*b*) this expression can take the form

$$(-\gamma_\mu \partial_\mu + m)_{\alpha\beta}(2\pi)^{-3} \int_{k_0=\omega} \frac{d^3k}{2\omega} e^{ik(x-x')},$$

and by referring to the definition of $\Delta^+(x-x')$, we finally obtain, as can be easily seen,

$$\langle 0 \,|\, \psi_\alpha(x)\bar{\psi}_\beta(x') \,|\, 0 \rangle = -i(\gamma_\mu \partial_\mu - m)_{\alpha\beta}\Delta^+(x-x') = -iS_{\alpha\beta}^+(x-x'). \tag{6.1.109a}$$

Similarly,

$$\langle 0 \,|\, \bar{\psi}_\alpha(x)\psi_\beta(x') \,|\, 0 \rangle = -iS_{\bar{\beta}\alpha}(x'-x). \tag{6.1.109b}$$

† It is extremely important to note that a straightforward evaluation of the operator

$$P_0 = -4 \int \psi(x)\gamma_4\, \partial_4\psi(x)\, d^3\boldsymbol{x}$$

in terms of the creation and annihilation operators gives the *nondefinite* operator

$$P_0 = \int_{k_0=\omega} d^3\boldsymbol{k}k_0[\tilde{a}_r^-(\boldsymbol{k})a_r^+(\boldsymbol{k})a - \tilde{a}_r^+(\boldsymbol{k})_r^-(\boldsymbol{k})]$$

in sharp contradistinction to the scalar field where P_0 is expressed with a + sign. This is the main reason why it is necessary to quantize such a field by anticommutators in order to obtain a positive definite P_0 operator and such that $\langle 0 \,|\, P_0 \,|\, 0 \rangle = 0$. (See footnote on p. 192.)

Another expectation value in the vacuum state is worth computing: the so-called "fermion propagator":

$$\varepsilon(x-x')\langle 0|P\{\psi_\alpha(x)\bar{\psi}_\beta(x')\}|0\rangle \tag{6.1.110a}$$

which we can also express [cf. (9.84)] in the form:

$$\langle 0|T\{\psi_\alpha(x)\bar{\psi}_\beta(x')\}|0\rangle,$$

an expression which can be compared with (5.84*a*), where the chronological operator $P\{\ldots\}$ is defined. We therefore obtain

$$\varepsilon(x-x')\langle 0|P\{\psi_\alpha(x)\bar{\psi}_\beta(x')\}|0\rangle \tag{6.1.110b}$$
$$= \begin{cases} \langle 0|\psi_\alpha(x)\bar{\psi}_\beta(x')|0\rangle & \text{if} \quad x_0 > x_0', \\ -\langle 0|\bar{\psi}_\beta(x')\psi_\alpha(x)|0\rangle & \text{if} \quad x_0 < x_0'. \end{cases}$$

The expressions of the right sides are given by formulae (6.1.109); we finally have

$$\varepsilon(x-x')\langle 0|P\{\psi_\alpha(x)\bar{\psi}_\beta(x')\}|0\rangle \tag{6.1.111}$$
$$= \begin{cases} -iS^+_{\alpha\beta}(x-x') & \text{if} \quad x_0 > x_0' \\ +iS^-_{\alpha\beta}(x-x') & \text{if} \quad x_0 < x_0' \end{cases}$$
$$= -\frac{1}{2}\, S^{(c)}_{\alpha\beta}(x-x') = \lim_{\varepsilon=0} \frac{i}{(2\pi)^4}\int d^4k e^{ik(x-x')}\frac{(i\gamma k-m)_{\alpha\beta}}{k^2+m^2-i\varepsilon},$$

in accordance with the table on p. 236.

We should find in similar fashion that

$$\langle 0|[\psi_\alpha(x), \bar{\psi}_\beta(x')]_-|0\rangle = -S^{(1)}_{\alpha\beta}(x-x'). \tag{6.1.112}$$

Note finally that we obtain the set of relations

$$\left.\begin{aligned} [\psi^+_\alpha(x), \psi^-_\beta(x')]_+ &= [\bar{\psi}^-_\alpha(x), \bar{\psi}^+_\beta(x')]_+ = 0, \\ [\bar{\psi}^+_\alpha(x), \psi^+_\beta(x')]_+ &= [\bar{\psi}^-_\alpha(x), \psi^-_\beta(x')]_+ = 0, \\ [\psi^-_\alpha(x), \bar{\psi}^+_\beta(x')]_+ &= -iS^-_{\alpha\beta}(x-x'). \end{aligned}\right\} \tag{6.1.113}$$

15. Field of spin $\frac{1}{2}$ and mass zero. Non-conservation of the parity

The equation of a Dirac field corresponding to particles of zero mass can be obtained very simply by making $m=0$ in the Dirac equation; it thus takes the form

$$\gamma\,\partial\psi = 0 \tag{6.1.114}$$

and is clearly invariant under space symmetry, time reversal and charge conjugation.

Experiments on β-decay tend to prove that the corresponding interactions, together with a number of interactions known as weak interactions, are not invariant under space symmetries. This equation must therefore be replaced by another one which is not invariant under space reflections.

Such an equation, which had already been investigated by Weyl and Pauli, was reintroduced into quantum field theory by Lee and Yang. Let us choose the representation (6.1.33) of the γ-matrices and define the two wave functions

$$\psi_{\mathrm{I}}(x) = \begin{pmatrix} \psi_1(x) \\ \psi_2(x) \end{pmatrix}, \qquad \psi_{\mathrm{II}}(x) = \begin{pmatrix} \psi_3(x) \\ \psi_4(x) \end{pmatrix},$$

where $\psi_1, \psi_2, \psi_3, \psi_4$ are the components of the Dirac wave function $\psi(x)$. Using the fact that the four matrices γ_μ are antidiagonal, it is easy to see that ψ_{I} and ψ_{II} obey the equations

$$i\,\frac{\partial \psi_{\mathrm{I}}(x)}{\partial t} = -i\boldsymbol{\sigma}\cdot\nabla\psi_{\mathrm{I}}(x), \tag{6.1.115a}$$

$$i\,\frac{\partial \psi_{\mathrm{II}}(x)}{\partial t} = i\boldsymbol{\sigma}\cdot\nabla\psi_{\mathrm{II}}(x), \tag{6.1.115b}$$

where $\boldsymbol{\sigma}$ has as its components the Pauli matrices $\sigma_1, \sigma_2, \sigma_3$ given by (6.1.19).

If we look for plane wave solutions, namely

$$\psi_{\mathrm{I}}(x) = A_{\mathrm{I}}(p)e^{ipx}, \qquad \psi_{\mathrm{II}}(x) = A_{\mathrm{II}}(p)e^{ipx}, \tag{6.1.116}$$

we find that $A_{\mathrm{I}}(p)$ and $A_{\mathrm{II}}(p)$ must obey the algebraic equations:

$$p_0A_{\mathrm{I}}(p) = \sigma_jp_jA_{\mathrm{I}}(p) \equiv H_{\mathrm{I}}(\boldsymbol{p})A_{\mathrm{I}}(p), \tag{6.1.117a}$$

$$p_0A_{\mathrm{II}}(p) = -\sigma_jp_jA_{\mathrm{II}}(p) \equiv H_{\mathrm{II}}(\boldsymbol{p})A_{\mathrm{II}}(p). \tag{6.1.117b}$$

Consider the spin operator in wave mechanics; according to (6.1.9) and (5.1.34), we have

$$S_k = -iJ_{lm} = \frac{1}{2}\begin{pmatrix} \sigma_k & 0 \\ 0 & \sigma_k \end{pmatrix},$$

consequently the polarization operator $\Sigma(\boldsymbol{p})$ in the direction of the momentum

$$\Sigma_p = \frac{\boldsymbol{S}\cdot\boldsymbol{p}}{|\boldsymbol{p}|}$$

is such that

$$H_{\mathrm{I}}(\boldsymbol{p}) = 2|\boldsymbol{p}|\Sigma_p, \tag{6.1.118a}$$

$$H_{\mathrm{II}}(\boldsymbol{p}) = -2|\boldsymbol{p}|\Sigma_p. \tag{6.1.118b}$$

To describe the neutrino we shall choose eqn. (6.1.117*b*); the energy is therefore given by (6.1.118*b*): we are dealing with a particle if Σ is negative, that is to say if the spin has an opposite direction to that of the momentum, otherwise we are dealing with a particle with negative energy, the corresponding "hole" of which is described by eqn. (6.1.117*a*). We can therefore say that the neutrino is a particle with left-helicity, while the antineutrino is a particle with right-helicity. The neutrino is not therefore a Majorana particle defined by the identity of the particle and of the antiparticle.

This theory does not conserve the parity: indeed a space symmetry transforms the vectors $\boldsymbol{p}$ into $-\boldsymbol{p}$, leaves the pseudo-vector $\boldsymbol{S}$ invariant and consequently exchanges particles and antiparticles.

Similarly, this theory is not invariant under charge conjugation. This operation does in fact transform a particle into an antiparticle but it changes neither its momentum nor its spin.

We can readily verify the fact that the theory is invariant under a time reversal and the product of the charge conjugation by a space symmetry.

It is therefore invariant under the product of these three operations, and thus provides an example of an application of the TCP theorem discussed on p. 212.

We can easily write down the wave equation of the neutrino with the help of the γ-matrices by using the projection operators $\frac{1\pm\gamma_5}{2}$. We shall not go into details or discuss the quantization of this field (which anyway presents no special difficulties); readers interested in these matters may consult the articles quoted below.†

PART 2

MAXWELL FIELD (FREE PHOTONS)

It is in the investigation of the electromagnetic field or photon field whose quanta are zero mass neutral particles, described by a vector field, and of the interactions of this field with the electron field, that quantum field theory has obtained its most decisive results, since the theoretical forecasts are in complete agreement with the most delicate experimental results (Lamb and Retherford shift; anomalous magnetic momentum of the electron). The electromagnetic field was, moreover, the first field to be quantized and it has served as a guide for the quantization of other fields such as the meson field.

We are therefore going to investigate the free photon field in some detail. Despite the fact that the formalism of the quantization of this field follows fairly closely that of the neutral scalar field, the zero mass of the photon and the Lorentz condition introduce some rather subtle procedures into the theory, as we shall be seeing below.

† T. D. Lee and C. N. Yang, *Phys. Rev.*, **105**, 1671 (1957); L. Landau, *Nucl. Phys.*, **3**, 127 (1957); A. Salam, *Nuovo Cimento*, **5**, 299 (1957). For the quantization of this field, cf. H. Umezawa and A. Visconti, *Nucl. Phys.*, **4**, 224 (1957).

16. Classical Maxwell equations†

Consider first the non-quantized Maxwell equations in the presence of a given current density $\boldsymbol{J}(\boldsymbol{x}, t)$ and charge density $\varrho(\boldsymbol{x}, t)$.

The electric field $\boldsymbol{E}$ and the magnetic field $\boldsymbol{H}$ are generally described by the system of Maxwell equations:

$$\left.\begin{aligned} &\text{I.} \quad \nabla \wedge \boldsymbol{H} = \frac{\partial \boldsymbol{E}}{\partial t} + \boldsymbol{J}, \\ &\text{II.} \quad \nabla \wedge \boldsymbol{E} = -\frac{\partial \boldsymbol{H}}{\partial t}, \\ &\text{III.} \quad \nabla \cdot \boldsymbol{E} = \varrho, \\ &\text{IV.} \quad \nabla \cdot \boldsymbol{H} = 0. \end{aligned}\right\} \tag{6.2.1}$$

By eliminating $\boldsymbol{E}$ and $\boldsymbol{H}$ between eqns. (II) and (III), we obtain the continuity equation:

$$\nabla \cdot \boldsymbol{J} + \frac{\partial \varrho}{\partial t} = 0. \tag{6.2.2}$$

Equation (IV) shows that $\boldsymbol{H}$ is of the form

$$\boldsymbol{H} = \nabla \wedge \boldsymbol{A}. \tag{6.2.3a}$$

By carrying this expression of $\boldsymbol{H}$ into (II), we get

$$\boldsymbol{E} = -\frac{\partial \boldsymbol{A}}{\partial t} - \nabla \Phi, \tag{6.2.3b}$$

where Φ is a function of x which can be determined as follows. Carry the expressions of $\boldsymbol{E}$ and $\boldsymbol{H}$ into eqn. (I)†

$$\nabla \wedge \boldsymbol{H} = \nabla \wedge \nabla \wedge \boldsymbol{A} = \nabla(\nabla \cdot \boldsymbol{A}) - \nabla^2 \cdot \boldsymbol{A} = -\frac{\partial^2 \boldsymbol{A}}{\partial t^2} - \nabla \frac{\partial \Phi}{\partial t} + \boldsymbol{J}$$

or, alternatively,

$$\frac{\partial^2 \boldsymbol{A}}{\partial t^2} - \nabla^2 \cdot \boldsymbol{A} + \nabla \left\{ \nabla \cdot \boldsymbol{A} + \frac{\partial \Phi}{\partial t} \right\} = \boldsymbol{J}. \tag{6.2.4}$$

If therefore we make Φ obey the Lorentz condition:

$$\nabla \cdot \boldsymbol{A} + \frac{\partial \Phi}{\partial t} = 0, \tag{6.2.5}$$

the vector-potential $\boldsymbol{A}$ then obeys the equation

$$\Box \boldsymbol{A} = -\boldsymbol{J}, \tag{6.2.6a}$$

while the potential Φ obeys

$$\Box \Phi = -\varrho, \tag{6.2.6b}$$

† For a summary of the classical theory of electromagnetic radiation, cf. W. Heitler, *Quantum Theory of Radiation* (Chap. I), Oxford, 1954.

an equation obtained by combining eqns. (6.2.1 III) and (6.2.3*b*). Note furthermore that conversely, if we start from the system of equations formed by (I), (III), (6.2.3*a*), (6.2.3*b*) [instead of from the system (6.2.1)], we obtain the remaining Maxwell equations, namely eqns. (II) and (IV).

The Maxwell equations have to be completed by the law of force expressing the action of the electromagnetic field on a current and a charge. A density of force is defined:

$$\boldsymbol{f} = \varrho \boldsymbol{E} + \boldsymbol{J} \wedge \boldsymbol{H}, \tag{6.2.7a}$$

so that the force acting upon a volume V of matter is

$$\boldsymbol{F} = \int_V \boldsymbol{f}(\boldsymbol{x})\, d^3\boldsymbol{x}. \tag{6.2.7b}$$

The Poynting vector

$$\boldsymbol{S} = \boldsymbol{E} \wedge \boldsymbol{H} \tag{6.2.8}$$

allows the diminution of the energy

$$W = \frac{1}{2} \int_V (\boldsymbol{E}^2 + \boldsymbol{H}^2)\, d^3\boldsymbol{x} \tag{6.2.9a}$$

enclosed within a volume V limited by a surface σ, per unit of time, to be expressed:

$$\frac{dW}{dt} = -\oint_\sigma \boldsymbol{S} \cdot \boldsymbol{n}\, d\sigma; \tag{6.2.9b}$$

$\boldsymbol{n}$ is the unit normal to σ pointing outwards.

It can moreover be proved that to the radiation defined by the vector $\boldsymbol{S}$ there corresponds the following total momentum:

$$\boldsymbol{P} = \int_V \boldsymbol{S}\, d^3\boldsymbol{x}. \tag{6.2.10}$$

The Lienard and Wiechert formulae give the expression of the vector potential at a point $\boldsymbol{x}$ when this potential is produced by a point charge with a density $e\,\delta(\boldsymbol{x}-\boldsymbol{x}')$ moving at a speed $\boldsymbol{v}$.† By introducing

$$s = r - \boldsymbol{v} \cdot \boldsymbol{r}, \quad r = |\boldsymbol{x} - \boldsymbol{x}'|,$$

these formulae can be written:

$$\Phi(\boldsymbol{x}, t) = \frac{e}{4\pi} \frac{1}{s(\tau)}, \tag{6.2.11a}$$

$$\boldsymbol{A}(\boldsymbol{x}, t) = \frac{e}{4\pi} \frac{\boldsymbol{v}(\tau)}{s(\tau)}. \quad \tau = t - r. \tag{6.2.11b}$$

† See also H. Bacry, *Annales de Phys.*, **8**, 197 (1963).

We then obtain the fields **E** and **H** by means of formulae (6.2.3):

$$\boldsymbol{E} = \frac{e}{4\pi}\frac{1-\beta^2}{s^3(\tau)}(\boldsymbol{r}-r\boldsymbol{v})+\frac{e}{4\pi s^3}\,\boldsymbol{r}\wedge\left[(\boldsymbol{r}-r\boldsymbol{v})\wedge\frac{d\boldsymbol{v}}{dt}\right], \qquad (6.2.12a)$$

$$\boldsymbol{H} = -\frac{e}{4\pi}\frac{1}{r}\boldsymbol{E}\wedge\boldsymbol{r}, \quad \beta = |\boldsymbol{v}| \qquad (6.2.12b)$$

where all the quantities contained in these formulae are to be considered at the time $\tau = t-r\ (c=1)$.

It is well known that the electric field can be analysed into two parts: the first part depends on $\boldsymbol{v}$ $\left(\text{and } not \text{ on } \frac{d\boldsymbol{v}}{dt}\right)$, behaves like $\frac{1}{r^2}$ at infinity and is reduced to the Coulomb field when $\boldsymbol{v} = 0$.

The second part is proportional to $\frac{d\boldsymbol{v}}{dt}$, behaves like $\frac{1}{r}$ at infinity and is orthogonal to $\boldsymbol{x}$. It represents the transversal field and the region of space in which it predominates is called the "wave zone".

Three-vector and four-dimensional formulations of the Maxwell equations

Three-vector formulation	Four-dimensional formulation
	$x_\mu = \begin{cases}\boldsymbol{x}\\ x_4 = ix_0\end{cases}$
	$A_\mu = \begin{cases}\boldsymbol{A}\\ i\Phi\end{cases}$
	$H_1 = F_{23}, \quad H_2 = F_{31}, \quad H_3 = F_{12}$
	$iE_j = F_{4j}$
	$J_\mu = \begin{cases}\varrho\boldsymbol{v}\\ i\varrho\end{cases}$
$\boldsymbol{E} = -\frac{\partial \boldsymbol{A}}{\partial t}-\nabla\Phi$	$F_{\mu\nu} = A_{\nu,\mu}-A_{\mu,\nu}$
$\boldsymbol{H} = \nabla\wedge\boldsymbol{A}$	$F_{\varrho\sigma}F_{\varrho\sigma} = 2(\boldsymbol{H}^2-\boldsymbol{E}^2)$
(IV) $\nabla\cdot\boldsymbol{H} = 0$	$F_{\alpha\beta,\gamma}+F_{\beta\gamma,\alpha}+F_{\gamma\alpha,\beta} = 0$
(II) $\nabla\wedge\boldsymbol{E}+\frac{\partial \boldsymbol{H}}{\partial t} = 0$	or $\varepsilon_{\lambda\mu\nu\varrho}F_{\mu\nu,\varrho} = 0$
(III) $\nabla\cdot\boldsymbol{E} = \varrho$, (I) $\nabla\wedge\boldsymbol{H} = \frac{\partial \boldsymbol{E}}{\partial t}+\boldsymbol{J}$	$F_{\mu\nu,\nu} = J_\mu$
	$\Box A_\mu = -J_\mu$
$\Box\boldsymbol{A} = -\boldsymbol{J}$, $\Box\Phi = -\varrho\left(\Box\equiv\nabla^2-\frac{\partial^2}{\partial t^2}\right)$	

Three-vector formulation	Four-dimensional formulation
$\nabla\cdot\boldsymbol{J}+\dfrac{\partial\varrho}{\partial t}=0$	$J_{\mu,\mu}=0$
$\nabla\cdot\boldsymbol{A}+\dfrac{\partial\Phi}{\partial t}=0$	$A_{\mu,\mu}=0$
$\boldsymbol{f}=\varrho\boldsymbol{E}+\boldsymbol{J}\wedge\boldsymbol{H}$	$f_\mu=F_{\mu\nu}J_\nu \quad (f_4=i\varrho\boldsymbol{E}\cdot\boldsymbol{v})$ $=T_{\mu\nu,\nu}$
	$T_{\mu\nu}=F_{\varrho\mu}F_{\varrho\nu}-\dfrac{1}{4}\delta_{\mu\nu}F_{\varrho\sigma}F_{\varrho\sigma}$
$\boldsymbol{S}=\boldsymbol{E}\wedge\boldsymbol{H}$ $W=\dfrac{1}{2}\int(\boldsymbol{E}^2+\boldsymbol{H}^2)\,d^3\boldsymbol{x}$	$T_{\mu\nu}=\begin{pmatrix} T_{11} & T_{12} & T_{13} & iS_1 \\ T_{21} & T_{22} & T_{23} & iS_2 \\ T_{31} & T_{32} & T_{33} & iS_3 \\ \underbrace{iS_1 \quad iS_2 \quad iS_3}_{i\boldsymbol{S}} & & & -\frac{1}{2}(\boldsymbol{E}^2+\boldsymbol{H}^2) \end{pmatrix}$
$\boldsymbol{P}=\int\boldsymbol{S}\,d^3\boldsymbol{x}$	$P_\mu=\int T_{\mu\nu}\,d\sigma_\nu$
	$=-i\int T_{\mu4}\,d^3\boldsymbol{x}=\begin{cases}\int_V \boldsymbol{E}\wedge\boldsymbol{H}\,d^3\boldsymbol{x}=\boldsymbol{P}\\ i\int_V \dfrac{\boldsymbol{E}^2+\boldsymbol{H}^2}{2}\,d^3\boldsymbol{x}=P_4.\end{cases}$

It can finally be proved that a point charge radiates the energy

$$W=\frac{2}{3}\frac{e^2}{4\pi}\left(\frac{d\boldsymbol{v}}{dt}\right)^2. \tag{6.2.13}$$

Since this radiation occurs at the expense of the kinetic energy of the particle, it corresponds to a "field reaction". This reaction is in fact a result of the force exerted by a charged particle upon itself (proper force). Indeed, let de and de' be two charges taken at two different points of a particle: the proper force is obtained by integrating the action exerted by one charge upon the other. By means of certain assumptions made for the purpose of simplification (rigid and spherical distribution of charges with a radius $\bar{r}_0$, etc.), we find that

$$\boldsymbol{F}_{(S)}=-\frac{2}{3}\frac{d\boldsymbol{v}}{dt}\int\frac{de\,de'}{4\pi r}+\frac{2}{3}\frac{e^2}{4\pi}\frac{d^2\boldsymbol{v}}{dt^2}; \tag{6.2.14}$$

the term of higher order to be added to this formula is of the form $e^2\bar{r}_0 \dfrac{d^3v}{dt^3}$ and cancels when the radius of the particle becomes zero.

When the particle is a uniformly charged sphere, the first term of (6.2.14) contains the factor:

$$\frac{1}{2}\int \frac{de\, de'}{4\pi r} = \frac{e^2}{4\pi \bar{r}_0^2}$$

and the proper force becomes infinite for a point particle. We shall be meeting this conclusion again in quantum field theory in the problem of infinite proper energies which we shall discuss in detail further on.

In the table on p. 250, we have arranged all the formulae of Maxwell theory, expressed either in three-vector or four-dimensional formulation.

17. Gauge transformation

Let us now return to the free photon field. We shall write the Maxwell equations (6.2.1) in covariant form, by introducing the skew-symmetrical tensor

$$F_{\mu\nu}(x) = -F_{\nu\mu}(x), \tag{6.2.14a}$$

$$\left.\begin{aligned} F_{ij} &= H_k \quad (i, j, k, \text{ even permutation of the numbers } 1, 2, 3),\\ F_{4j} &= iE_j. \end{aligned}\right\} \tag{6.2.14b}$$

These equations take the form

$$F_{\mu\nu,\,\lambda} + F_{\nu\lambda,\,\mu} + F_{\lambda\mu,\,\nu} = 0, \tag{6.2.14c}$$

$$F_{\mu\nu,\,\nu} = 0. \tag{6.2.14d}$$

We infer from them that each component $F_{\mu\nu}$ obeys the equation

$$\Box F_{\mu\nu} = 0. \tag{6.2.15}$$

We then introduce a four-vector A_μ, such that

$$F_{\mu\nu}(x) = A_{\nu,\,\mu}(x) - A_{\mu,\,\nu}(x). \tag{6.2.16}$$

Equations (6.2.14*a* and *c*) are identically obeyed; eqn. (6.2.14*d*) takes the form

$$\Box A_\mu - \frac{\partial \chi(x)}{\partial x_\mu} = 0 \tag{6.2.17a}$$

with

$$\chi(x) \equiv A_{\mu,\,\mu}(x). \tag{6.2.17b}$$

Equations (6.2.6) ($\Box A_\mu = 0$) can therefore only be obtained when the A_μ's are made to satisfy the Lorentz condition:

$$\chi(x) = 0, \tag{6.2.18}$$

a covariant form of (6.2.5).

It is interesting to note that eqns. (6.2.6) can be obtained from (6.2.17) by defining a special gauge for the A_μ's.

It is, indeed, easy to verify that the system of eqns. (6.2.14) is invariant under the "Gauge transformation":

$$A_\mu(x) \to A'_\mu(x) = A_\mu(x) + \frac{\partial \Lambda(x)}{\partial x_\mu}, \tag{6.2.19}$$

where $\Lambda(x)$ is an arbitrary function.

Consider next a particular form of Λ: $\Lambda \equiv \varphi$ such that

$$\Box\varphi + \chi(x) = 0; \tag{6.2.20}$$

the field $A'_\mu = \mathcal{A}_\mu$ corresponding to this particular gauge obeys in that case the following relations:

$$\mathcal{A}_{\mu,\mu} = 0, \tag{6.2.21a}$$

$$\Box\mathcal{A}_\mu = \Box A_\mu + \partial_\mu \Box\varphi = 0; \tag{6.2.21b}$$

this field does in fact satisfy the Lorentz condition and eqn. (6.2.6).

We can therefore start from the following system of equations:

$$F_{\mu\nu} = A_{\nu,\mu} - A_{\mu,\nu}, \tag{6.2.22a}$$

$$\Box A_\mu = 0, \tag{6.2.22b}$$

$$A_{\mu,\mu} = 0 \tag{6.2.22c}$$

obtained from the system formed by eqns. (6.2.3), (6.2.6) and (6.2.5). We can readily see that this system is indeed equivalent to the Maxwell equations (6.2.1) since, as may be verified, the last two equations involve $F_{\mu\nu,\nu} = 0$. Finally the system (6.2.22) is formed by eqns. (6.2.3 I and III); this system is equivalent to (6.2.1) by virtue of the remark following formula (6.2.6); on the other hand, it is only invariant under the restricted gauge group given by:

$$\Box\varphi(x) = 0. \tag{6.2.22d}$$

Let us define, in conclusion, the Coulomb gauge: it is expressed in a particular frame, that in which

$$\nabla \cdot \boldsymbol{A} = 0. \tag{6.2.23}$$

This condition replaces the Lorentz condition and the vector potential then obeys the equations

$$-\Box\boldsymbol{A} + \nabla\frac{\partial\Phi}{\partial t} = \boldsymbol{J}, \tag{6.2.24a}$$

$$\nabla^2\Phi = -\varrho. \tag{6.2.24b}$$

The first equation follows directly from (6.2.4) and the second is obtained by combining (III) with (6.2.3*b*). The scalar potential $\Phi(x)$ obeys in that case the well-known Poisson equation.

18. Elementary solutions

We can consider these solutions as particular cases of the equation

$$\{\Box - m^2\}\Delta^{(e)}(x) = \dots$$

for $m = 0$. This is in fact what we did in the table on p. 167, where these new functions were denoted by

$$D^{(e)}(x) = \Delta^{(e)}(x, m = 0).$$

We verify in particular that

$$D(x) = -\frac{1}{(2\pi)^3}\int e^{i\boldsymbol{k}.\boldsymbol{x}}\frac{\sin \omega x_0}{\omega}\, d^3\boldsymbol{k} = -\frac{1}{2\pi}\,\varepsilon(x)\,\delta(x_\mu x_\mu), \quad (6.2.25)$$

$$D^{(1)}(x) = \frac{1}{(2\pi)^3}\int e^{i\boldsymbol{k}.\boldsymbol{x}}\frac{\cos \omega x_0}{\omega}\, d^3\boldsymbol{k} = \frac{1}{2\pi^2}\,P\frac{1}{x_\mu x_\mu}. \quad (6.2.26)$$

19. Lagrangian formalism and quantization

Let us start from the Lagrangian density:

$$\mathcal{L} = -\frac{1}{4}F_{\mu\nu}F_{\mu\nu} - \frac{1}{2}(A_{\mu,\,\mu})^2. \quad (6.2.27)$$

Since the variations are carried out with respect to the A_μ's, the field equations take the form

$$F_{\mu\nu,\,\nu} - \frac{\partial\chi(x)}{\partial x_\mu} = 0 \quad (6.2.28a)$$

and by introducing the potential four-vector A_μ, the preceding equation is transformed into

$$\Box A_\mu(x) = 0. \quad (6.2.28b)$$

Finally, differentiating (6.2.28*b*) with respect to μ and summing over μ, we also get

$$\Box\chi(x) = 0. \quad (6.2.28c)$$

The canonically conjugate variable to the $A_\mu(x)$'s can be easily computed

$$\Pi_\mu = n_\varrho\frac{\partial\mathcal{L}}{\partial A_{\mu,\,\varrho}} = n_\varrho F_{\mu\varrho} - n_\mu\chi \quad (6.2.29)$$

and reduce for the hyperplane t = constant ($n_\mu = 0, 0, 0, -i$) to

$$\Pi_j = iF_{4j}, \quad (6.2.30a)$$

$$\Pi_4 = i\chi. \quad (6.2.30b)$$

Equations (6.2.28*a*) differ from the Maxwell equations by the presence of the term $\frac{\partial \chi}{\partial x_\mu}$. But if we introduce the Lorentz condition:

$$\chi(x) = 0 \tag{6.2.31}$$

in order to obtain these equations, it follows in accordance with (6.2.30*b*) that Π_4 is also identically zero; it is therefore no longer possible to satisfy the commutation rule between A_4 and Π_4.

Fermi suggested replacing (6.2.31) by the following equation, known as the "Fermi condition",†

$$\chi(x)\Psi = 0, \tag{6.2.32}$$

where Ψ is the state vector of the system (Heisenberg picture). In other words, the Maxwell equations are only valid in a subspace of the original space $\mathcal{R}$ (in which the operators A_μ are defined). We can also say that these equations are only valid for the expectation values

$$\partial_\nu \langle \Psi | F_{\mu\nu} | \Psi \rangle = 0. \tag{6.2.33}$$

The existence of the vector Ψ, an eigenvector of the operator $\chi(x)$ corresponding to the eigenvalue 0, raises a number of difficulties which will be discussed in the next section; let us note for the time being that $\chi(x)$, which obeys eqn. (6.2.28*c*), is a linear functional of $\chi(\boldsymbol{x}, 0)$ and of $\frac{\partial \chi(x)}{\partial t}$ for $t = 0$; in order to realize the Fermi condition, it is therefore sufficient to make the state vector obey the relations

$$\chi(\boldsymbol{x}, 0)\Psi = 0, \quad \lim_{t=0} \frac{\partial \chi(x)}{\partial t} \Psi = 0.$$

The condition $\chi(x) = 0$ must therefore be considered as an auxiliary condition and field equations then take the form:

$$\Box A_\mu = 0. \tag{6.2.34}$$

Note also that by starting from the Lagrangian density:

$$\mathcal{L} = -\frac{1}{4} F_{\mu\nu} F_{\mu\nu}, \tag{6.2.35a}$$

† E. Fermi, *Rev. Mod. Phys.*, **4**, 87 (1932): it should also be noted that eqn. (6.2.32) has to be obeyed for any x. In other words, the continuous set of operators $\chi(x)$ has to have a common eigenvector Ψ corresponding to the eigenvalue O. If therefore x and x' are two different points:

$$[\chi(x), \chi(x')] = 0.$$

We easily verify that relation is obeyed; we obtain, indeed, making use of the commutation rules (6.2.36) of the A_μ's:

$$[\chi(x), \chi(x')] = i\Box D(x-x') = 0.$$

the field equations are $F_{\mu\nu,\nu} = 0$. An investigation of the theory inferred from this Lagrangian shows that the Hamilton equations are not independent; the picture is not canonical.

We can also use the following density:

$$\mathcal{L} = -\frac{1}{2} A_{\mu,\nu} A_{\mu,\nu}, \tag{6.2.35b}$$

expressed exclusively in terms of the potentials. The field equations are then $\Box A_\mu = 0$, and the Lorentz condition has to be postulated. Although the discussion of the quantization is a simpler matter, the tensor $T_{\mu\nu}$ is then expressed by means of the potentials $A_\mu(x)$ alone (and not of the $F_{\mu\nu}$'s, which are gauge invariant observables) and the gauge invariance of this tensor requires some calculations. In spite of these disadvantages, we shall be using it further on [formulae (6.2.97) *et seq.*] for the computation of certain observables.

The commutation rules are very simple since $\mathcal{D}(\partial) = I$; they are

$$[A_\mu(x), A_\nu(x')] = i\,\delta_{\mu\nu} D(x-x'). \tag{6.2.36}$$

The methods described in the last chapter will enable us to split A_μ into parts with positive and negative frequencies [(5.55) *et seq.*]:

$$A_\mu(x) = (2\pi)^{-\frac{3}{2}} \int_{k_0=K} \frac{d^3\boldsymbol{k}}{\sqrt{2K}} \left(a_\mu(\boldsymbol{k})e^{ikx} + a_\mu^*(\boldsymbol{k})e^{-ikx}\right), K = |\boldsymbol{k}|. \tag{6.2.37}$$

It should be noted that the sign * denotes here neither the complex conjugate nor its Hermitian conjugate: true, the photon field is indeed a neutral field, but the fourth component $A_4(x)$ is not Hermitian but anti-Hermitian:

$$\tilde{A}_4(x) = (i\Phi(x))^\sim = -A_4(x), \tag{6.2.38}$$

we therefore get

$$a_j^*(\boldsymbol{k}) = \tilde{a}_j(\boldsymbol{k}), \quad a_4^*(\boldsymbol{k}) = -\tilde{a}_4(\boldsymbol{k}). \tag{6.2.39}$$

Moreover, the interpretation of the a^* and a as creation and annihilation operators calls for certain precautions. Making use of the commutation rules between the reduced variables

$$[a_\mu(\boldsymbol{k}), a_\nu^*(\boldsymbol{k}')] = \delta_{\mu\nu}\,\delta(\boldsymbol{k}-\boldsymbol{k}'), \tag{6.2.40}$$

we see that

$$\left.\begin{array}{ll} a_j(\boldsymbol{k}),\ \tilde{a}_4(\boldsymbol{k}) & \text{are annihilation operators,} \\ \tilde{a}_j(\boldsymbol{k}),\ a_4(\boldsymbol{k}) & \text{are creation operators.} \end{array}\right\} \tag{6.2.41}$$

Before we discuss the results of this property of the above operators, it will be advisable to introduce into the vector space of the index μ or polarization space† a complete orthonormal set of vectors $e^{(1)}(k)$, $e^{(2)}(k)$, $e^{(3)}(k)$, $e^{(4)}(k)$ such that:

$$e^{(\lambda)}e^{(\lambda')} \equiv \sum_{\mu} e_{\mu}^{(\lambda)}e_{\mu}^{(\lambda')} = \delta_{\lambda\lambda'}, \tag{6.2.42a}$$

$$\Big\{\sum_{\lambda} e^{(\lambda)}e^{(\lambda)}\Big\}_{\mu\nu} \equiv \sum_{\lambda} e_{\mu}^{(\lambda)}e_{\nu}^{(\lambda)} = \delta_{\mu\nu}. \tag{6.2.42b}$$

As an example of such a system, we can consider the following:

$$\left.\begin{array}{llll} \boldsymbol{k}\cdot\boldsymbol{e}^{(1)}(\boldsymbol{k}) = 0, & e_4^{(1)} = 0, & \boldsymbol{k}\cdot\boldsymbol{e}^{(2)}(\boldsymbol{k}) = 0, & e_4^{(2)} = 0, \\ \boldsymbol{e}^{(3)}(\boldsymbol{k}) = \dfrac{\boldsymbol{k}}{K} & e_4^{(3)} = 0, & \boldsymbol{e}^{(4)}(\boldsymbol{k}) = 0, & e_4^{(4)} = 1. \end{array}\right\} \tag{6.2.43}$$

Split the four-vector $a(\boldsymbol{k})$ with respect to the vectors $e^{(\lambda)}$:

$$\left.\begin{aligned} a\ (\boldsymbol{k}) &= \sum_{\lambda} c_{\lambda}e^{(\lambda)}(\boldsymbol{k}), \\ a^*(\boldsymbol{k}) &= \sum_{\lambda} c_{\lambda}^{*}e^{(\lambda)}(\boldsymbol{k}). \end{aligned}\right\} \tag{6.2.44a}$$

† Let x and y be the components along $\mathbf{e}^{(1)}$ and $\mathbf{e}^{(2)}$ (transverse components, p. 258); they are of the form

$$x = a\cos\omega t; \quad y = b\cos(\omega t-\varphi);$$

the point (x, y) describes an ellipse of semi-axes A and B. By setting $B/A = \tan\psi$, we obtain for the angle α between Ox and the large axis of this ellipse:

$$\tan 2\psi = 2ab(a^4+b^4+2a^2b^2\cos 2\varphi)^{-\frac{1}{2}}\sin\varphi$$
$$\tan 2\alpha = 2ab(b^2-a^2)^{-1}\cos\varphi.$$

Consider the vector u of a two-dimensional complex space the components of which are:

$$u_1 = \frac{a}{2}\exp(i\omega t), \quad u_2 = \frac{b}{2}\exp i\{\omega t-\varphi\},$$

and define the Stokes parameters:

$$s_0 = \langle u\,|\,I\,|\,u\rangle = \frac{1}{4}|a|^2+\frac{1}{4}|b|^2$$

$$s_1 = \langle u\,|\,\sigma_3\,|\,u\rangle = \frac{1}{4}|a|^2-\frac{1}{4}|b|^2$$

$$s_2 = \langle u\,|\,\sigma_1\,|\,u\rangle = \frac{1}{2}ab\cos\varphi$$

$$s_3 = \langle u\,|\,\sigma_2\,|\,u\rangle = -\frac{1}{2}ab\sin\varphi.$$

We see that all the elements characterizing the transversal polarization are given by these parameters; the intensity is equal to $4s_0$, and:

$$\tan 2\alpha = s_2/s_1 \qquad \tan 2\psi = -s_3(s_2^2+s_1^2)^{-\frac{1}{2}}.$$

(Cf. also footnote, p. 224.)

It follows that

$$c_\lambda(\boldsymbol{k}) = a(\boldsymbol{k})e^{(\lambda)}(\boldsymbol{k}) \equiv \sum_\mu a_\mu(\boldsymbol{k})e_\mu^{(\lambda)}(\boldsymbol{k}) \tag{6.2.44b}$$

and a similar expression is obtained for c_λ^*. The relations (6.2.39) supply compatibility relations for the operators c [satisfied for the choice (6.2.43), cf. (6.2.47*a*)].

Finally, with the aid of the vectors $e^{(\lambda)}(\boldsymbol{k})$, the vector potential $A(x)$ can be written:

$$\begin{aligned} A_\mu(x) &= \sum_\lambda (2\pi)^{-\frac{3}{2}} \int_{k_0=K} \frac{d^3\boldsymbol{k}}{\sqrt{2K}} e_\mu^{(\lambda)}(\boldsymbol{k}) \left(c_\lambda(\boldsymbol{k})e^{ikx} + c_\lambda^*(\boldsymbol{k})e^{-ikx}\right) \\ &\equiv \sum_\lambda A_\mu^{(\lambda)}(x). \end{aligned} \tag{6.2.45}$$

On the other hand, if we carry into (6.2.40) the expression of the $a(\boldsymbol{k})$'s in terms of the $e^{(\lambda)}(\boldsymbol{k})$'s the orthogonality of the preceding vectors enables us to formulate commutation relations for the operation $c_\lambda(\boldsymbol{k})$:

$$[c_\lambda(\boldsymbol{k}), c_{\lambda'}^*(\boldsymbol{k}')] = \delta_{\lambda\lambda'}\delta(\boldsymbol{k}-\boldsymbol{k}') \tag{6.2.46}$$

and by adopting the frame of reference (6.2.43) we find

$$c_j^*(\boldsymbol{k}) = \tilde{c}_j(\boldsymbol{k}), \quad c_4^*(\boldsymbol{k}) = -\tilde{c}_4(\boldsymbol{k}), \tag{6.2.47a}$$

which leads to the following interpretation [cf. (6.2.101) and (6.2.102)]:

$$\left.\begin{array}{lll} c_j(\boldsymbol{k}), & \tilde{c}_4(\boldsymbol{k}) & \text{are annihilation operators,} \\ \tilde{c}_j(\boldsymbol{k}), & c_4(\boldsymbol{k}) & \text{are creation operators.} \end{array}\right\} \tag{6.2.47b}$$

With the representation (6.2.43) of the $e^{(\lambda)}$'s the vectors $A^{(1)}(x)$ and $A^{(2)}(x)$ correspond to a transversally polarized radiation, whereas $A^{(3)}(x)$ corresponds to a longitudinally polarized radiation; $A^{(4)}(x)$ is said to correspond to a scalar polarization. The terms "longitudinal" and "scalar" photons are often used for the particles described by $A^{(3)}$ and $A^{(4)}$. In the system (6.2.43) the four vectors $A^{(\lambda)}$ can be expressed explicitly as follows:

$$\left.\begin{aligned} A_l^{(r)}(x) &= (2\pi)^{-\frac{3}{2}} \int_{k_0=K} \frac{d^3\boldsymbol{k}}{\sqrt{2K}} e_l^{(r)}\left(c_r(\boldsymbol{k})e^{ikx} + c_r^*(\boldsymbol{k})e^{-ikx}\right), \\ A_4^{(r)}(x) &= 0, \quad K = |\boldsymbol{k}| \end{aligned}\right\} \tag{6.2.48}$$

where there is no summation over $r = 1, 2$; similarly, we obtain

$$\left.\begin{aligned} A_l^{(3)}(x) &= (2\pi)^{-\frac{3}{2}} \int_{k_0=K} \frac{d^3\boldsymbol{k}}{\sqrt{2K}} \frac{k_l}{K} \left(c_3(k)e^{ikx} + c_3^*(\boldsymbol{k})e^{-ikx}\right), \\ A_4^{(3)}(x) &= 0, \end{aligned}\right\}$$

and also

$$\left.\begin{aligned} A_j^{(4)}(x) &= 0, \\ A_4^{(4)}(x) &= (2\pi)^{-\frac{3}{2}} \int_{k_0=K} \frac{d^3\boldsymbol{k}}{\sqrt{2K}} \left(c_4(\boldsymbol{k})e^{ikx} + c_4^*(\boldsymbol{k})e^{-ikx}\right). \end{aligned}\right\}$$

20. Fermi condition. Elimination of the longitudinal and scalar components

We now intend to investigate in detail the Fermi condition. We shall see that it makes it possible to eliminate the longitudinal and scalar photons, but necessitates a number of precautions when it comes to defining the state vector. The general expression of $\chi(x)$ can be written from (6.2.45):

$$\chi(x) = i\sum_{\lambda,\mu}(2\pi)^{-\frac{3}{2}}\int_{k_0=K}\frac{d^3\boldsymbol{k}}{\sqrt{2K}}k_\mu e_\mu^{(\lambda)}\left(c_\lambda(\boldsymbol{k})e^{ikx}-c_\lambda^*(\boldsymbol{k})e^{-ikx}\right) \quad (6.2.49a)$$

and for the particular choice (6.2.43) of the $e^{(\lambda)}$'s the above expression is reduced to

$$\chi(x) = i(2\pi)^{-\frac{3}{2}}\int_{k_0=K}\frac{d^3\boldsymbol{k}}{\sqrt{2K}}K\left[\left(c_3(\boldsymbol{k})+ic_4(\boldsymbol{k})\right)e^{ikx}-\left(c_3^*(\boldsymbol{k})+ic_4^*(\boldsymbol{k})\right)e^{-ikx}\right]. \quad (6.2.49b)$$

Consider now the state vector Ψ; it can be split into a tensor product of vectors $\Psi(\boldsymbol{k})$:†

$$\Psi = \prod_{\boldsymbol{k}}\Psi(\boldsymbol{k}) \quad (\textstyle\prod \text{ indicates a tensor product}) \quad (6.2.50a)$$

such that only the operators $c(\boldsymbol{k})$ and $c^*(\boldsymbol{k})$ depending on the momentum $\boldsymbol{k}$ act on the vector $\Psi(\boldsymbol{k})$. Each vector $\Psi(\boldsymbol{k})$ can be separately written as follows:

$$\Psi(\boldsymbol{k}) = \Psi_{\text{tr}}(\boldsymbol{k})\otimes\Psi_{\text{ls}}(\boldsymbol{k}), \quad (6.2.50b)$$

where $\Psi_{\text{tr}}(\boldsymbol{k})$ is the vector on which the transversal operators c_r and c_r^* alone act, with $r=1, 2$, while only the longitudinal and scalar operators $c_{r'}$ and $c_{r'}^*$ with $r'=3, 4$ act on Ψ_{ls}.

The Fermi condition for the vector $\Psi_{\text{ls}}(\boldsymbol{k})$ is then reflected by the two equations:

$$\left.\begin{aligned}\left[c_3(\boldsymbol{k})+ic_4(\boldsymbol{k})\right]\Psi_{\text{ls}}(\boldsymbol{k}) &= 0,\\ \left[c_3^*(\boldsymbol{k})+ic_4^*(\boldsymbol{k})\right]\Psi_{\text{ls}}(\boldsymbol{k}) &= 0.\end{aligned}\right\} \quad (6.2.51)$$

† The state vector Ψ is, indeed, an eigenvector of observables which are given by sums of bilinear products in the variables $c_\mu(\boldsymbol{k})$, such as (6.2.101) and (6.2.102). It can also be seen at this point that the Fermi condition eliminates the longitudinal and scalar components. Indeed, multiply the first eqn. (6.2.51) by $c_3^*(\boldsymbol{k})-ic_4^*(\boldsymbol{k})$, the second one by $c_3(\boldsymbol{k})-ic_4(\boldsymbol{k})$ and add, one gets:

$$\begin{aligned}\{[c_3^*(\boldsymbol{k}), c_3(\boldsymbol{k})]_+ + [c_4^*(\boldsymbol{k}), c_4(\boldsymbol{k})]_+ + i[c_3^*(\boldsymbol{k}), c_4(\boldsymbol{k})]\\ + i[c_3(\boldsymbol{k}), c_4^*(\boldsymbol{k})]\}\,\Psi_{\text{ls}}(\boldsymbol{k}) = 0.\end{aligned}$$

From the relations (6.2.46) it follows that the commutators are 0; there remains:

$$\{c_3^*(\boldsymbol{k})c_3(\boldsymbol{k})+c_3(\boldsymbol{k})c_3^*(\boldsymbol{k})+c_4^*(\boldsymbol{k})c_4(\boldsymbol{k})+c_4(\boldsymbol{k})c_4^*(\boldsymbol{k})\}\,\Psi_{\text{ls}}(\boldsymbol{k}) = 0.$$

Considering the energy operator given by (6.2.102), it follows that only the transverse components contribute (see p. 269).

Let us, furthermore, suppose the system to be enclosed within a cube with side L (so as to avoid all difficulties in the normalization of the states) and let us write the general expression of the vector $\Psi_{ls}(\boldsymbol{k})$ by making the longitudinal and scalar creation operators act on the vacuum (Chap. V, § 15):

$$\Psi_{ls}(\boldsymbol{k}) = \sum_{n,\, m} \frac{C_{n,\, m}}{\sqrt{n!\, m!}} [c_3^*(\boldsymbol{k})]^n [c_4(\boldsymbol{k})]^m |0\rangle. \tag{6.2.52}$$

Carrying (6.2.52) into (6.2.51), we obtain the linear homogeneous system in $C_{n,\, m}$†

$$\left.\begin{aligned} \sqrt{n}\, C_{n,\, m} + i\sqrt{m}\, C_{n-1,\, m-1} &= 0, \\ \sqrt{m}\, C_{n,\, m} + i\sqrt{n}\, C_{n-1,\, m-1} &= 0, \end{aligned}\right\} \tag{6.2.53}$$

which is only compatible if $n = m$ and in that case

$$C_{n,\, m} = (-i)^n \delta_{nm}, \tag{6.2.54}$$

so that the state vector $\Psi_{ls}(\boldsymbol{k})$ then takes the form:

$$\Psi_{ls}(\boldsymbol{k}) = \sum_{n=0} \frac{(-i)^n}{n!} (c_3^*(\boldsymbol{k})\, c_4(\boldsymbol{k}))^n |0\rangle = e^{-ic_3^*(\boldsymbol{k})\, c_4(\boldsymbol{k})} |0\rangle. \tag{6.2.55}$$

This vector cannot, however, be normalized: its norm is infinite. Indeed, set

$$\langle \Psi_{ls}(\boldsymbol{k}) | \Psi_{ls}(\boldsymbol{k}) \rangle = \sum_{n,\, m} \frac{(-i)^{n+m}}{n!\, m!} \langle 0 | (c_3(\boldsymbol{k}) c_4^*(\boldsymbol{k}))^n (c_3^*(\boldsymbol{k})\, c_4(\boldsymbol{k}))^m |0\rangle. \tag{6.2.56a}$$

Now, by virtue of the commutation rules,

$$\langle 0 | (c_3 c_4^*)^n (c_3^* c_4)^m | 0 \rangle = (-1)^n (n!)^2 \delta_{nm}, \tag{6.2.56b}$$

so that actually the norm of the vector $\Psi_{ls}(\boldsymbol{k})$ is infinite and this is the fundamental difficulty introduced by the Fermi condition.

There are three methods of overcoming this difficulty: the first two, which will be discussed in this section and the following one, are purely mathematical. The third method which is the basis of de Broglie's wave mechanics of the photon is far more physical: it consists in treating the photon as a vector particle with mass μ and in passing to the limit $\mu = 0$ once all calculations have been effected. This last method will be discussed in the following section.

† To obtain (6.2.53) use the commutation rules and such relations as

$$\begin{aligned} c_3(c_3^*)^n &= n(c_3^*)^{n-1} + (c_3^*)^n c_3, \\ c_4^*(c_4)^n &= -n(c_4)^{n-1} + (c_4)^n c_4^*. \end{aligned}$$

The first method consists in introducing a parameter λ into the definition of the state vector Ψ_{ls} in such a way that the state vector is normalizable and obeys the Fermi condition when this parameter converges to 1.† Let us therefore set

$$\Psi_{\text{ls}}^{(\lambda)}(\boldsymbol{k}) = \sqrt{1-\lambda^2} \sum_{n=0} \frac{(-i\lambda)^n}{n!} (c_3^*(\boldsymbol{k})\, c_4(\boldsymbol{k}))^n |0\rangle. \qquad (6.2.57)$$

We verify that we indeed obtain

$$\langle \Psi_{\text{ls}}^{(\lambda)}(\boldsymbol{k}) | \Psi_{\text{ls}}^{(\lambda)}(\boldsymbol{k}) \rangle = (1-\lambda^2) \sum_{n=0} \lambda^{2n} = 1. \qquad (6.2.58)$$

Equations (6.2.51) are obviously no longer satisfied; they become

$$\left.\begin{aligned} [c_3(\boldsymbol{k})+ic_4(\boldsymbol{k})]\Psi_{\text{ls}}^{(\lambda)}(\boldsymbol{k}) &= \alpha(\lambda, \boldsymbol{k}) = i(1-\lambda)c_4(\boldsymbol{k})\Psi_{\text{ls}}^{(\lambda)}(\boldsymbol{k}),\\ [c_3^*(\boldsymbol{k})+ic_4^*(\boldsymbol{k})]\Psi_{\text{ls}}^{(\lambda)}(\boldsymbol{k}) &= \beta(\lambda, \boldsymbol{k}) = (1-\lambda)c_3^*(\boldsymbol{k})\Psi_{\text{ls}}^{(\lambda)}(\boldsymbol{k}). \end{aligned}\right\} \qquad (6.2.59)$$

The norms of the vectors $\alpha(\lambda, k)$ and $\beta(\lambda, k)$ (in $\mathcal{R}$-space) are easy to compute: we verify that they are proportional to

$$\frac{1-\lambda}{1+\lambda}.$$

Since these norms tend to zero when λ becomes 1, we can therefore say that the Fermi condition is only satisfied at this limit. We can now compute the expectation values of the products of longitudinal and scalar creation and annihilation operators. Compute, for instance, $\langle \Psi_{\text{ls}}^{(\lambda)}(\boldsymbol{k}) | c_3^* (\boldsymbol{k}) c_3(\boldsymbol{k}) | \Psi_{\text{ls}}^{(\lambda)}(\boldsymbol{k}) \rangle$; we find that this expectation value is given by the expression

$$(1-\lambda^2) \sum \frac{(-i\lambda)^{m+n}}{m!\, n-1!} \langle 0 | (c_4^* c_3)^m (c_3^* c_4)^n | 0 \rangle = (1-\lambda^2) \sum n(\lambda^2)^n = \frac{\lambda^2}{1-\lambda^2}.$$

Similarly, we obtain

$$\langle \Psi_{\text{ls}}^{(\lambda)}(\boldsymbol{k}) | c_3(\boldsymbol{k}) c_3^*(\boldsymbol{k}) | \Psi_{\text{ls}}^{(\lambda)}(\boldsymbol{k}) \rangle = \frac{1}{1-\lambda^2},$$

together with the following formulae (which the reader can verify for himself):

$$\left.\begin{aligned} \langle \Psi_{\text{ls}}^{(\lambda)}(\boldsymbol{k}) | c_4 c_4^* | \Psi_{\text{ls}}^{(\lambda)}(\boldsymbol{k}) \rangle &= \langle \Psi_{\text{ls}}^{(\lambda)} | c_3^* c_3 | \Psi_{\text{ls}}^{(\lambda)} \rangle = -\frac{\lambda^2}{1-\lambda^2},\\ \langle \Psi_{\text{ls}}^{(\lambda)}(\boldsymbol{k}) | c_4^* c_4 | \Psi_{\text{ls}}^{(\lambda)}(\boldsymbol{k}) \rangle &= \langle \Psi_{\text{ls}}^{(\lambda)} | c_3 c_3^* | \Psi_{\text{ls}}^{(\lambda)} \rangle = -\frac{1}{1-\lambda^2}. \end{aligned}\right\} \qquad (6.2.60)$$

An important result can be inferred from these formulae, namely that the longitudinal and scalar photons give no contribution to the expecta-

† R. Utiyama, T. Imamura, S. Sanakawa, T. Dodo, *Prog. Theor. Phys.*, **6**, 587 (1951).

tion values of the observables. Consider the operator $\mathcal{N}(\boldsymbol{k})$ as introduced in (6.2.101); separate in its expression the respective contributions of the longitudinal and scalar photons: $\mathcal{N}_j(\boldsymbol{k})+\mathcal{N}_s(\boldsymbol{k})$; in accordance with the preceding formulae, we have†

$$\langle \Psi_{\mathrm{ls}}^{(\lambda)}(\boldsymbol{k}) \,|\, \mathcal{N}_1(\boldsymbol{k})+\mathcal{N}_s(\boldsymbol{k}) \,|\, \Psi_{\mathrm{ls}}^{(\lambda)}(\boldsymbol{k}) \rangle \tag{6.2.61}$$
$$= \frac{1}{2} \langle \Psi_{\mathrm{ls}}^{(\lambda)} \,|\, c_3 c_3^* + c_3^* c_3 + c_4 c_4^* + c_4^* c_4 \,|\, \Psi_{\mathrm{ls}}^{(\lambda)} \rangle = 0.$$

The property referred to is therefore proved since any other observable can be expressed with the aid of the operator $\mathcal{N}(\boldsymbol{k})$.

In many applications, the longitudinal and scalar components may therefore be simply omitted. We shall be showing, moreover, that the gauge invariance of the theory is not affected by this elimination, by investigating the method of quantization of the Maxwell equations with an indefinite metric.

21. Quantization in an indefinite metric

The infinite norm of the state vector describing a system of free photons is therefore the main difficulty we have come across in the foregoing survey of the Maxwell field. This infinite norm is a consequence of the fact that the Fermi condition is reflected by the two equations (6.2.51) in each of which creation and annihilation operators are found to be mixed. In a theory in which c_3 and c_4 would be annihilation operators and c_3^* and c_4^* creation operators such a difficulty might not arise. A theory on these lines has been obtained independently by Gupta and Bleuler; it is based on the definition of a vector space with indefinite norm.‡ We shall begin by investigating a few characteristics of such a space.

We are therefore interested in defining a vector space with a metric such that the norm of each vector be real, but not positive definite. Denoting by $\langle u \mid v \rangle_\eta$ the new scalar product we propose to obtain

$$\|f\|^2 = \langle f|f\rangle_\eta = \langle f|f\rangle_\eta^*. \tag{6.2.62}$$

Let η be then an operator of this space, known as "metric operator"; we define the scalar product as follows:

$$\langle f|g\rangle_\eta \equiv \langle f|\eta|g\rangle. \tag{6.2.63}$$

The following property of the scalar product

$$\langle f|g\rangle_\eta = \langle g|f\rangle_\eta^* \tag{6.2.64}$$

† We apologize for referring to a formula which will not be proved until later. The proof of this formula is independent of the above considerations.

‡ S. N. Gupta, *Proc. Phys. Soc.*, **64**, 850 (1951); K. Bleuler, *Helv. Phys. Acta*, **23**, 567 (1950).

implies, furthermore, that the metric operator is Hermitian; we shall be assuming this hereafter.

By virtue of the definition of the scalar product, the matrix element of an operator A is therefore

$$\langle f | A | g \rangle_\eta \equiv \langle f | \eta A | g \rangle \tag{6.2.65}$$

and we shall adopt as a definition of the adjoint operator $\bar{A}$ the one which conforms to the usual definition:

$$\langle f | A | g \rangle_\eta = \langle g | \bar{A} | f \rangle_\eta^*. \tag{6.2.66}$$

We then have the following set of propositions:

COROLLARY 1. *Between the generalized adjoint operators $\bar{A}$ and the adjoint operator there exists the following relation:*

$$\bar{A} = \eta^{-1} \tilde{A} \eta. \tag{6.2.67}$$

Indeed, on the one hand we have:

$$\langle f | A | g \rangle_\eta = \langle f | \eta A | g \rangle = \langle g | \tilde{A} \eta | f \rangle^*$$

and on the other:

$$\langle g | \bar{A} | f \rangle_\eta^* = \langle g | \eta \bar{A} | f \rangle^*,$$

which leads us to (6.2.67).

COROLLARY 2. *We have*

$$(\bar{A})^- = A. \tag{6.2.68}$$

COROLLARY 3. *If*

$$[\eta, A] = 0, \tag{6.2.69a}$$

then

$$\bar{A} = \tilde{A}. \tag{6.2.69b}$$

The proof of these two corollaries is straightforward.

An operator A will be called a generalized Hermitian operator if

$$\bar{A} = A. \tag{6.2.70}$$

We then obtain the following two theorems:

THEOREM 1. *The expectation value of a generalized Hermitian operator is real.*

Indeed,

$$\langle f | A | f \rangle_\eta = \langle f | \bar{A} | f \rangle_\eta = \langle f | A | f \rangle_\eta^*. \tag{6.2.71}$$

THEOREM 2. *If A is a Hermitian operator and if*

$$[\eta, A] = 0,$$

the operator A is a generalized Hermitian operator.

If A is an anti-Hermitian operator and if

$$[\eta, A]_+ = 0,$$

the operator A is a generalized Hermitian operator.

The above theorem can be readily proved, making use of the preceding corollaries.

Finally, a generalized unitary operator is defined as follows:

$$A^{-1} = \bar{A}. \tag{6.2.72}$$

The canonical transformations generated by such an operator turn a generalized Hermitian operator into another one of the same kind, as may be easily seen.

The key to this quantization method which we now intend to investigate, is given by theorem 2: we have to define an operator which obeys the relations

$$[\eta, A_j(x)] = 0, \tag{6.2.73a}$$

$$[\eta, A_4(x)]_+ = 0, \tag{6.2.73b}$$

from which it follows that the A_μ's are generalized Hermitian operators.

The operator η can be defined by an infinite tensor product of operators relating to each component and to each momentum $\boldsymbol{k}$. We shall choose the factors relating to the space components $A_j(x)$ equal to the operator identity, and the factor relating to A_4 and to the momentum $\boldsymbol{k}$,† namely,

$$\eta(\boldsymbol{k}) = (-1)^{\mathcal{N}_4(\boldsymbol{k})}, \tag{6.2.74a}$$

with

$$\mathcal{N}_4(\boldsymbol{k}) = c_4^*(\boldsymbol{k})c_4(\boldsymbol{k}). \tag{6.2.74b}$$

We have, indeed,

$$[\eta(\boldsymbol{k}), c_4(\boldsymbol{k})]_+ = 0, \qquad [\eta(\boldsymbol{k}), c_4^*(\boldsymbol{k})]_+ = 0, \tag{6.2.75}$$

since if we consider the complete orthonormal set of the eigenvectors of $\mathcal{N}_4$

$$\mathcal{N}_4 \,|\, n_4 \rangle = n_4 \,|\, n_4 \rangle, \tag{6.2.76}$$

$$c_4 \,|\, n_4 \rangle = \sqrt{n_4+1}\,|\, n_4+1 \rangle, \qquad c_4^* \,|\, n_4 \rangle = \sqrt{n_4}\,|\, n_4-1 \rangle,$$

it is easy to verify the validity of eqns. (6.2.75) by making them act on each of the vectors of this set. These equations lead to eqns. (6.2.73) and consequently the choice (6.2.74) of η is the one required.

† $\mathcal{N}_4(\boldsymbol{k})$ in (6.2.74*b*) is one of the terms of the operator defined in (6.2.101) and which can be defined as the number operator.

In the metric η, the expansion of A_μ is written, as may be easily verified:

$$A_\mu(x) = (2\pi)^{-\frac{3}{2}} \int_{k_0=K} \frac{d^3\boldsymbol{k}}{\sqrt{2K}} (a_\mu(\boldsymbol{k})e^{ikx} + \bar{a}_\mu(\boldsymbol{k})e^{-ikx})$$

$$= \sum_\lambda (2\pi)^{-\frac{3}{2}} \int_{k_0=K} \frac{d^3k}{\sqrt{2K}} e_\mu^{(\lambda)}(c_\lambda(\boldsymbol{k})e^{ikx} + \bar{c}_\lambda(\boldsymbol{k})e^{-ikx}) \tag{6.2.77}$$

and we obtain the following commutation rules

$$[c_\mu(\boldsymbol{k}), \bar{c}_\nu(\boldsymbol{k}')] = \delta_{\mu\nu}\,\delta(\boldsymbol{k}-\boldsymbol{k}') \tag{6.2.78}$$

and the Fermi condition:

$$(c_3+ic_4)\,|\,\Psi\rangle_\eta = 0, \qquad (\bar{c}_3+i\bar{c}_4)\,|\,\Psi\rangle_\eta = 0. \tag{6.2.79}$$

Consider, as in the preceding section, the state vector Ψ as the tensor product of a state vector describing the transversal photons and of a state vector Ψ_{ls} describing the longitudinal and scalar photons. The vector Ψ_{ls} can be represented by an infinite tensor product of vectors $\Psi_{\text{ls}}(\boldsymbol{k})$; split each of these vectors Ψ into its components in the orthonormal set of eigenvectors common to $\mathcal{N}_3(\boldsymbol{k})$ and $\mathcal{N}_4(\boldsymbol{k})$ as follows:†

$$\Psi_{\text{ls}}(\boldsymbol{k}) = \sum C_{n_3, n_4}\,|\,n_3, n_4\rangle, \tag{6.2.80a}$$

with

$$\langle n_3, n_4\,|\,n_3', n_4'\rangle_\eta = \langle n_3, n_4\,|\,(-1)^{\mathcal{N}_4}\,|\,n_3', n_4'\rangle = (-1)^{n_4}\,\delta_{n_3, n_3'}\,\delta_{n_4, n_4'}. \tag{6.2.80b}$$

It is then easy to verify that the two relations (6.2.79) of the Fermi condition reduce to the following one:

$$\sqrt{n_3+1}\,C_{n_3+1, n_4} + i\sqrt{n_4+1}\,C_{n_3, n_4+1} = 0, \tag{6.2.81}$$

where the coefficients $C_{n,m}$ are zero if one of the indices is negative.

From the basis $|\,n_3, n_4\rangle$, a new basis can be defined

$$\left.\begin{aligned} &|\,\Psi_{\text{ls}}^{(0)}\rangle = |\,0, 0\rangle, \\ &|\,\Psi_{\text{ls}}^{(1)}\rangle = |\,1, 0\rangle + i\,|\,0, 1\rangle, \\ &\dots\dots\dots\dots\dots\dots\dots\dots, \\ &|\,\Psi_{\text{ls}}^{(n)}\rangle = \sum_{r=0}^{n} (i)^r \sqrt{C_n^r}\,|\,n-r, r\rangle, \end{aligned}\right\} \tag{6.2.82}$$

where the label n of $\Psi_{\text{ls}}^{(n)}$ denotes the total number of longitudinal and scalar photons; it can be easily verified that the vectors of this basis obey the following relations:

$$\left.\begin{aligned} &\langle\,\Psi_{\text{ls}}^{(0)}\,|\,\Psi_{\text{ls}}^{(n)}\rangle_\eta = \delta_{n0}, && \text{provided that } n \text{ or } n' \neq 0; \\ &\langle\,\Psi_{\text{ls}}^{(n)}\,|\,\Psi_{\text{ls}}^{(n)}\rangle_\eta = 0 && n \neq n' \text{ or } n = n'. \end{aligned}\right\} \tag{6.2.83}$$

† Cf. (6.2.101).

Thus the norm in the η metric of all the $|\Psi_{ls}^{(n)}\rangle$ vanishes except when $n=0$ in which case it is positive. Any negative norm is therefore excluded with this basis: a state represented by a linear combination of $|\Psi_{ls}^{(0)}\rangle$ and $\Psi_{ls}^{(n)}\rangle$ is a normalizable vector with norm I.

Furthermore, if we return to the space x, the Fermi condition (6.2.79), (6.2.81) now only affects the parts of $A_\mu(x)$ with positive frequencies; this condition can be replaced by the following weaker condition:

$$\frac{\partial A_\mu^+(x)}{\partial x_\mu}|\Psi\rangle = 0. \tag{6.2.84}$$

We are now going to prove that the states characterized respectively by the vectors $|\Psi_{ls}^{(0)}\rangle$ and $|\Psi_{ls}^{(0)}\rangle + \Sigma c_n |\Psi_{ls}^{(n)}\rangle$ correspond to the same physical state by showing that the expectation values of $F_{\mu\nu}$ and A_μ (expressed in the metric η) differ in these two cases by a gauge transformation.

Therefore set

$$|\Psi_{ls}\rangle = |\Psi_{ls}^{(0)}\rangle + \sum c_n |\Psi_{ls}^{(n)}\rangle, \tag{6.2.85}$$

where the numbers c_n are arbitrary and calculate the expectation value of

$$\langle\Psi_{ls}|F_{\mu\nu}|\Psi_{ls}\rangle_\eta \equiv \langle\Psi_{ls}|A_{\nu,\mu} - A_{\mu,\nu}|\Psi_{ls}\rangle_\eta. \tag{6.2.86}$$

We can easily verify from the expression (6.2.77) of $A_\mu(x)$ that we have the following commutation relation:

$$\left[F_{\mu\nu}(x), \frac{\partial A_\lambda^+(x)}{\partial x_\lambda}\right] = 0,$$

so that by virtue of (6.2.84)

$$\frac{\partial A_\lambda^+(x)}{\partial x_\lambda} F_{\mu\nu}(x)|\Psi_{ls}\rangle = 0. \tag{6.2.87}$$

The vector $F_{\mu\nu}|\Psi_{ls}\rangle$ does therefore represent a state of the system of photons since it obeys the Fermi condition. We can, moreover, verify that in the system (6.2.82), the vector $F_{\mu\nu}|\Psi_{ls}^{(n)}\rangle$, $n \neq 0$ is a vector of the space spanned by all the vectors of this system except the vector $|\Psi_{ls}^0\rangle$. For this purpose it is sufficient to write down the longitudinal and scalar components of $F_{\mu\nu}$ and to note that they cannot diminish any of the number n_3 and n_4 of longitudinal and scalar photons. Insert (6.2.85) into (6.2.86) and use the orthonormality relations (6.2.83); bearing in mind the remark we have just made, we obtain

$$\langle\Psi_{ls}|F_{\mu\nu}|\Psi_{ls}\rangle = \langle\Psi_{ls}^{(0)}|F_{\mu\nu}|\Psi_{ls}^{(0)}\rangle, \tag{6.2.88}$$

in other words, "the expectation values of the electric and magnetic fields, observable and gauge invariant quantities, depend exclusively on the vector $|0,0\rangle$ representing a state without either longitudinal or scalar photons".

The elimination of these two kinds of photons from all the observable and gauge invariant quantities is thus achieved and we have by virtue of the first of eqns. (6.2.83):

$$\langle \Psi_{1s} | F_{\mu\nu} | \Psi_{1s} \rangle_\eta = \langle \Psi_{1s}^0 | F_{\mu\nu} | \Psi_{1s}^0 \rangle.$$

Consider now the expectation values of the potentials

$$\langle \Psi_{1s} | A_\mu | \Psi_{1s} \rangle_\eta = \langle \Psi_{1s}^{(0)} | A_\mu | \Psi_{1s}^{(0)} \rangle_\eta + \langle A_\mu \rangle, \tag{6.2.89}$$

where $\langle A_\mu \rangle$ represents the contribution of all the components $\Psi_{1s}^{(n)}$.

From these expectation values, we can infer those of the $F_{\mu\nu}$'s; indeed:

$$\langle \Psi_{1s} | F_{\mu\nu} | \Psi_{1s} \rangle = \langle \Psi_{1s}^{(0)} | A_{\nu,\mu} - A_{\mu,\nu} | \Psi_{1s}^{(0)} \rangle_\eta + \frac{\partial \langle A_\nu \rangle}{\partial x_\mu} - \frac{\partial \langle A_\mu \rangle}{\partial x_\nu} \tag{6.2.90}$$

so that by virtue of (6.2.88):

$$\frac{\partial \langle A_\nu \rangle}{\partial x_\mu} - \frac{\partial \langle A_\mu \rangle}{\partial x_\nu} = 0.$$

The latter equality shows that the field $\langle A_\mu \rangle$ derives from a four-gradient:

$$\langle A_\nu \rangle = \frac{\partial \Phi(x)}{\partial x_\nu}. \tag{6.2.91}$$

Consequently,

$$\langle \Psi_{1s} | A_\mu | \Psi_{1s} \rangle_\eta = \langle \Psi_{1s}^{(0)} | A_\mu(x) | \Psi_{1s}^{(0)} \rangle_\eta + \frac{\partial \Phi(x)}{\partial x_\mu}. \tag{6.2.92}$$

It is apparent that Φ (like A_μ) obeys

$$\Box \Phi(x) = 0.$$

We thus see finally that $\Psi_{1s} | A_\mu | \Psi_{1s} \rangle_\eta$ differs from $\langle \Psi_{1s}^{(0)} | A_\mu | \Psi_{1s}^{(0)} \rangle_\eta$ in a simple change of gauge.

22. Observables of the electromagnetic field

The operator $T_{\mu\nu}$ which is derived from the Lagrangian density (6.2.27) is written:

$$T_{\mu\nu} = -\frac{\partial \mathcal{L}}{\partial A_{\varrho,\nu}} A_{\varrho,\mu} + \mathcal{L}\,\delta_{\mu\nu} = -F_{\varrho\nu} A_{\varrho,\mu} + \chi(x) A_{\nu,\mu} + \mathcal{L}\,\delta_{\mu\nu}. \tag{6.2.93}$$

This tensor is neither symmetrical nor gauge invariant. The tensor which has these properties is the tensor $\Theta_{\mu\nu}$ which necessitates the calculation of the density of the spin angular momentum $\mathfrak{M}_{\mu\nu\lambda}(x)$.

For this purpose, note first that since A_μ is a four-vector, an infinitesimal Lorentz transformation

$$A'_\mu(x') = A_\mu(x) + \varepsilon_{\mu\nu} A_\nu(x) = A_\mu(x) + \frac{1}{2}\, \varepsilon_{\sigma\tau} J_{\sigma\tau\mu\nu} A_\nu(x)$$

makes it possible to write the six matrices $J_{\sigma\tau}$:

$$J_{\sigma\tau\mu\nu} = \delta_{\sigma\mu}\,\delta_{\tau\nu} - \delta_{\sigma\nu}\,\delta_{\tau\mu}. \tag{6.2.94}$$

The density of the spin tensor is therefore expressed as follows [cf. (3.49)]:

$$\mathfrak{M}_{\mu\nu\lambda} = \frac{\partial \mathcal{L}}{\partial A_{\alpha,\,\lambda}}\, J_{\mu\nu\alpha\beta} A_\beta(x) \tag{6.2.95}$$

$$= F_{\mu\lambda} A_\nu - F_{\nu\lambda} A_\mu - \delta_{\mu\lambda} \chi(x) A_\nu(x) + \delta_{\nu\lambda} \chi_\mu(x) A_\mu(x)$$

and the tensor $\Theta_{\mu\nu}$ takes the following form:

$$\Theta_{\mu\nu} = \frac{1}{2}\,(F_{\lambda\mu} F_{\lambda\nu} + F_{\lambda\nu} F_{\lambda\mu}) - \frac{1}{4}\,\delta_{\mu\nu}(F_{\sigma\tau} F_{\sigma\tau}) + \Theta'_{\mu\nu}, \tag{6.2.96}$$

where $\Theta'_{\mu\nu}$ is a symmetrical tensor the terms of which are proportional to $\chi(x)$, to its derivatives and to its square, such that in accordance with the Fermi condition:

$$\langle \Psi | \Theta'_{\mu\nu} | \Psi \rangle = 0.$$

$\Theta'_{\mu\nu}$ is therefore a tensor with zero expectation value for any physical state and can be neglected. $\Theta_{\mu\nu}$ then becomes the classical energy–momentum tensor which is expressed in the table on p. 250.

To introduce the main observables, it is simpler to start from the Lagrangian density (6.2.35*b*):

$$\mathcal{L} = -\frac{1}{2}\, A_{\mu,\,\nu} A_{\mu,\,\nu} \tag{6.2.97}$$

which differs from the expression (6.2.27) only by terms proportional to χ, to its derivatives and by four-dimensional divergences; we obtain, indeed,

$$\frac{1}{4}\, F_{\mu\nu} F_{\mu\nu} - \frac{1}{2}\,(\chi(x))^2 \tag{6.2.98}$$

$$= -\left(\frac{1}{2}\, A_{\mu,\,\nu} A_{\mu,\,\nu} + \frac{1}{2}\, A_\mu \frac{\partial \chi(x)}{\partial x_\mu} - \frac{1}{2}\frac{\partial}{\partial x_\nu}\{A_\mu A_{\nu,\,\mu}\}\right) - \frac{1}{2}\,(\chi(x))^2.$$

The energy–momentum tensor is expressed in the simple form:

$$T_{\mu\nu} = A_{\lambda,\,\mu} A_{\lambda,\,\nu} - \frac{1}{2}\,\delta_{\mu\nu}(A_{\sigma,\,\tau} A_{\sigma,\,\tau}). \tag{6.2.99}$$

From this symmetrical tensor, we infer the momentum

$$P_\mu = -i \int \left\{ A_{\lambda,\mu} A_{\lambda,4} - \frac{1}{2}\, \delta_{\mu 4} (A_{\sigma,\tau} A_{\sigma,\tau}) \right\} d^3x \qquad (6.2.100)$$

or by introducing the operators c_λ and c_λ^*:

$$P_\mu = \int_{k_0 = K} d^3k k_\mu \frac{1}{2} (c_\lambda(\boldsymbol{k}) c_\lambda^*(\boldsymbol{k}) + c_\lambda^*(\boldsymbol{k}) c_\lambda(\boldsymbol{k})) \qquad (6.2.101)$$

$$= \int_{k_0 = K} d^3k k_\mu \sum_\lambda \mathcal{N}_\lambda(\boldsymbol{k}).$$

We see in particular that the energy operator

$$P_0 = \int_{k_0 = K} d^3k K \frac{1}{2} (c_j(\boldsymbol{k}) \tilde{c}_j(\boldsymbol{k}) + \tilde{c}_j(\boldsymbol{k}) c_j(\boldsymbol{k}) - c_4(\boldsymbol{k}) \tilde{c}_4(\boldsymbol{k}) + c_4(\boldsymbol{k}) \tilde{c}_4(\boldsymbol{k})) \qquad (6.2.102)$$

has positive expectation values solely as a result of the Fermi condition by which the negative definite term is eliminated [cf. (6.2.61)]. Finally, all the observables are expressed by means of the two operators

$$\mathcal{N}_r(\boldsymbol{k}) = \frac{1}{2} (c_r(\boldsymbol{k}) \tilde{c}_r(\boldsymbol{k}) + \tilde{c}_r(\boldsymbol{k}) c_r(\boldsymbol{k})) \qquad (r = 1, 2) \qquad (6.2.103)$$

up to an infinite term.

The problem of the infinite vacuum-expectation values can be dealt with as we have already done by simple subtraction or by the introduction of ordered products.

It will be noted, finally, that the number of photons with momentum $\boldsymbol{k}$ and polarization state r contribute to the characterization of a state. On the other hand, the zero value of the mass of the photon indicates that its polarization in a given direction [cf. (4.5*c*)] can only take two values, although the spin of the photon is—by virtue of the structure of the field equations—equal to 1.

The above property is expressed by stating that there are only two possibilities for the spin of the photon with respect to a given direction: the spin is parallel or antiparallel to that direction.

This interpretation can be confirmed by a direct investigation of the spin operator [in accordance with (6.2.95), for instance].

23. Measurement of the electric and magnetic fields at two different space–time points

From the commutation rules of the A_μ's we can easily go over to the commutation rules obeyed by the $F_{\mu\nu}$'s; a simple calculation shows that

$$[F_{\mu\nu}(x), F_{\varrho\lambda}(x')] = i\{\delta_{\mu\lambda}\partial_\nu\partial_\varrho - \delta_{\mu\varrho}\partial_\nu\partial_\lambda + \delta_{\nu\varrho}\partial_\mu\partial_\lambda - \delta_{\nu\lambda}\partial_\mu\partial_\varrho\} D(x-x'). \tag{6.2.104}$$

It follows therefore that at two different space–time points the operators $\boldsymbol{E}(x)$ and $\boldsymbol{H}(x)$ associated with the electric and magnetic fields cannot be measured simultaneously, contrary to what occurs in the classical electrodynamics theory. It is interesting to compare the results that may be inferred from the above formula with the uncertainty relations between two canonically conjugate variables. These anologies can be brought out very clearly if we note that what is being measured is not the field $F_{\mu\nu}(x)$ at a well-defined point x, but the expectation value of this field on a particular space–time volume $\mathcal{V}$:

$$f_{\mu\nu}[\mathcal{V}] = \frac{1}{\mathcal{V}} \int_{\mathcal{V}} F_{\mu\nu}(x)\, d^4x.$$

A subtle discussion initiated by Bohr and Rosenfeld proves that the $f_{\mu\nu}$'s obey all the conditions to which we can legitimately subject them. These authors go even further and show that experiments can be imagined on macroscopic test-bodies which would enable one to verify the consequences of the commutation rules obeyed by the operators $f_{\mu\nu}[\mathcal{V}]$ and even those which can be inferred from the commutation rules of currents $J_\mu(x)$.

It would be impossible to recapitulate this discussion without entering on a subtle and therefore somewhat long analysis; we can do no more than refer readers to the articles quoted below.†

PART 3

PROCA–DE BROGLIE FIELD‡

The Proca–de Broglie field describes particles with non-zero mass and spin 1. Its wave function can be defined either by a four-vector $\Phi_\mu(x)$ or within the framework of the fusion theory by the tensor product of two

† W. Pauli (Ed.), *Niels Bohr and the Development of Physics*, p. 70 *et seq.*, Pergamon Press, London. For the original articles: N. Bohr and L. Rosenfeld, *Dan. Mat. fys. Medd.*, **12**, 8 (1933); *Phys. Rev.*, **78**, 794 (1950). L. Rosenfeld, *Physica*, **19**, 859 (1953); E. Cornaldesi, *Supl. Nuovo Cimento*, **10**, 83 (1953); B. Feretti, *Nuovo Cimento*, **12**, 585 (1954).

‡ A. L. Proca, *Journal Phys.*, **7**, 347 (1936); L. de Broglie, *Théorie générale des particles à spin*, Gauthier-Villars, 1943,

spinors (this last wave function then describes a mixture of particles of spin 0 and 1).

The particles of this field have been called "vector mesons"; it should be added that they have not been observed as free particles, but as resonances. They must in any case be distinguished from the pseudo-scalar mesons which describe nuclear forces.

We may also try to describe the photon as a vector particle, of spin 1, with vanishing mass. This is the fundamental idea in de Broglie's wave mechanics of the photon and it raises, as we shall be seeing, problems of a certain complexity.

24. Wave equations

We are therefore interested in describing this field by a four-vector with components $\Phi_\mu(x)$, each of which by definition obeys the equation

$$\{\Box - m^2\}\Phi_\mu(x) = 0. \tag{6.3.1a}$$

In order to restrict its spin to 1, we shall also make it satisfy the Lorentz condition (which must be considered as an equation and not as a condition concerning the state vector)

$$\frac{\partial \Phi_\mu(x)}{\partial x_\mu} = 0, \tag{6.3.1b}$$

we therefore obtain finally only three independent components $\Phi_\mu(x)$.

Because of the purely imaginary character of $\dfrac{\partial}{\partial x_4}$, we do indeed obtain

$$(\Box - m^2)\tilde{\Phi}_\mu(x) = 0,$$

but $\partial_\mu \tilde{\Phi}_\mu(x)$ is no longer zero. The field which obeys the same equations (6.3.1) is, as may be easily verified,

$$\Phi^*(x) = \begin{pmatrix} \tilde{\Phi}_1(x) \\ \tilde{\Phi}_2(x) \\ \tilde{\Phi}_3(x) \\ -\tilde{\Phi}_4(x) \end{pmatrix} \tag{6.3.2}$$

and the asterisk * as superscript to the Φ_μ's does not denote the complex-conjugation.

Introduce the skew-symmetrical tensor

$$\left.\begin{aligned} f_{\mu\nu} &= \Phi_{\nu,\,\mu} - \Phi_{\mu,\,\nu}, \\ f^*_{\mu\nu} &= \Phi^*_{\nu,\,\mu} - \Phi^*_{\mu,\,\nu}. \end{aligned}\right\} \tag{6.3.3}$$

Equations (6.3.1) take the form

$$\{\Box - m^2\}\Phi_\mu = \partial_\varrho \partial_\varrho \Phi_\mu - \partial_\mu \partial_\varrho \Phi_\varrho - m^2 \Phi_\mu = f_{\varrho\mu,\,\varrho} - m^2 \Phi_\mu = 0,$$

which leads to another form of the wave equation

$$f_{\mu\varrho,\varrho}+m^2\Phi_\mu = 0, \tag{6.3.4}$$

where the part played by the mass m of the particle can be clearly seen.

Returning to the variables Φ_μ, eqn. (6.3.4) can also take the form

$$\{\partial_\mu\partial_\varrho-\{\Box-m^2\}\,\delta_{\mu\varrho}\}\Phi_\varrho(x) = 0. \tag{6.3.5}$$

Conversely, the equivalent equations (6.3.4) and (6.3.5) lead to the wave equation (6.3.1a) and the Lorentz condition (6.3.1b). To see this, one has only to differentiate (6.3.4) with respect to x_μ: the term $f_{\mu\varrho,\varrho,\mu}$ is zero since $f_{\mu\varrho}$ is skew-symmetrical, and, since m is non-zero, we infer from it (6.3.1b).

25. Lagrangian formalism and quantization

The field equations (6.3.4) are inferred from the following Lagrangian density:

$$\mathcal{L} = -\frac{1}{2}f^*_{\mu\nu}f_{\mu\nu}-m^2\Phi^*_\lambda\Phi_\lambda, \tag{6.3.6}$$

as may be easily verified.

Using the field equations in their form (6.3.5), a straightforward calculation gives the differential operator $\mathcal{D}(\partial)$:

$$\mathcal{D}_{\mu\nu}(\partial) = \delta_{\mu\nu}-\frac{1}{m^2}\,\partial_\mu\partial_\nu. \tag{6.3.7}$$

The commutation rules of the field variables are therefore

$$[\Phi_\mu(x), \tilde{\Phi}_\nu(x')] = i\left\{\delta_{\mu\nu}-\frac{1}{m^2}\,\partial_\mu\partial_\nu\right\}\Delta(x-x'). \tag{6.3.8}$$

It is easy to go over to the reduced variables (creation and annihilation operators); let us write the matrix elements $D(ik)$ (5.110) in the following form:

$$D_{\mu\nu}(ik) = c_{\mu\varrho}e^{(\varrho)}_\nu, \tag{6.3.9}$$

where $e^{(\varrho)}$, $\varrho = 1, 2, 3, 4$ is the set of vectors (6.2.43) introduced for the Maxwell field. The coefficients $c_{\mu\varrho}$ must be determined from eqn. (5.113), which takes the form

$$c_{\mu\varrho}c^*_{\mu\varrho'}e^{(\varrho)}_\alpha e^{(\varrho')}_\beta = \delta_{\alpha\beta}+\frac{k_\alpha k_\beta}{m^2}$$

and we finally obtain the following expression of the matrix $D(ik)$:

$$D_{jr} = e^{(r)}_j \quad (r = 1, 2); \qquad D_{j3} = \frac{\omega}{m}\,e^{(3)}_j; \qquad D_{j4} = 0;$$

$$D_{4r} = 0 \quad (r = 1, 2); \qquad D_{43} = \frac{|\boldsymbol{k}|}{m}; \qquad D_{44} = 0.$$

The decomposition of the field variables can be easily written:

$$\Phi_j(x) = (2\pi)^{-\frac{3}{2}} \int_{k_0=\omega} \frac{d^3\boldsymbol{k}}{\sqrt{2\omega}}$$
$$\times\left[\left(\sum_{r=1}^{3} e_j^{(r)}\varphi_r(\boldsymbol{k})+\frac{\omega}{m}\, e_j^{(3)}\varphi_3(\boldsymbol{k})\right) e^{ikx}\right.$$
$$\left.+\left(\sum_{r=1}^{2} e_j^{(r)}\tilde{\varphi}_r(\boldsymbol{k})+\frac{\omega}{m}\, e_j^{(3)}\tilde{\varphi}_3(\boldsymbol{k})\right) e^{-ikx}\right], \tag{6.3.10}$$
$$\Phi_0(x) = (2\pi)^{-\frac{3}{2}} \int_{k_0=\omega} \frac{d^3\boldsymbol{k}}{\sqrt{2\omega}}\, \frac{|\boldsymbol{k}|}{m}\, (\varphi_3(\boldsymbol{k})e^{ikx}+\tilde{\varphi}_3(\boldsymbol{k})e^{-ikx})$$

and between these reduced variables there exist the following commutation rules:

$$[\varphi_j(\mathrm{k}),\, \tilde{\varphi}_l(\boldsymbol{k}')] = \delta_{jl}\, \delta(\boldsymbol{k}-\boldsymbol{k}'), \tag{6.3.11}$$

from which we can define the creation and annihilation operators.

26. Observables of the field

Starting from well-known formulae, we obtain successively

$$T_{\mu\nu} = f^*_{\nu\varrho}\Phi_{\varrho,\mu}+\Phi^*_{\varrho,\mu}f_{\nu\varrho}+\mathcal{L}\,\delta_{\mu\nu}, \tag{6.3.12}$$
$$\mathfrak{M}_{\mu\nu\lambda} = f^*_{\mu\lambda}\Phi_\nu-f_{\nu\lambda}\Phi_\mu+\Phi^*_\nu f_{\mu\nu}-\Phi^*_\mu f_{\nu\lambda}, \tag{6.3.13}$$

with $J_{\mu\nu\lambda\varrho}$ given by (6.2.94). The symmetrical tensor $\Theta_{\mu\nu}$ is then written:

$$\Theta_{\mu\nu} = f^*_{\nu\varrho}f_{\mu\varrho}+f^*_{\mu\varrho}f_{\nu\varrho}-m^2(\Phi^*_\nu\Phi_\mu+\Phi^*_\mu\Phi_\nu)+\mathcal{L}\,\delta_{\mu\nu} \tag{6.3.14}$$

and the current is expressed in the form:

$$J_\mu(x) = ie(f^*_{\varrho\mu}\Phi_\varrho-\Phi^*_\varrho f_{\varrho\mu}). \tag{6.3.15}$$

We leave it to the reader to write down P_μ, $\mathcal{N}$, . . ., in terms of the creation and annihilation operators.

27. Stueckelberg formulation

By introducing a fifth component which has no effect on the observables, Stueckelberg expressed the Lagrangian formalism of the vector meson in a form very close to that of quantum electrodynamics.†

Consider a neutral vector field, described by Hermitian and anti-Hermitian Φ_μ operators and an independent scalar field $B(x)$. If we start from the Lagrangian density

$$\mathcal{L} \equiv -\frac{1}{2}(\Phi_{\mu,\nu}\Phi_{\mu,\nu}+m^2\Phi_\mu\Phi_\mu)-\frac{1}{2}(B_{,\mu}\, B_{,\mu}+m^2B^2), \tag{6.3.16}$$

† E. C. G. Stueckelberg, *Helv. Phys. Acta*, **11**, 225 (1938).

the wave equations are then

$$\left.\begin{aligned}\{\Box - m^2\}\,\Phi_\mu(x) &= 0,\\ \{\Box - m^2\}\,B(x) &= 0,\end{aligned}\right\} \tag{6.3.17}$$

to which we shall add the auxiliary condition (a condition on the state vector)

$$\chi(x) \equiv \frac{\partial \Phi_\mu}{\partial x_\mu} + mB(x) = 0. \tag{6.3.18}$$

Equations (6.3.17) then take the form

$$f_{\mu\nu,\,\nu} + m^2\Phi_\mu = -m\,\frac{\partial B}{\partial x_\mu} \tag{6.3.19}$$

and we come back to eqn. (6.3.4) when $B(x) = 0$.

These equations are invariant under a continuous group analogous to the gauge group of electrodynamics. Indeed, let there be a scalar $\Lambda(x)$, a solution of the equation

$$(\Box - m^2)\,\Lambda(x) = 0. \tag{6.3.20}$$

If we carry out the transformation

$$\Phi'_\mu = \Phi_\mu + \frac{\partial \Lambda}{\partial x_\mu}, \qquad B' = B - m\,\Lambda(x), \tag{6.3.21}$$

wave equations and the auxiliary condition keep their form, as may be easily verified. With the following choice of the gauge $\Lambda(x)$:

$$B(x) - m\,\Lambda(x) = 0 \tag{6.3.22}$$

[a choice possible by virtue of (6.3.17)], the new field variables $U_\mu(x)$ are

$$U_\mu(x) = \Phi_\mu(x) + \frac{1}{m}\frac{\partial B}{\partial x_\mu}. \tag{6.3.23a}$$

They obey the equation of the vector meson field

$$\frac{\partial}{\partial x_\nu}\{U_{\nu,\,\mu} - U_{\mu,\,\nu}\} + m^2 U_\mu = 0 \tag{6.3.23b}$$

and the auxiliary condition (6.3.18) takes the form

$$\chi(x) - U_{\mu,\,\mu} = 0. \tag{6.3.23c}$$

Let us now consider the commutation rules; since the fields $\Phi_\mu(x)$ and $B(x)$ obey the equations of the scalar meson, we can set

$$\left.\begin{aligned}[\Phi_\mu(x), \Phi_\nu(x')] &= i\,\delta_{\mu\nu}\,\Delta(x-x'),\\ [B(x), B(x')] &= i\,\Delta(x-x'),\\ [\Phi_\mu(x), B(x')] &= 0.\end{aligned}\right\} \tag{6.3.24}$$

We verify, furthermore, that

$$[\chi(x), \chi(x')] = 0, \tag{6.3.25}$$

so that the auxiliary condition (6.3.18) is valid for any x. It is easy to infer the commutation rules of the U_μ's from those of the Φ_μ's; we get

$$[U_\mu(x), U_\nu(x')] = [\Phi_\mu(x), \Phi_\nu(x')] - \frac{1}{m^2}\frac{\partial^2}{\partial x_\mu \, \partial x_\nu}[B(x), B(x')] \tag{6.3.26}$$

$$= i\left\{\delta_{\mu\nu} - \frac{1}{m^2}\frac{\partial^2}{\partial x_\mu \, \partial x_\nu}\right\}\Delta(x-x')$$

and the variables $U_\mu(x)$ indeed satisfy the commutation rules of the vector field.

The physical quantities can be expressed very easily. Note for instance that the expression of $T_{\mu\nu}$ is

$$T_{\mu\nu} = \Phi_{\sigma,\,\mu}\Phi_{\sigma,\,\nu} + B_{,\,\mu}B_{,\,\nu} + \mathcal{L}\,\delta_{\mu\nu} \tag{6.3.27}$$

and that after the gauge transformation (6.3.20) $T_{\mu\nu}$ becomes

$$T_{\mu\nu} \rightarrow T'_{\mu\nu} = T_{\mu\nu} - \Lambda_{,\mu}\chi_{,\nu} - \Lambda_{,\nu}\chi_{,\mu} + (\Lambda_{,\varrho}\chi_{,\varrho} + m^2\,\Lambda\chi)\,\delta_{\mu\nu}$$

and consequently, by virtue of the auxiliary condition, the expectation values of $T_{\mu\nu}$ and $T'_{\mu\nu}$ are equal.

28. Wave mechanics of the photon and vector field with zero mass

When elementary particles are classified by the methods of the fusion theory (*op. cit.* in Chap. IV), one is led to consider a particle of spin 1 as a fusion of two particles of spin $\frac{1}{2}$ and non-zero masses. The question may then arise to what extent such particles can be associated with the electromagnetic field, or in other words: "Can a non-zero mass be attributed to the photon?" De Broglie has carried out an exhaustive analysis of the experimental results which cause a mass-zero to be attributed to the photon and he has concluded that a mass in the neighbourhood of 10^{-65} g could not be distinguished experimentally from the mass zero.†

In that case, the photon would be a vector particle, of spin 1, with a comparatively small mass, and in the formalism of the wave mechanics of the photon we can find a confirmation of the essential results of quantum field theory as described above.

Without going into details, we shall confine ourselves to investigating a formal aspect of this problem by raising the question: to what extent can the passing to the limit $m = 0$ which enables us to describe the photon be

† L. de Broglie, *Mécanique ondulatoire du photon et Théorie quantique des champs*, Gauthier-Villars, 1949.

justified? If, for instance, the photon were described directly by the vector field $\Phi_\mu(x)$ in the formalism of the first section of this part, the passing to the limit $m=0$ would render the commutation rules (6.3.8) meaningless. The Stueckelberg formulation palliates these difficulties to a certain degree: it is apparent in the first place that when m becomes zero, the Stueckelberg field $\Phi_\mu(x)$ converges to $A_\mu(x)$ [eqns. (6.3.17) and (6.3.18)]. But in that case the Fermi condition (6.3.18) would raise the same difficulties we ran up against in the case of the Maxwell field. The technique adopted consists in introducing the field $U_\mu(x)$ of the preceding section and then proving that the condition (6.3.23c) leads to a normalizable vector state and that we obtain the Fermi condition when $m = 0$. We shall therefore compute all the expectation values with the help of the state vector describing the field $U_\mu(x)$, then we shall pass to the limit $m = 0$.†

Using the notation of the preceding section, let us consider the two systems of field variables:

(a) the fields $\Phi_\mu(x)$ and $B(x)$ obeying eqns. (6.3.17) and (6.3.18), which we can rewrite:

$$\{\Box - m^2\}\,\Phi_\mu(x) = 0, \tag{6.3.28a}$$

$$\{\Box - m^2\}\;B(x) = 0, \tag{6.3.28b}$$

$$B(x) = -\frac{1}{m}\frac{\partial \Phi_\mu}{\partial x_\mu} \tag{6.3.28c}$$

and we include the auxiliary condition in the wave equations;

(b) the field $U_\mu(x)$ defined by

$$U_\mu(x) = \Phi_\mu(x) + \frac{1}{m}\frac{\partial B}{\partial x_\mu}, \tag{6.3.29}$$

which obeys the commutation rules (6.3.26) and has a Fourier transform analogous to (6.3.10):

$$U_j(x) = (2\pi)^{-\frac{3}{2}} \int_{k_0=\omega} \frac{d^3\boldsymbol{k}}{\sqrt{2\omega}} \times\Big[\Big(\sum_r e_j^{(r)} u_r(\boldsymbol{k}) + \frac{\omega}{m}\, e_j^{(3)} u_3(\boldsymbol{k})\Big) e^{ikx} + \Big(\sum_r e_j^{(r)} \tilde{u}_r(\boldsymbol{k}) + \frac{\omega}{m}\, e_j^{(3)} \tilde{u}_r(\boldsymbol{k})\Big) e^{-ikx}\Big], \tag{6.3.30}$$

$$U_0(x) = (2\pi)^{-\frac{3}{2}} \int_{k_0=\omega} \frac{d^3\boldsymbol{k}}{\sqrt{2\omega}}\, \frac{|\boldsymbol{k}|}{m}\, [u_3(\boldsymbol{k}) e^{ikx} + \tilde{u}_3(\boldsymbol{k}) e^{-ikx}],$$

† F. Coester, *Phys. Rev.*, **83**, 798 (1951); R. T. Glauber, *Prog. Theor. Phys.*, **9**, 295 (1953). For a different formalism, cf. F. J. Belinfante, *Phys. Rev.*, **84**, 644 (1951); J. G. Valatin, *Dan. Mat. fys. Medd.*, **26**, no. 13 (1951).

with $r=1, 2$ and

$$[u_l(\boldsymbol{k}), \tilde{u}_{l'}(\boldsymbol{k}')] = \delta_{ll'}\delta(\boldsymbol{k}-\boldsymbol{k}'). \tag{6.3.31}$$

As for the field $B(x)$, it can be written

$$B(x) = (2\pi)^{-\frac{3}{2}} \int_{k_0=\omega} \frac{d^3\boldsymbol{k}}{\sqrt{2\omega}} [b(\boldsymbol{k})e^{ikx} + \tilde{b}(\boldsymbol{k})e^{-ikx}], \tag{6.3.32}$$

with

$$[b(\boldsymbol{k}), \tilde{b}(\boldsymbol{k}')] = \delta(\boldsymbol{k}-\boldsymbol{k}'). \tag{6.3.33}$$

Note first that if Ψ is the state vector, we obtain by virtue of (6.3.28),

$$\langle \Psi | \frac{\partial \Phi_\mu}{\partial x_\mu} + mB(x) | \Psi \rangle = 0; \tag{6.3.34}$$

if therefore we make the auxiliary field $B(x)$ describe particles which provide no contribution to the results observed, in other words if we assume

$$\langle \Psi | B(x) | \Psi \rangle = 0, \tag{6.3.35}$$

the Fermi condition for the field Φ_μ is found to be satisfied. Let us now further suppose that the particles described by the longitudinal component of $U(x)$ do not exist in nature, in other words that the only possible states are those which obey

$$U_3(x) | \Psi \rangle = 0; \tag{6.3.36a}$$

an explicit computation proves, indeed, that these longitudinal particles do not react with the electrons. It may further be noted that equation (6.3.35) also gives

$$\tilde{b}(\boldsymbol{k}) | \Psi \rangle = 0. \tag{6.3.36b}$$

Insert into eqns. (6.3.28*c*) and (6.3.29) the expressions (6.3.30) of U_μ, (6.3.10) of Φ_μ and (6.3.32) of $B(x)$; we get

$$b(\boldsymbol{k}) = -\frac{i}{m}[|\boldsymbol{k}|\varphi_3(\boldsymbol{k}) + i\omega\varphi_4(\boldsymbol{k})]; \tag{6.3.37}$$

$$u^{(r)}(\boldsymbol{k}) = \varphi_r(\boldsymbol{k}) \quad (r = 1, 2),$$

$$u^{(3)}(\boldsymbol{k}) = \frac{m}{\omega}\varphi_3(\boldsymbol{k}) + \frac{i}{\omega}|\boldsymbol{k}|b(\boldsymbol{k}) = \frac{1}{m}[\omega\varphi_3(\boldsymbol{k}) + i|\boldsymbol{k}|\varphi_4(\boldsymbol{k})].$$

Finally, formulae (6.3.36) applied to the transversal component of the vector $\Psi(\boldsymbol{k})$ [cf. (6.2.50*b*)] yield

$$\left.\begin{aligned} (\omega\varphi_3(\boldsymbol{k}) + i|\boldsymbol{k}|\varphi_4(\boldsymbol{k}))|\Psi_{1s}(\boldsymbol{k})\rangle &= 0, \\ (|\boldsymbol{k}|\tilde{\varphi}_3(\boldsymbol{k}) + i\omega\tilde{\varphi}_4(\boldsymbol{k}))|\Psi_{1s}(\boldsymbol{k})\rangle &= 0. \end{aligned}\right\} \tag{6.3.38}$$

By making $m=0$, we find ourselves back at the expressions (6.2.51) of electrodynamics. But, and this is the important point, the vector $\Psi_{ls}(\boldsymbol{k})$ can be normalized; we have

$$|\Psi_{ls}(\boldsymbol{k})\rangle = \frac{m}{\omega} \sum_{n_3, n_4} \left(-i\frac{\boldsymbol{k}}{\omega}\right)^{n_3} \delta_{n_3, n_4} |n_3, n_4\rangle \tag{6.3.39}$$

where the notation of formulae (6.2.80) has been adopted.

To sum up the technique with which this new method provides us let us suppose that we wish to compute an expectation value formed with the field variables $A_\mu(x)$ of electrodynamics. We begin by replacing them by Φ_μ's which we express with the aid of the U_μ's, then we pass to the limit $m = 0$. Although the U_μ's themselves have no definite limit values, the expectation values investigated converge to well-defined limits and the state vector does satisfy the auxiliary condition.

To give one example, let us compute

$$\langle 0| \frac{\partial A_\mu(x)}{\partial x_\mu} \frac{\partial A_\nu(x)}{\partial x_\nu} |0\rangle;$$

we write this expression in the form

$$\lim_{m=0} \langle 0| \frac{\partial \Phi_\mu(x)}{\partial x_\mu} \frac{\partial \Phi_\nu(x)}{\partial x_\nu} |0\rangle$$

$$= \lim m^2 \langle 0| B(x)B(x')|0\rangle = -i \lim_{m=0} m^2 \Delta^+(x-x') = 0.$$

29. Petiau–Duffin–Kemmer equation†

We propose to express the equations of the vector particle (therefore spin 1) in the form

$$\{\beta_\mu \partial_\mu + m\} Q(x) = 0 \tag{6.3.40}$$

and we have seen in Chap. IV (4.28*b*) that one has to take β_μ's which obey the Petiau–Duffin–Kemmer relations.

$$\beta_\mu \beta_\nu \beta_\lambda + \beta_\lambda \beta_\nu \beta_\mu = \beta_\mu \delta_{\lambda\nu} + \beta_\lambda \delta_{\mu\nu}. \tag{6.3.41}$$

It is convenient to introduce the matrix

$$\eta_\mu = 2\beta_\mu^2 - I, \tag{6.3.42}$$

which obeys the following relations:

$$\left.\begin{aligned} \eta_\mu^2 = I, \quad [\eta_\mu, \eta_\nu]_- &= 0; \\ [\beta_\mu, \eta_\nu]_+ &= 0 \quad (\mu \neq \nu), \\ \beta_\mu = \beta_\mu \eta_\mu = \eta_\mu \beta_\mu & \quad \text{(no summation over } \mu). \end{aligned}\right\} \tag{6.3.43}$$

† Cf. bibliographical note on p. 145.

The matrix η of the general theory can be chosen equal to η_4:

$$\bar{Q}(x) = \tilde{Q}(x)\eta_4, \tag{6.3.44}$$

since with the representation of the β_μ's we shall be giving at the end of the section, the β_μ's (and consequently the η_μ's) can be chosen Hermitian. We thus obtain

$$\bar{Q}(x)\{\beta_\mu \overleftarrow{\partial}_\mu - m\} = 0, \tag{6.3.45}$$

and we can verify, making use of (6.3.41), that by defining

$$J_{\mu\nu} = [\beta_\mu, \beta_\nu], \tag{6.3.46}$$

$J_{\mu\nu}$ indeed obeys the characteristic relation (4.17).

Equation (6.3.40) describes particles of spin 0 and 1; we can see this in the following way.

Consider first the matrices

$$P^{(0)} = \beta_1^2 \ldots \beta_4^2, \qquad P_\mu^{(0)} = P^{(0)}\beta_\mu, \tag{6.3.47}$$

then set

$$P^{(0)}Q(x) = \Phi(x), \quad P_\mu Q(x) = \Phi_\mu(x); \tag{6.3.48}$$

we verify that we have

$$\partial_\mu \Phi_\mu + m\Phi = 0, \quad \partial_\mu \Phi + m\Phi_\mu = 0 \tag{6.3.49}$$

and that we obtain the wave equation of the scalar particles.

$$\partial_\mu \partial_\mu \Phi - m^2\Phi = 0.$$

Consider next the following matrices:

$$\left.\begin{aligned} P_j^{(1)} &= -\beta_1^2\beta_2^2\beta_3^2\beta_j\beta_4, \\ P_4^{(1)} &= \beta_1^2\beta_2^2\beta_3^2(1-\beta_4^2), \\ P_{\mu\nu} &= P_\mu^{(1)}\beta_\nu; \end{aligned}\right\} \tag{6.3.50}$$

we verify that we have

$$P_{\mu\nu} = -P_{\nu\mu}, \quad P_{\mu\nu}^{(1)}\beta_\nu\beta_\sigma = \delta_{\nu\sigma}P_\mu^{(1)} - \delta_{\mu\sigma}P_\nu^{(1)}. \tag{6.3.51}$$

Setting

$$P_\mu^{(1)}Q(x) = m\Phi_\mu(x), \quad P_{\mu\nu}Q(x) = F_{\mu\nu}(x), \tag{6.3.52}$$

we obtain the equations of a vector particle

$$P_{\mu\nu,\,\mu} - m^2\Phi_\nu = 0, \quad F_{\mu\nu} = \Phi_{\nu,\,\mu} - \Phi_{\mu,\,\nu}. \tag{6.3.53}$$

Below we give representations of the β_μ's and leave it to the reader to verify that they obey (6.3.41).

Setting aside the case $\beta_1 = 0, \ldots, \beta_4 = 0$, consider the case of the spin 0; the matrices β_μ take the form

$$\beta_1 = \left[\begin{array}{c:c} 0 & \begin{matrix}0\\1\\0\\0\end{matrix} \\ \hdashline \begin{matrix}0&1&0&0\end{matrix} & 0 \end{array}\right], \quad \beta_2 = \left[\begin{array}{c:c} 0 & \begin{matrix}0\\0\\1\\0\end{matrix} \\ \hdashline \begin{matrix}0&0&1&0\end{matrix} & 0 \end{array}\right],$$

$$\beta_3 = \left[\begin{array}{c:c} 0 & \begin{matrix}0\\0\\0\\1\end{matrix} \\ \hdashline \begin{matrix}0&0&0&1\end{matrix} & 0 \end{array}\right], \quad \beta_4 = \left[\begin{array}{c:c} 0 & \begin{matrix}-i\\0\\0\\0\end{matrix} \\ \hdashline \begin{matrix}i&0&0&0\end{matrix} & 0 \end{array}\right]$$

and correspond to a wave function $Q(x)$ with five components formed by the four Φ_μ's and Φ of (6.3.48).

Turning to the particles of spin 1, the matrices are now:

$$\beta_1 = \left[\begin{array}{c:c:c:c} 0 & 0 & 0 & \begin{matrix}-1\\0\\0\end{matrix} \\ \hdashline 0 & 0 & \begin{matrix}0&0&0\\0&0&-1\\0&1&0\end{matrix} & \begin{matrix}0\\0\\0\end{matrix} \\ \hdashline 0 & \begin{matrix}0&0&0\\0&0&1\\0&-1&0\end{matrix} & 0 & \begin{matrix}0\\0\\0\end{matrix} \\ \hdashline \begin{matrix}-1&0&0\end{matrix} & \begin{matrix}0&0&0\end{matrix} & \begin{matrix}0&0&0\end{matrix} & 0 \end{array}\right],$$

$$\beta_2 = \left[\begin{array}{c:c:c:c} 0 & 0 & 0 & \begin{matrix}0\\-1\\0\end{matrix} \\ \hdashline 0 & 0 & \begin{matrix}0&0&1\\0&0&0\\-1&0&0\end{matrix} & \begin{matrix}0\\0\\0\end{matrix} \\ \hdashline 0 & \begin{matrix}0&0&-1\\0&0&0\\1&0&0\end{matrix} & 0 & \begin{matrix}0\\0\\0\end{matrix} \\ \hdashline \begin{matrix}0&-1&0\end{matrix} & \begin{matrix}0&0&0\end{matrix} & \begin{matrix}0&0&0\end{matrix} & 0 \end{array}\right],$$

$$
\beta_3 = \left[\begin{array}{c|c|c|c}
0 & 0 & 0 & \begin{matrix} 0 \\ 0 \\ -1 \end{matrix} \\ \hline
0 & 0 & \begin{matrix} 0 & -1 & 0 \\ 1 & 0 & 0 \\ 0 & 0 & 0 \end{matrix} & \begin{matrix} 0 \\ 0 \\ 0 \end{matrix} \\ \hline
0 & \begin{matrix} 0 & 1 & 0 \\ -1 & 0 & 0 \\ 0 & 0 & 0 \end{matrix} & 0 & \begin{matrix} 0 \\ 0 \\ 0 \end{matrix} \\ \hline
\begin{matrix} 0 & 0 & -1 \end{matrix} & \begin{matrix} 0 & 0 & 0 \end{matrix} & \begin{matrix} 0 & 0 & 0 \end{matrix} & 0
\end{array}\right],
$$

$$
\beta_4 = \left[\begin{array}{c|c|c|c}
0 & 0 & \begin{matrix} -i & 0 & 0 \\ 0 & -i & 0 \\ 0 & 0 & -i \end{matrix} & \begin{matrix} 0 \\ 0 \\ 0 \end{matrix} \\ \hline
0 & 0 & 0 & \begin{matrix} 0 \\ 0 \\ 0 \end{matrix} \\ \hline
\begin{matrix} i & 0 & 0 \\ 0 & i & 0 \\ 0 & 0 & i \end{matrix} & 0 & 0 & \begin{matrix} 0 \\ 0 \\ 0 \end{matrix} \\ \hline
\begin{matrix} 0 & 0 & 0 \end{matrix} & \begin{matrix} 0 & 0 & 0 \end{matrix} & \begin{matrix} 0 & 0 & 0 \end{matrix} & 0
\end{array}\right].
$$

These matrices correspond to a wave function $Q(x)$ with ten components formed by the six $F_{\mu\nu}$ and the four Φ_μ of (6.3.52).†

† For a fuller study, cf. H. Umezawa, *Quantum Field Theory*, North Holland, 1956, pp. 85–91.

Index